ROCK | WATER | LIFE

ROCK |

Duke University Press
Durham and London 2020

WATER | LIFE

ecology & humanities
for a decolonial south africa

LESLEY GREEN

Library of Congress Cataloging-in-Publication Data
Names: Green, Lesley, [date] author.
Title: Rock / water / life : ecology and humanities for a decolonial South
    Africa / Lesley Green.
Description: Durham : Duke University Press, 2020. | In title, "[/]" is
    expressed as a vertical bar; in table of contents, "[a]" is expressed as the
    phonetic schwa symbol (upside-down e). | Includes bibliographical
    references and index.
Identifiers: LCCN 2019041853 (print)
LCCN 2019041854 (ebook)
ISBN 9781478003694 (hardcover)
ISBN 9781478003991 (paperback)
ISBN 9781478004615 (ebook)
Subjects: LCSH: Environmental sciences—Political aspects—South Africa. |
    Environmental policy—South Africa. | Environmental justice—
    South Africa. | Racism—Environmental aspects—South Africa. |
    Science and the humanities.
Classification: LCC GE190.S6 G74 2020 (print) | LCC GE190. S6 (ebook) |
    DDC 304.20968—dc23
LC record available at https://lccn.loc.gov/2019041853
LC ebook record available at https://lccn.loc.gov/2019041854

Cover art: *Fire and Ice*, 2017. Digital print on archival paper 1000 x 438 mm.
    © Simon Sephton. Courtesy of the artist.

for Nazeer, Susanna, Sinegugu, Nonhle, Bazooka,
Selene, Serge, Jenni, Davine, Maryam, Noora, Caron
connectors of rock, water, life

Whether we like it or not there is
at the present time an eminent, tentacular civilization.
Because it is clear that now we have entered the era of the finite world. . . .
Still more,
with modern European thought was born a new process . . .
a process of reification, that is,
the thingification of the world. . . .
The consequence you know,
it is the appearance of the mechanized world,
the world of efficiency
but also the world in which people themselves become things.
In short
we are facing a gradual devaluation of the world,
which leads quite naturally to an inhuman world
on whose trajectory lies
contempt, war, exploitation of humans by humans.

—Aimé Césaire, speaking at the
Festival mondial des arts nègres in Dakar, 1966

Here in South Africa, we are always in the crucible.
There are never any shortcuts. All we can do is to be present.

—Jennifer Ferguson

How should we construct our question
so that it has a chance of interesting those to whom we ask it
and a chance of receiving interesting answers?

—Isabelle Stengers and Vinciane Despret

CONTENTS

FOREWORD

The most adequate rendering of what Lesley Green's book asks her readers to accept feeling, thinking, and imagining might well be expressed by her quotation of singer and poet Jennifer Ferguson: "Here in South Africa, we are always in the crucible. There are never any shortcuts. All we can do is to be present."

As an academic working at the University of Cape Town, Lesley Green was indeed present in the spring of 2015, when activist students and staff obtained the removal of the statue of Cecil John Rhodes that dominated the campus; and again in October 2015, when the #FeesMustFall protest followed the #RhodesMustFall one, and, ten months later resurged with the demand for a free, decolonised education for all. To be present, however, does not mean simply to be there. It means allowing oneself to be affected; to give up one's reasons, however legitimate; to pick and choose; or to object on this point or the other. It means feeling that there are times when such reasons create a distance, the luxury of an aloofness only "whiteness" allows.

During my stay in Cape Town, just after the 2015 protests, Lesley and I talked a lot about whiteness and the terrible feeling that the trust in the possibility of a "Rainbow Nation" had dissipated as a dream—back into the crucible as the quasi-miraculous shortcut willed by Nelson Mandela was falling apart. Donna Haraway's motto, "staying with the trouble," was all the more haunting.

I had come to Cape Town to meet with Lesley and work with her students on the particular situation of environmental humanities in South Africa. I understood from her writing that environmental questions in the post-apartheid context are folded in explicit conflictual entanglements that make the possibility of consensus, that is, of "sensing together," a fully developed challenge. Here nothing is innocent. You cannot forget, when you are concerned by the devastation of the environment, that this concern was, and is still, the justification for protection that has meant fencing away marauders and poachers, defending "nature" against the menace of humans who happen to be poor, and are thus invariably black. But here also, disempowered fishers had felt authorised to examine and learn collectively how to question the ways in which, in the name of science, they were excluded from decision-making processes that concerned them. Lesley's chapter about the politics associated with the fixing of fishing quotas offers an inspiring example of how disempowered people were able to contest the rationale of a so-called objective (managerial) scientific argumentation in Parliament.

What could a country where the need to decolonise ways of thinking, imagining, and doing is taken seriously teach us all? Could post-apartheid South Africa endorse the challenge of environmental justice? Could this place where injustice once ruled supreme, and where the reference to dignity now had a political power of its own, be a fostering place for the invention of a social ecology of knowledges and experiences? The question was all the more important because South Africa had already experienced the perils of simply placing "white" and "indigenous" knowledges in opposition—the mistake made by former president Thabo Mbeki when, as Lesley concludes about the tragic bungling of the AIDS crisis, he tried to change the players on a chess board, instead of changing the game itself.

During my stay, Lesley and her students made me understand that my hopeful interest was, if not misplaced, at least hard to foster at a time when the very legacy of Mandela's Rainbow Nation was sinking into the anger and disillusionment of the younger generation, the feeling that "whites" had cheated both them and their future. But they did so in a way that taught me what it took to attend to this anger rather than to "correct" it with the certainly true, but oh-so-easy argument: that what reduced the Rainbow Nation to an impossible dream was first of all the acceptance of the "neoliberal" promise, the reification of economic growth as the only way to bring general prosperity and consensus.

It would have been easy, and true, to plead that if anger targets whiteness, it should acknowledge that the meaning of whiteness has changed. It is no longer synonymous with discrimination but rather with "whitewashing"—that is, with wiping away any active memory of whatever would empower resistance: those responsible have no particular colour, nor ideological stance—ideology is a thing of the past. We are all now constrained to serve anonymous necessity, or, as Lesley writes, the three interlocked gods of reason: technical efficiency, economic productivity, scientific objectivity. In other words, such a proposition offers to share a sad, impotent lucidity—"you" should not mistake us for your enemies; "we" are all in the same boat, suffering the same hegemony.

Berthold Brecht famously wrote that we speak about the violence of the river when it drags everything with it, but we never call violent the banks that channel it. As an inhabitant of academia, I must recognise the way it channels successive generations of students into thoughtlessness. Even critical lucidity, as cultivated by academics tirelessly analysing and denouncing the tentacular power of the neoliberal domination, does not help very much when it is a question of imagining how to escape the hold of the three gods mentioned above. Indeed, academic critique seldom creates ways of sustaining the imaginative and practical resources liable to be reclaimed against domination. Rather, it prioritises the danger of being seduced into thinking that those resources could be up to the task. The thesis that it is easier to anticipate the end of the world than the end of capitalism has been repeated as a manner of academic mantra. Engraining the fear of being duped—or being shown by colleagues to be duped—in the minds of generations of students may be seen as a free gift from the critical academy to capitalism.

Lesley's book engages this issue in an intensively situated manner, fully accepting that as a daughter to a family where the fear of shame in nonconformity with whiteness was massaged into her skin, she cannot accept the comfort of academic distancing. It is no longer sufficient to remember that as a student, she became an active part of the anti-apartheid protest movement. Facing an anger that identifies her as a "white" academic, she has chosen to abandon the protection of academic conformity, but not to side with the anger. She has accepted learning, thinking, and suffering with what anger risks forgetting—the land and its rocks, waters, and living inhabitants. She has transformed the writing of this book, rich in knowledges and documentation, into a journey into feeling, opening herself to the ghosts, to the voiceless presence of those, humans and non-humans, who were, and go on being, victims of the gods of reason.

What readers will discover is an entangled double rendering: of haunted landscapes, and of a researcher who turns "reflexivity" into the art of letting herself be affected by past and present socio-ecological devastation in order to become able to stay with the trouble; to refuse analytical understanding of its power to distance. She does not go beyond the facts. There is no beyond when one tells about destruction and mutilation. She remembers, remembers, rearticulates what was done to this country, and is not transmitted to the students of her university—or only as deplorable but cold cases. Yet she cannot participate in the angry decolonising slogans of those students against "Science," because she knows that refusing white knowledge *en bloc* is a trap. Their enemies play at being horrified but are really delighted when listening on the internet to a student claiming that "science as a whole is a product of Western modernity, and the whole thing should be, like, scratched off"—a video that went viral. But she can learn to feel in the very fibres of her being how right they are to contest the kind of unthinking, unfeeling objectivity they are taught to conform to.

The succeeding "Fallist" movements were born in universities, places which are today all over the world under the yoke of neomanagerial governance and the unrelenting pressure for "breakthroughs" ensuring competitive innovation. As such, their task is to address continuously renewed generations of students, whom they have to convince over and over again that the way they are trained makes sense. Most of the time it works, and the disappointed and recalcitrant ones simply disappear. But from time to time, the institution's bluff is called, and the river overflows its banks. Even if such events are transient, they are not the expression of some utopian dream; rather, they are an awakening from a sleepwalking routine by a gust of vital questions—questions that make our definitions of knowledge stagger because they open to an outside these definitions ignore. Coming back to the challenge of decolonising universities and the kind of knowledges they produce and transmit, we all ignore what this challenge entails and demands, and students do not know better. What they know instead is that what they are presented with has the dubious effect of making them forget the question.

Lesley has allowed herself to be haunted by the question, to stay with the trouble, while the academically safe position would have been to analyse the incapacity of the institution to answer the decolonising challenge. She works with this challenge, works for a future that would include her children together with those students and staff who refuse the bitter reality that is stealing their own future. Her stance reminds me of John Dewey's call to his social sciences colleagues, when he wrote in 1922: "Be the evils what they may, the experiment is not yet played out. The United States are not yet made; they are not a finished fact to be categorically assessed."[1] Today, the call to resist matter-of-fact assessment may reverberate everywhere on this earth, as everywhere what is categorically assessed is the way we participate, eyes wide open, in the ever-accelerating unravelling of the earth's entangled socio-ecological worlds. But Dewey's call has a special meaning in South Africa. Here it is a question of reclaiming the legacy of an actually experimental weaving, of refusing to betray the trust that made this now-unravelling dream possible.

Reclaiming means reacquainting oneself with generative resources, resources that sustain and inspire. Lesley proposes that her readers remember that the trust in the possibility of weaving together different voices and perspectives was born of the African soil, this soil from which colonisation has cut people off. The African "dilemma tales" do not honour confrontational truth, as cherished by our academic tradition, with authors rivalling for authority. They stage situations as multi-authored ones, gathering human and non-human protagonists, experiencing them with their own perspective, and participating in them in their own ways. Against the blind and brutal shortcuts proposed by the gods of reason, such tales demand from those who claim to care for a situation—teachers and researchers, for instance—the capacity to cultivate thinking and imagination that would not be about what this situation should conform to, but that enable them to stay alongside as it unfolds.

Learning to think alongside more and more troubled and troubling situations may be a vital challenge, and a demanding relevant substitute for the conquering machine that has been called reason. Lesley's journey into the South African crucible shows us a path for reclaiming an exercise of reason worth keeping in the hard times which are coming.

—Isabelle Stengers, Université Libre de Bruxelles

ACKNOWLEDGMENTS

So many thanks are due. Isabelle Stengers read every chapter as it emerged and sent back responses that took my thinking further. Editor Helen Moffett's generosity was a joy. Elizabeth Ault and the team at Duke kept the finish line attainable. Robert Morrell and Marilet Sienaert approved the small grant that seeded the book in 2012.

I thank colleagues affiliated with Environmental Humanities South—Frank Matose, Nikiwe Solomon, Michelle Pressend, Hedley Twidle, Virginia MacKenny, Bodhisattva Kar, Ian Rijsdijk, Lance van Sittert, Shadreck Chirikure, Horman Chitonge, Maano Ramutsindela, Carla Tsampiras, Philip Aghoghovwia, Tania Katzschner, Andre Goodrich, Oliver Mtapuri, and Artwell Nhemachena—and all of the master's and PhD students in Environmental Humanities South 2015–19 and in Contested Ecologies 2010–15, for their commitment to research in this field.

Marisol de la Cadena's work and friendship have been an inspiration, and likewise those of Donna Haraway, Anna Tsing, Mario Blaser, Eduardo Viveiros de Castro, Deborah Danowski, Beth Povinelli, Kristina Lyons, Jenny Reardon, Karen Barad, Elaine Gan, Beth Stevens, TJ Demos, Heather Swanson, Maria Puig de la Bellacasa, Pedro Neves Marques, Helen Verran, Bruno Latour, Bronislaw Szerzynski, Ruth Kansky, Hylton White, Diana Gibson, Jess Auerbach, Dominic Boyer, Cymene Howe, Rob Nixon, Astrida Neimanis, Elspeth Probyn, Kate Johnston, Fiona Probyn-Rapsey, Manuel Tironi, Thom van Dooren, Thomas Cousins, and Steven Robins. In his capacity as humanities dean, Sakhela Buhlungu offered immense interest, along with support when he could and good humour when he couldn't.

Exceptional administrative support came from Vivienne Toleni. I would not have completed the manuscript had it not been for her daily assistance, offered with patience, persistence, laughter, and kindness.

Suzanne Fraser's Cape Town sabbatical brought our household many laughs and rich conversations. Trish Strydom, Ester Haumann, Jeanette Walker, Margaret Dyasi, Jax Westraad, Lynne Simpson, Tony Simpson, Ilsa Boswell, Alison Moultrie, Annie Holmes, Vanessa Farr: so many of our conversations are reflected on these pages in ways big and small; thank you. "Die Biervrouens"—Tracy Konstant, Helene van der Merwe—dankie julle.

I especially want to acknowledge the tenacity and courage of University of Cape Town colleagues in every Faculty who had the courage to "stay with the trouble" through difficult times. You know who you are.

In Mpumalanga, thank you to David Bunn at Wits Rural Facility for generous hospitality. At SANParks Kruger, Emile Smidt, Danie Pienaar, and Louise Swemmer helped me to broaden my Cape-based thinking, as did Wendy Annecke in the Cape SANParks research division.

Environmental activists whose determination to imagine and begin to build a different world has taught me so much include Caron von Zeil of Reclaim Camissa; Davine Cloete of Lutzville; Gathuru Mburu of the Institute for Cultural Ecology in Kenya; Jenni Trethowan of the Baboon Matters Trust; Makgoma Lekalakala and Liz McDaid, who took on the proposed nuclear development; Maryam Salie and Noora Salie of the Sandvlei United Community Organisation; Nazeer Sondhay and Susanna Coleman of the Philippi Horticultural Area Campaign; Nnimmo Bassey of the Health of Mother Earth Foundation; Nonhle Mbuthuma of the Amadiba Crisis Committee at Xolobeni; Sinegugu Zukulu for his courage and tenacity in defending Mpondoland; Selene Smith of Coastal Links in Langebaan; the extraordinary environmental attorneys Cormac Cullinan and Samuel Nguiffo; and the team at the Centre for Environmental Rights. *Phambili!*

———

Research funds are gratefully acknowledged from the National Research Foundation (SeaChange, 2010–12; Human-Social Dynamics (The Making of an Environmental Public), 2015–17; National Research Foundation Ratings 2015–18), the National Institute of Humanities and Social Sciences, the Andrew W. Mellon Foundation, and the University of Cape Town via the Programme for Enhancement of Research Capacity and the annual University of Cape Town University Research Committee grants, as well as support for Isabelle Stengers's visit in 2015.

This work has its beginnings in the Sawyer Seminar on Knowledges and Ways of Knowing in the PostColonial University funded by the Andrew W. Mellon Foundation (2009–11), in which the Contested Ecologies project grew. The Mellon Foundation's South Africa Programme funded the launch of the Environmental Humanities South graduate programme (2014–17) and supported its transformation into a fully fledged research center at the University of Cape Town (2017–21). The support and interest of Stuart Saunders and Saleem Badat, at Andrew W. Mellon, are deeply appreciated.

Completion of the book was made possible by a Fulbright Fellowship for a residency at the Science and Justice Research Center of the University of California at Santa Cruz (UCSC), January to June 2018. Sincere thanks to Kim Williams and her team at the United States Consulate in Cape Town for their unstinting support, and at UCSC, Anna Tsing, Kristina Lyons, Donna Haraway, Karen Barad, Jenny Reardon, and Colleen Massengale.

———

Each individual chapter has had a long history, and some sections appeared in prior publications.

An early version of the introduction was invited by Deborah Posel for a workshop at the University of Cape Town with Rob Nixon, and that event kicked off the entire Environmental Humanities South project. Some parts of that paper appeared in the journal *Resilience* (volume 1, issue 2) in 2014.

Chapter 1: In 2012, the African Climate and Development Initiative's Bergrivier workshop team started me thinking about rivers and water, and the Berg River supply to Cape Town. Caron von Zeil pointed me to the streams and the importance of their history. Henrik Ernstson invited me to share an early draft with his urban natures team. Nick Shepherd organised the walking workshop in the Table Mountain National Park. The team at Kaskazi Kayaks got me paddling in Table Bay, which provided insight into the relationships of bay, city, mountain, streams, and history.

Chapter 2 began on a committee for the University of Cape Town's Shale Gas Working Group, initiated by Marilet Sienaert and chaired by Dean Francis Petersen. Eduardo Viveiros de Castro suggested I present some of the Karoo material at the Thousand Names of Gaia workshop in Rio in 2014, and thought with me at length about cement. A section of the chapter was published by Jan Glazewski and Surina Esterhuyse in 2016. Thank you to Leonie Joubert, Ben Schoeman, Isa Marques, Nathan Honey, Lawrence van der Merwe, and Miriam Waltz for the off-season trip to the AfrikaBurn installation on fracking; and to Pedro Neves Marques for the invitation to write the Changing of the Gods of Reason piece, which responded to the work of the Rhodes Must Fall collective at the University of Cape Town. Thank you to all who commented on the piece at the Ethnographic Engagements seminar at the University of California at Santa Cruz; and to Dominic Boyer, who hosted its discussion at a Centre for Energy Humanities seminar at Rice University, Houston, in 2018.

Chapter 3 began as a project with the ABC team in 2010—Robert Morrell, Timm Hoffman, David Gammon, Helen Verran, Millie Hilgart, Nicola Wheat, and Josh Cohen. It grew from there in discussions with Kristina Lyon, Karen Barad, Jenny Reardon, Donna Haraway, and the Science and Justice Research Center at Santa Cruz. Isabelle Stengers kept gently suggesting that this material ought to be in this book, and the invitation from the Science Faculty Engagements students at the University of Cape Town to share it enabled a dialogue to begin. Some of this chapter was published in the *South African Journal of Science*.

Chapter 4: Thanks to Isabelle Stengers and Alison Moultrie and the Cologne Anthropology and Global Studies programme for critical responses to earlier versions. In the development of the chapter, I acknowledge particularly the work of Michelle Pressend, Nazeer Sondhay, and Susanna Coleman; Davine Cloete; Nonhle Mbuthuma; Maria Puig de la Bellacasa; Kristina Lyons; Gathuru Mburu; Chris Mabeza; Claire Pentecost, Anna Tsing, Elaine Gan, and the wider team at the HKW Workshop in Berlin in 2016. At UCT, colleagues thinking about soil include Virginia MacKenny, Zayaan Khan, Nikiwe Solomon, Nteboheng Phakisi, Deanna Polic, Benjamin Klein, and Leonie Joubert, and comments from the audience at the session at the Knowledge, Cultures, Ecologies conference in Santiago 2017, and the Human Natures lecture in Sydney in 2019. My aunt Lynne Simpson introduced me to the work of Nathaniel Morgan—thank you. Frank Matose, Maano Ramutsindela, Tafadzwa Mushonga, Marieke Norton, Giulio Brandi, and James Murombedzi offered extended conversations on the militarisation of conservation. Thank you to Rob Nixon for hosting the discussion of this chapter at the Princeton Environmental Institute. An early version of the soil argument was presented at the "Agroecology for the 21st Century" conference in Cape Town, which was hosted by Rachel Wynberg, and published in the *Daily Maverick* in February 2019.

Chapter 5 began as an anthropology class field trip to see people encountering baboons at the side of the road in 2013, and grew from there. Thanks to Jenni Trethowan and Ruth Kansky for teaching me what you did about baboons; to Donna Haraway for her extensive comments on an earlier version; to Gregory and Lindiwe Mthembu-Salter for hosting me in Scarborough; and to Danny Solomon for sharing his PhD material. Early versions were presented at the Spaces of Attunement workshop at Cardiff University in Wales; also at the Indexing the Human series of Stellenbosch University's

Sociology and Anthropology Department; and at the university formerly known as Rhodes at the Association of University English Teachers of South Africa.

Chapter 6 has a long history in a fisheries anthropology project dating back to 2010. Sincere thanks to those who thought with me as this particular chapter developed: Lance van Sittert; Barbara Paterson, Marieke Norton, Kelsey Draper, and the fisheries anthropology team; Bruno Latour for his responses to earlier papers in this field and the invitation to present the lobster material at one of the An Inquiry into Modes of Existence (AIME) workshops in Paris in 2014, where I was grateful for comments from Martin Girardeau, Vincent Lepinay, Christophe Leclerq, and Isabelle Stengers. Jo Barnes, Leslie Petrik, Melissa Zackon, Nikiwe Solomon, Cecilia Sanusi, Adeola Abegunde, Mark Jackson, Neil Overy, and Jean Tresfon continue to courageously document river and marine pollutants that end up in Cape Town's urban oceans. Thanks also to Serge Raemaekers and the Abalobi team; Selene Smith for her time in Langebaan; Jackie Sunde and Kim Prochazka for comments on earlier versions; Tracy Fincham and the team at Kaskazi Kayaks on the issue of clean seawater; and Coleen Moloney for trusting that the work would go somewhere useful. The final chapter took form in response to Elspeth Probyn's invitation to present this at the Sustaining the Seas conference in Sydney in 2017. Parts of this chapter have been published in *Reset Modernity!* and in *Catalyst*.

The conclusion—"Coda"—developed from extensive discussions in the graduate courses "Science, Nature, Democracy" and "Researching the Anthropocene" from 2015 to 2019. I am grateful for all the comments on its presentation from participants in the EHS seminar at the University of Cape Town in August 2019.

The bibliography owes its finessing to Thomas Cartwright.

Finally: my sons, Jordan and Jonathan, were willing (and sometimes unwilling) participants in seven years of research and writing trips, and celebrated with me when it was all done. Thank you: I treasure all that we have been able to learn together.

Near Cape Point. Photo: Lesley Green

DIFFERENT QUESTIONS, DIFFERENT ANSWERS

Baboons. Porcupines. Otters. Lynx. African genet cats. Crayfish. Sharks.
Dusky dolphins. Killer whales. Southern right whales.
Seals.
Owls. Fish eagles. Black eagles. Sugarbirds. Sunbirds. Oystercatchers.
African penguins. Black-shouldered kites. Rock kestrels.
Harlequin snakes. Puff adders. Rinkhals. Cape cobras. Mole snakes.
Olive house snakes.
Bloukopkoggelmanders.
Tortoises. Baboon spiders. Scorpions.
Stick insects. Cicadas. Praying mantis.
Duikers. Steenbokkies.
Copper blue butterflies.

These are some of the 351 air-breathing creatures that traverse the edges of
Cape Town, South Africa, amid the suburban islands of the south peninsula
around which the Indian Ocean swirls into the Atlantic. A fence crosses from
one ocean to the other, marking the edge of Cape Point Nature Reserve. The
fence stops the eland, the bontebok, the rooibokke, the ostriches, and the
law-abiding. To pass through the gate into the reserve, I need an annual Wild
Card that costs me more than a ten-year US visa, plus an extra card for my
bicycle, and extra for snorkelling or fishing or staying overnight. When I applied for my Wild Card, I was also invited to marry a staffer of South African
National Parks (SANParks), since the system had no variable for a solo par-
ent with children. The staff member on the line from Pretoria suggested that
I put in the identity number of the desk attendant under "spouse." I declined
the offer of nuptials, however generous, so according to SANParks records,
I'm married to my sister.

But I'm getting away from my story, and in any case, today the Wild Card has been forgotten at home, and I'm at Cape Point with my Cannondale that leaps forward like a Lamborghini when I put pedal to metal. Notwithstanding my offer of every identity card that I have in my possession, the only visa that SANParks will accept is my Visa credit card. The morning sun dazzles the Indian Ocean while my card and the card machine chatter away about my bank balance. A slip prints, and I sign. The woman at the toll till calls over my shoulder to her colleague behind me: "The Russians are here." Her tone is flat. The same words would have scrambled the South African Navy in Simonstown twenty years ago, but today it is just business. Black Mercs, Audis, BMWs sweep past the official roadway via a side entrance, gratis: no Wild Cards, visas, or Visa cards needed. They are here on BRICS business, no doubt: to negotiate for nature in the form of uranium, methane, undersea oil and gas, elsewhere in South Africa, for the Brazil–Russia–India–China–South Africa alliance. It must have been their day out to come and see where the two oceans meet, take in a bit of fenced-off nature—something different from discussions about GDP, nuclear power, and the BRICS Bank.

Their convoy leaves, and my cycling companion and I work our way around the hill in the still morning air, along the miles of roadway that is mercifully tour-bus-free this early on a Saturday morning. At the end of the road there's the Cape Point lighthouse and the blessed coffee shop. We lean our bikes on a bronze baboon statue as the first tour bus comes in with a whoosh of brakes. A clutch of Germans is ushered in by a tour guide with bottle-red hair, a gold necklace, and a sunbed tan. Heeled in Nike, crowned in Ray-Bans, and hung with Nikon, the visitors wear the dust-free, sweat-proof raiments of the duty-free perfumed classes—khaki here, apricot there, a dash of white and gold; hints of leopard; a whisper of zebra. They hover like bumblebees around their sushi and champagne when out of nowhere a baboon with a baby dangling from her belly darts in and makes off with a fistful of sugar sachets. Mayhem erupts. With brooms and mops, the staff charge the German line. Mother and baby hop over the electric fence. A ranger with popping buttons puffs up the koppie (small hill) to give chase. Amid the general hilarity re Close Encounter with Baboons, the cappuccino sippers open conversations across the customs attending table borders. A South African at the table next to me says hi. I learn he is a helicopter pilot who exports agricultural materials to India. He and his brother bought a farm in the province of Mpumalanga near the Kruger National Park three ears ago, but they're still waiting for their licence to quarry coal. "The prob-
not the extraction," he says. "It's what comes after. You have to restore
ᴛ."

As I'm cycling home, there's not much to see—the tour buses are out, so the animals are not—but beyond the nature reserve fence there are baboon monitors in the streets of Scarborough, their paintball guns covered. Dressed in bright yellow traffic vests atop bush-green uniforms, they're the wildlife riot police, aka Human Wildlife Solutions.

I'm curious. When I arrive home, I look up their website and find in their annual report the names of the local baboon troops: the Waterfall troop, the Misty Cliffs troop, the Da Gama troop, the Groot Olifantsbos troop. The map of troop turf backs onto Da Gama Park, Red Hill shack settlement, Simonstown, and Smitswinkel Bay; their land goes into Cape Point, where there are another four troops, touching the residential areas called Scarborough, Misty Cliffs, Ocean View, Masiphumelele, Capri. I'm intrigued, and scope the area out on Google Earth. A wider range of South African publics is hard to imagine along any one road. How do different ideas about baboons and other creatures play out in everyday life, I wonder: do some neighbourhoods love the wildlife in the region? Is there a bushmeat trade? Are children safe from the rangers' paintballs? What about baby baboons and juveniles? I pore over Google Earth, wondering what baboon troops and human neighbours think of one another across the edges of the city and the wild.

A few days later, I'm back in a car to drive the circle of roads here that ring the south peninsula section of Table Mountain National Park, starting at Da Gama Park, with its abandoned military buildings and its ship-shattering cannon pointing out over Simonstown. Its concrete fence posts are slowly splitting open as the steel inside rusts in a slow-motion argument between chemistry and property. The top road leads over into Welcome Glen, where hard times announce themselves in patches of rust and you can buy sacred crystals at bargain prices from the Scratch Patch, where agates and quartzes are tumbled and polished and sometimes dyed turquoise. Next door is the old Marine Oil Refinery—bulldozed and awaiting a mall, but still bearing the name that was a polite term for the business of boiling down southern right whales. Around the bend and over the hill is Capri, home to middle classes, with wind-shredded exotic palms in streets with names like "Bermuda Way." Across the road is Masiphumelele, where over the past few years, a child in my son's class has lost his home three times in shack fires.

Some weeks later I join a hike, walking Cape Point to Kommetjie for two days, with a group of researchers who are there to think about Table Mountain and its many natures.

Day 1: I'm astonished at how hard it is to walk on the mountain, clambering over rocks. My gear is perfect: a light pack, great boots. The problem is in my centre. I feel like a Picasso painting. I'm used to roads, pavements, stairs, where I don't have to think about my feet. To walk here, I need to feel my core; but I can't—I am walking feet topped by thinking head, fending off images of an ice-cream cone about to lose its scoop. Most of my energy goes to balancing; trying to feel that body core that has gone AWOL after twelve years of child-rearing, a divorce, and an academic day-and-night job.

Day 2: The wind has moved in. Great gusts of icy south-easterly winds blow up from the Antarctic. My backpack is like a sail. I bob around the mountain trail like an insect on a car aerial. My legs are sore, and I'm the slowpoke at the back. Keeping going is the focus. A rhythm finally settles in, from my feet to my being: a refrain that keeps me walking. It's a song—a rap—that rises from nowhere, but I can feel it in my belly, and, most important, I can walk it. Greet—the earth—with e-very footfall. Greet—the earth—with e-very footfall. Greet—the earth—with e-very footfall. Greetings become caresses—gratitude for the gift of a secure step. I walk more gently; wondering what would be different if greeting the earth with every footfall was how I lived all the time . . . or perhaps if that was how everyone lived. Still focused on balancing and not slip-sliding away, I clutch onto the mountain with my feet the way a toddler clutches the side of a cot. The earth owes me nothing. I owe it everything.

The trail takes us past the back of Ocean View, home to apartheid-displaced fishers and their families, next to a farm that used to feed the sailors who arrived in Simonstown in the 1700s. It's warmer now. The sun is higher, and west of the ridge, the southeaster no longer rips. The view across Ocean View, Noordhoek, and Hout Bay is a tiny vista of the immensity of humans being together in a city whose edges form a wild space like no other. Multiple publics; multiple species; living despite the earth, despite each other, all navigating high walls, electric fences, security gates, and security systems. Here on the Cape Peninsula—my home for thirteen years—earth still gives life to all, even if our electricity comes from the sun's energy stored in the geological era called the Carboniferous, before the Jurassic, via coal quarried from farmlands in Mpumalanga, making South Africans among the worst per capita polluters on the planet. The Russian discussions on our energy futures are based on rocks that date back to the days when Africa was part of Pangea the supercontinent, and the Karoo was at the South Pole. Their business plan appears to have been drawn up despite warnings by earth scientists and climatologists that fossil fuels and nuclear radiation have changed the planet enough to warrant naming our time a new geological era.

The earth scientists call it the Anthropocene, a time when "global human"—Anthropos—has changed the earth's system of energy and chemical flows. They are wrong, of course: it is not universal humanity that has done this, but the societies caught up in a globalised economic system based on natural resource extraction and capital accumulation. My colleagues in the social sciences and humanities prefer the term "Capitalocene" for that reason. The irony! Russians and liberated South Africans are advancing the Capitalocene: negotiating the extraction of uranium despite the fragile economy, despite our fragile ecology. Political time—the five-year election cycle—is now also geological time, a period that changes the planet, forever; leaves a stratigraphy in the rocks, in the ice shelves, on the sea levels, on species that live or die. Law has a geological effect. So does philosophy. The social contract that undergirds modern democracies globally has produced a new age of extinctions: the loss of 53 percent of animal species since 1970, even in the age of conservation fences and national parks. Perhaps future archaeologists will read ours as the Age of the Angry Earth: a time of failed fences on an angry planet; a time of social contracts and constitutions that are cutting off from the tree of life the branch on which humans are living with barely a thought for negotiations over the future of our multispecies companions and our geological soulmates—rocks, lakes, atmospheres, oceans—in our humans-only parliaments.

What would it take to negotiate a truce with the earth, in the South Africa that entered global history because here at the Cape of Storms was a mountain that yielded springs of fresh water all year long?

Do we need the idea of the wild, of green, to save the planet?

Negotiating a truce with planetary systems and local ecologies has failed. The ideas of our times—environment, Wild Cards, ecosystem services— have not provided more than a few ecological zones amid a permanent war on ecology. And those zones are overwhelmingly preserves of elites: whites like myself who own Cannondale bicycles and would one day like to take the kids camping there, in a privately owned SUV.

Cycling around the peninsula, weekend after weekend, taught me that what I have been taught to see, and what I expect to see, and what I have learned to name and connect, did not give me the tools to "think" the connections that my bicycle was making, slow spoke by slow spoke. Pedalling in the early morning crisp air out on the peninsula, from the shanty settlements at Masiphumelele and Ocean View to the extreme wealth at Misty Cliffs, and across regional histories and geological times and municipal elections and cups of coffee, was a weekly provocation for about a year: surely, it seemed to dawn on me, there was another way of thinking; another way of working together; another way of living with the earth . . . another way of "thinking Cape Town."

The available options for thinking about the environment seemed to be these: an impossible romanticism, evident in so many greenie projects, where nature is paradise; or a modernist idyll in which Nature is where Natives come from, and They Have Nature But We Do Not (unless we have that SUV). Paradises Lost. Paradises Found. Paradises built with Parking Lots outside Cultural Villages in which humans play, for a fee, the-culture-of-humans-in-nature. A sign in the change-room of the outdoor sports shop: "Our company provides for the outdoor market and the outdoor aspirational market." Could living with Nature be not Extreme Sport or Wild Culture or Aspirational Market, but the Home version? The Greek word *oikos*, for "household," gives us the words "ecology," "economy," and "ecumene"—being together. Can we restore the planetary household *oikos*: in which earth, soil, species are together ecology, economy, and ecumene? Might our en-vironment become also our in-vironment, in which we recognise the geology that forms our bones and our legal system; our households and our food supply chain? Why is it so hard for us to imagine that future archaeologists will see in our bones the Big Farmer and Big Pharma that fed us the preservatives and pesticides and persistent organic pollutants that made us as sick as our soils?

Thinking of nature as something that belongs in a reserve is an idea that belongs in a museum. Notwithstanding all the environmental science we have, we're lost: we don't know where we are. We nature-lovers don't know how to live in the "nature-free" world of commodities that we've made on our planet.

When you're lost, you retrace your steps, as best you can. Go back to places you've seen before; exploring the routes in and through them again. Ask: How did we get here? What pasts are present, and what futures are forming? What connections exist that I didn't see before?

———

Contorted bodies. Pain etched in faces. Headless girls bending over backward, breasts to the sky. Mothers conjoined to babies. Endless images of entangled selves. Bizarre stone versions of the big beasts—elephant, buffalo, lion, giraffe, zebra, a massive hippo with shark teeth. Endless whale tails. I'm back on the road to Cape Point, in a car this time. I stopped at the stone carv-
· next to a muddy tour bus stop—always made a mental note to park and
¹ᵥ's the day. Wandered through them. They're higher than doors,
I look for something that speaks of a connection with the
onhood in pain, alongside bizarre Afro-kitsch versions of
.rnated in soapstone. The agonies of Frantz Fanon; of Aimé

Césaire's poetry; of traumatised selves; the pornography of Europe's gaze at Africa, cast in stone. I feel ill. Note to self: *Process this later*. A refugee amid scattered body parts turned to stone, I flee to the shop next door—Red Rock Tribal Art—and, from its name, expect a version of the cultural village. But I am surprised, meeting in it an owner and traveller who has an eye for artists whose representations speak of humanity | animality | earth that I recognise; that I could hope for.

Is it the white South African in me that feels such relief at finding a connection to nature that I can relate to? Is the pangolin that I buy, in the end, a romantic naïveté? I don't know anymore. Millipedes, carved chameleons, papier-mâché springbok heads designed to provoke an ironic guffaw at white hunters' fetish for stuffed heads on walls. Amid the Asante stools and kente cloths, a stone pond with algae and tadpoles and gorgeous succulents, I still feel ill. I'm staggering at the assault of trauma in the roadside soapstone. But the artists' nature assuages me, calms me. Restores my reality. I would sob with relief, if I could. But it wouldn't be normal, so I don't.

Shop owner Jules and I talk for a long time, about everyday things, about the environmental Greens, real and imagined, of Scarborough, about living ecologically. "The soil is our blood," she says. "I don't understand people who use industrial chemicals to clean their homes." My innards stop gasping. I take some pictures; take my leave. Go up to the restaurant next door—there's a guitarist whose music I love playing on Friday. I nod, I smile, I take pictures: landscapes with farmhouses, landscapes with shacks, landscapes with shacks and sea.

I drive up to the military base that's now the Table Mountain National Park Marine Protected Area Signals Division, behind tangled bursts of barbed wire that I step over to take pictures of the "Stop Crime" signboard in the dust. I photograph and photograph and photograph, taking 360-degree shots on a rock plinth, and stagger off at the last one. Across the road, I drive the short left down to Brooklands. A glade of blue agapanthus catches my eye. They're not indigenous here; must surely have been part of the old settlement. I see no house nearby, though, nothing to suggest this was once a garden. Carry on slowly down the perfectly tarred road, around a dad with four novice skateboarders age eight and up. Down to the Brooklands Water Scheme, and take the gravel road below the dam wall as far as I dare. U-turn, gingerly, a seven-point turn. Back down the tarred road, the blue catches my eye. Stop. Park. Walk. Carefully pick my way through the lawn grass that has gone wild and become mountainside undergrowth, checking for puff adder diamondbacks. Wouldn't want to get bitten here—no one would find me for a week. A few footfalls take me to plinths of concrete. They're made of broken bricks. One of them has bathroom tiles on it: white, with a royal navy trim that matches the grove of agapanthus. They're gorgeous. It's mid-December, and next year they will be remembered by those who once lived here, in Brooklands, who suffered the race-based forced removals of the 1970s, and who, *Groundup* magazine tells me, still meet here once a year on Heritage Day.[1]

Months later, after the statue of Cecil John Rhodes has been removed from the campus where I work, and I have read and heard black students speaking of their struggles with coloniality in the curriculum and in the everyday, I begin to understand the pain in the sculptures for tourists on safari who will put down money for a white gaze carved in ebony, not ivory.

———————

The Red Hill drive down is steep, a series of switchbacks that start at the camo-painted cannon near the entrance to Da Gama Park. It overlooks Simonstown naval harbour, where our national debt shudders in the waves. A cyclist has stopped, gasping at the last hairpin bend, standing on the tar astride his bike, leaning on his handlebars. He's too exhausted to look at me, or take in the "amper daar" ("almost there") that I want to offer him. I drive down slowly, looking for the faded old "bokkie" sign that warns against mountain fires. It's near the bottom, angled at those coming up, and covered with graffiti—but you can still see the Disney Bambi eyes in front of a mountain blaze, with the legend "Look What You've Done!" It's an accusation: *You're Guilty, Humanity*. It trumpets the attitude of those who believe that to save Table Mountain National Park's nature, we need to Keep All Non-paying Humans Out.

It's ironic, I think—or is it?—that the naval harbour below is a space permanently ready for war. The war in the harbour would be against foreigners in the name of the nation's people; but the war on the mountain is against non-paying humans in the name of the nation's nature, and what would have been a colonial war over regional resources two hundred years ago (even thirty years ago) is now a polite negotiation over finance managed by the people we elected to ensure liberation. Within this way of thinking, the people who throw up stumbling blocks are "unpatriotic." Ask the people of Xolobeni and Lutzville, who have been battling an Australian mining company which wants to mine their titanium-rich sands. Surely that's a version of civil war, in which government opposes its own people who want to live in their piece of the planet, and not off it?

Our national debt bobbing in the waves—the Corvettes and submarines—is core to the arms deal that crippled South Africa's polity in 1996, choking a newborn democracy with corruption: the birth (berth!) of "state capture" by corporate multinationals willing to grease palms. When I'd been at Navy Day with my sons, one of the vessels was already being used for spare parts—its name was plastered all over the equipment on the operations deck of its sister ship.

The cost of these vessels is responsible for the impoverishment of the naval dock as a whole: roads are in disrepair; barbed-wire fences are falling down. There's the palpable absence of a coast guard, in favour of a military that will fight imaginary wars that will make politicians great again, even though in 2012 South Africa's Exclusive Economic Zone was signed into international maritime law, giving the country exclusive access to the ocean's life and mineral resources for two hundred kilometres out to sea, along all three thousand kilometres of our coastline. As of that moment, South Africa had more marine area than land, but it could be patrolled only by a handful of multibillion-rand vessels of war that the navy could only send to sea for a few weeks in some financial years, according to submissions made to the corruption-investigating Arms Procurement Commission.[2] They are not the naval assets we need—and we can't afford their highly specialised maintenance.

The signs of an under-resourced everyday in the navy are all round in the Table Mountain National Park Signals Division, even in the peeling sign that greets me outside a fragile building on the mountaintop naval base. I photograph away, at the abandoned sentry post. No one stops me or asks any questions. I hear footsteps: a woman in a park ranger's uniform trudges across the disintegrating tarmac from one building to another. Her arm is strapped in a wrist guard. The occupational hazard in Simonstown's naval base in the electronics era is not going overboard, but going keyboard. The knowledge economy is a virtual world, and bodies don't do well in it.

Down at the docks, leisure yachts almost touch sides with the Corvettes. At the restaurant near Da Gama Park, there's a sign bearing information about dolphins and otters. A bit further on, some men fishing for supper. A diving shop. A kayak shop. I stop to start writing this at Just Sushi, overlooking the harbour where whale-watching charters dock and tuna charters come in to weigh their catches at the three-metre-high fish hoist, and the Salty Sea Dog makes the best hake and chips in the world.

In the weeks that follow, I see that uneasy alliance between the South African Defence Force and the South African nature conservationists mirrored in rhino management. The November 2014 cover of *Pop Mech*—the magazine more formally known as *Popular Mechanics*—shows a rhino made of the parts of a tank for a military hardware expo. A bumper sticker on a white Mercedes-Benz at my local petrol station in Newlands announces, "Save the rhino, hunt a poacher." The smugness of it makes even me—generally benign—want to crack the bumper. In what world do they live that makes it okay to speak of hunting people? Do they not know the same words were

Intro.1 | Bumper sticker, Newlands, Cape Town, 2016. Photo: Lesley Green.

used in the Cape only 125 years before, and one consequence was the genocide of the San? Can they not see that their "green" is the new white? That
people who once lived off the land and whose stories were full of accounts
of living with animals and learning from them were forced off the land in a
time when nature reserves were set up along with native reserves? And that
the high-budget game lodge that sponsored the sticker—Palala—is a hunting lodge targeting wealthy professionals? How has it been possible that conservation has come not only to speak of hunting people without a blink, but
to turn the phrase into a feel-good "that's-who-I-am-too" bumper sticker?

Violent conservation is not unique to South Africa. Rosaleen Duffy notes:

Conservationists increasingly talk in terms of a "war" to save species. International campaigns present a specific image: that parks agencies and
conservation NGOs are engaged in a continual battle to protect wildlife
from armies of highly organized criminal poachers who are financially
motivated. The war to save biodiversity is presented as a legitimate war

to save critically endangered species such as rhinos, tigers, gorillas and elephants. This is a significant shift in approach since the late 1990s, when community-based natural resource management (CBNRM) and participatory techniques were at their peak. Since the early 2000s there has been a . . . renewed interest in fortress conservation models to protect wildlife, including by military means.[3]

On the internet, "white green violence" is not hard to find. The new White Man's Burden—saving African nature from Africans—is stark in the work of VETPAW, aka Veterans Protecting African Wildlife, an NGO in which US military veterans get to put their theatre-of-war experiences in Iraq and Afghanistan to use on the African continent to protect African wildlife. It has set up "missions" in Kenya and South Africa. Many of the photos on its Facebook page would be at home in *Hustler* magazine, as their organisational diplomat is so often shot in a bikini and camo fatigues, draped in rifle-bullet necklaces. The television version is probably not far off.

What is the Black Man's Burden? For Joel Netshitenze, writing in 2016,

> Blackness cannot be defined by howls of pain in the face of a stubborn and all-encompassing racism. As during the struggle against apartheid colonialism, it should define its mission as being to resist, to persuade, to teach, to cajole and to lead in the name of an all-embracing humanity. . . . Blackness's "attitude of mind and way of life," to paraphrase the proponents of Black Consciousness, should turn grievance into strategy and action. The Black man's burden in today's South Africa should find expression in deliberate self-definition and self-assertion, in pursuit of excellence and acting as a force of example on what it means to be human and humane. Core to such an approach should be an ideal higher than pursuit of equality with whites. It should be about a new civilisation, "thoroughly spiritual and humanistic," which takes "its place . . . with other great human syntheses," "giving the world a more human face," to quote Seme, Luthuli and Biko, respectively.[4] Blackness should position itself as an integral and equal part of humanity, in dogged pursuit of excellence on a global scale.

Netshitenze's piece was greeted with outrage and dismissal by the student protesters whom he was addressing. His words were cerebral. When #Fallist students spoke or wrote, in 2015, they wrote of pain.

———

Where Kommetjie Road meets Slangkop Drive lies Ocean View: a place of forced removals, fisheries activism, and lobster poaching, whose trash was buried in the sand dunes up until the 1980s at a place known to surfers as the "Crayfish Factory." Across from Ocean View is Imhoff's Gift, the old farm that grew sailors' food, where now the monied can ride a camel, order another cappuccino, or feed peacocks. A horse farm has put up signs warning riders out on Long Beach to beware of the muggers in the sand dunes.

Closer to the Atlantic shoreline is Kommetjie: a place of surfers and fish traps from the earliest human settlements. The remains of some of them lie in boxes in the back stacks of the Iziko Museum, taken from caves and shelters all along the peninsula, where seashells can be found above the highest sea levels known in our lifetimes. Theirs is the legacy of an earlier, warmer era, and if archaeologists are correct, the perlemoen (abalone) they ate changed humans' evolutionary destiny.[5] Thirty killer whales beached near here a few summers ago, a little north of the Soetwater ("sweet water") campsite, where a few winters later, Nigerians and Somali refugees fleeing Cape Town's xenophobia were beached in tents at the sea's edge.

The road passes Misty Cliffs, where the sea mist hangs thick in the air all day; only sometimes burned off when the sun shows over the hills at eleven in the morning. Then there's Scarborough: where Scarborough Keepers is the name of the neighbourhood watch, and their warning board is not the usual suburban crosshairs of a gunsight or a crusader's sword and shield, but a circle of hand-holding humans.

I'm back on my bike, heading to Simonstown. It's midsummer. There's a shortcut over Red Hill at the informal settlement—shacks—and just past it there's that left turn to reach Brooklands, where dune moles excavate shards of porcelain teacups. The road curves around back to Da Gama Park, where 1970s prefab houses with Vibracrete walls and yellowing grass verges are interrupted by the same dune moles.

The hill above Simonstown is steep: grasses, rocks, and ferns amid shards of brown beer bottles and the occasional blue government-issue condom wrapper. A bit further to the left, there's a headland, and in its crevasse is a waterfall: home of the Waterfall baboon troop. The troop used to nest there, a Zimbabwean economic exile once told me. He makes a living selling teak fruit bowls made from Zimbabwean forest trees to the tourists passing the gates of Cape Point. Over the years of cycling here I've gotten to know his yellow bakkie, as small trucks are known in South Africa. One day I asked for him, and was led to a bush-green steel chest. Someone called his name and lifted the lid, and he climbed out, rubbing his eyes, extending his hand in greeting: "Hi, I'm Austin." He'd lived in the shacks on Red Hill since 1999. The baboons were terrible, he had said when I'd asked: they tore apart chickens, raided vegetable patches, trashed groceries. He said they were terrorising

the Red Hill shacks, ever since Nature Conservation cut down the Australian eucalyptus trees where they used to nest, to make way for indigenous
vegetation. The point is a contested one, I learned later from baboon observers, since the troop had several sleep sites. Why was it, I wondered, that
conservation officials were the target of this shack dweller's ire?

We chatted some more, then I cycled on, across the peninsula toward
Scarborough. The baboon monitors of Human Wildlife Solutions are like
bush fire-fighters holding a line above the vineyard. Or riot police. Most of
the humans here welcome the monitors. Do the wildlife?

That December I spend a few weeks living in Scarborough, to try to get a feel
for life with fence-crossing baboons. On Tuesdays the wheelie bins are out
for refuse collection. "The baboons are back," sings out a child to her mother
when she sees one. I follow her gaze to find the monitor, and then follow
his stare up the road. There's a young baboon greedily tearing apart a Pick n
Pay bag that peeps out of a shattered rubbish bin, its locks still intact. "You
just can't get ahead of them," my landlord tells me. "First we put locks on the
bins. Then the baboons learned to knock the bins over to pop the lids off. So
everyone laid their bins on their sides, but the baboons have learned that if
they jump on them long enough they split open. We just can't win."

As I continue on foot, my route takes me into the top road at Scarborough. "They've gone that way, behind the house," calls a woman in a lemon
chiffon dressing-gown from her balcony to the baboon monitor wearing
bush khaki. If she had curlers, it would complete the picture. The monitor
moves up the slopes, around the rocks, over a building site; his paintball gun
is covered with cloth so as not to upset their opponents here, who have listed
on their Facebook page the names and photographs of the terminated *Papio
ursinus*.

Finding a language in which to think and speak about "nature," "green," and
"environment," outside of the already written and the already said, is like
riding a bicycle through the bush instead of taking the road. A road would
be easier—but tarred ways only take you to what has already been mapped.
When you have already ridden down the written-down and not found a way
to answer the questions you have, you navigate the slow stuff, hoping that
there might be a different insight; different ways of seeing how the earth
relates to this age of the geological effects of humans that geologists call
the Anthropocene, in this strange and beautiful peninsula at the Cape c

Good Hope. When you are navigating an unthought trail, there are no auto-completes for words or thoughts.

Ours is a difficult moment in a difficult season. Telling a different story can begin, perhaps, from a self that has the courage to make—to scope—its land and sea and history in terms different to those with which we were raised. The question this book seeks to address is not how "we ought to do more" for the environment, but rather as an academic, a sixth-generation white South African, and a mother of two, how to find a way into ecological situations more thoughtfully than the auto-completes allow.

How to be more present to expunged pasts? How to imagine what others have felt in these places in other times, in other disciplines, as other species, as the earth itself? What is it to be present at the massive ecological destruction of our times, amid the pressing sense of the failure of "scientific nature" to find a voice in South African political life that can speak in voices other than the tones of "whiteliness" (to use Marilyn Frye's term): the expert, the judge, the martyr?[6] How to feel and think, and hold on to relationships that matter in a time of neoliberalism where all relations that matter in "the economy" have been translated into dollar values for "the market," while the rest have become invisible? How to enter into the life of the earth with those who share my space in this city, from my 1856 house that was built as a shop with the mountain's rocks and wood and clay to sell beer made from the mountain streams, six years before the last Cape Town lion was shot a mile away in 1862?

Sixteen decades later, that Westvoord Farm Shop that is now my home has been nipped and tucked by successive sell-offs into a suburban plot of 260 or so square metres, and the land that once pastured Khoe cattle, and then those of Van Riebeeck's and Rhodes's neighbours, now schools teenagers at Westerford High School. The ecologies of the city are fiercely contested: water, baboons, lobsters, streams, sanitation, air quality, alien plants, the daily discharge of the city's sewage into the ocean. Yet the night camera traps set up by my colleagues in biological sciences have yielded pictures of African genet cats prowling in the stream bed off Lemon Lane a few blocks away, along with Cape clawless otters. Radio collars on caracals find that six of them live above Rhodes Memorial, just a short walk from my house, a short left from the university, and a few hundred metres from Westbrook, the presidential mansion. And sometimes, at night, these caracals cross the M3 highway along which the good citizens of the southern suburbs crawl into work in their cars by the hundred thousand in the mornings and then home again in the evenings, perhaps in time to get their dogs and their Wild Cards, to go for a walk on the mountain's forested slopes.

Writing a book such as this began with a glimpse of a set of things that need rethinking together. That glimpse came as I cycled around the Cape Peninsula, encountering not only things and places known and protected by the environmental sciences, but also the sense that those things themselves—national parks, marine protected areas, baboon troops, streams, lobsters, fishers' struggles, urban farms—were also haunted. They had an existence in science and in environmental work; they also had another mode of existence that was unseen and unremarked. What's invisible are the stories in which they were once embedded in sets of relations, before they were reduced to words and objects, and the stories of their relations forgotten. The sense that concepts, in the world of natural science and environmentalism, are haunted is captured by Jacques Derrida when he plays with the word "ontologies"—the assumptions about what exists—and comes up with "hauntologies." Over time I began to get a sense that by understanding how environments became ghosted by their forgotten aspects, it was possible to find the place where some existences got lost, where they lost their ontological moorings, so to speak. In that moment, environmentalism became part of the era of expulsions, and of extractions driven by expulsions, and of the struggles against extinction in spaces of extraction.

How do you explain "ontologies" to a naturalist who thinks that his or her idea of "nature" is what is "just there," without thinking about how that idea of nature or environment came to be?

When I was a teenager, I was lost in a world of fundamentalism: the world of the incontrovertible Word, where the Word was God, and the Bible was literally true. It took me a long time to think my way out of that, and in consequence I'm somewhat allergic—or at least hypersensitive—to ways of making the world as if that way of world-making is the only way to make a world (or even worse, as if it is the only world).

Enter "Natural Science," stage right: a way of world-making with Things, as if no human mediations of those things have occurred. A piece of charcoal, for example, might be soil on a farm, or dirt in an office, but gold in a carbon-dating laboratory. It all depends on the relationships and technologies in which the charcoal is situated. I'm a great fan of science, but I'm not a fan of the political authority that "Science" (with a capital S) takes when it presents its findings in politics the same way as the Discovery Channel presents them on TV: uncontestable, heroic journeys into "the" singular and transcendent truth. I think the best scientists I have worked with are those who know science as a space of permanent doubt, permanent questioning and self-questioning, and who are open to rethinking their own situatedness and ontologies.

I was privileged to do some of that kind of work with chemist David Gammon and botanist Timm Hoffmann, in dialogue with anthropologist Joshua Cohen, exploring how healers in Namaqualand approach plant medicine, and how that differs from modernist approaches to medicine which have trained us to seek out molecules of warfare: antifungals, antibiotics, antivirals, anti-inflammatories, antihelminthics. At the end of that project (described in chapter 3), Cohen had enough material to argue that healers were primarily looking for plants that restored "krag," or energy/vitality/well-being. That didn't mean that healers were not using plants as anti-inflammatories or antihelminthics, but it did mean that it was wrong to reduce what they did to what we had always assumed about how plant medicine worked.

That's ontology, right there: looking at how you look; what you name; what matters to you. These are the "things" that science draws attention to in political life. But when a scientist or a profession starts to believe that their way of knowing something is the only existence that that thing has, that scientist is at risk of missing what might matter very much to others, or the relations that are really important to that thing.

"Colonisation is thingification," wrote Aimé Césaire, one of the founders of postcolonial thought.[7] For Césaire, the only way to think in a world that had been stupefied by colonial taxonomies and techniques of naming natural and human resources, and in so doing extracting them from their local relationships, was to turn to the surreal. Neither religious nor scientific words worked for him. Poetry was for him what bicycle riding was for me: a way of slowing down, questioning the connections that had been taught, and erased, and reduced to things. And making different connections.

Finding a place from which to reconnect scientific ontology with its hauntologies is the work of this book, because we can no longer reduce the ex-es of our time—extinctions, expulsions, extractions, existence—to individual problems to be addressed apart from one another. This book is in part a rebellion against the authoritarian claim to transcendent knowledge that is contained within many expressions of environmental sciences, environmental management, and environmental activism. In particular, it is a rebellion against the way in which "scientistic" approaches, in South Africa, have come to serve as an authorising space for white authority. Where "science" is understood to offer the ultimate truth about "nature," and politics is understood to be the opposite of science (and therefore parliamentary debate is to be confined to society, economy, and culture), a partial connection with racialised authority (always denied in the name of objectivity) has been inevitable.[8] Yet the South African scientific community has (with very few exceptions) been unwilling to acknowledge, much less explore, its connections to the racialised history of claims to authority in this country.

This book is therefore an attempt to reclaim the space of critical enquiry in the sciences of South African environmentalism, navigating a path that welcomes and celebrates scientific enquiry, scientific achievements, and integrative thinking, and questions scientific reductionism and transcendence, and associated forms of environmental authoritarianism.

It is difficult to cut such a path at a time when multiple assaults on science are part of political, legal, and corporate life—and indeed in the social sciences too, in endless claims that science is a matter of identity politics. This book, chapter by chapter, proposes a "resituating" of the "critical zone" as one that embraces the encounter of humans, technologies, and modes of doing politics, with this planet's planes of existence in rock, water, and life. That "critical zone"—itself a term borrowed from soil scientists who are thinking integratively—is vital to the future of sciences, politics, and universities. If environmental science and management are to mobilise an environmental public in South Africa, they need a different approach.

———

The writing of this book began on my bicycle, trying to process the dis-ease I felt traversing nature and city, wealth and poverty, navy and conservation. As the idea of writing a book on several different Western Cape environmental struggles evolved, it became more and more apparent that it was not possible to think about nature without attending to the nature that people constitute, and our bodily entanglements with the places in which we live and have our being. For that reason, the opening chapters offer historical and contemporary studies of two places—Table Mountain and the Karoo. As those studies took form, I realised that I was seeing the changing relations of rock, water, and life in South African history—and that tracing these was a way to challenge the illusion that nature and society are separate. The knowledge of them, and their governance, neither could nor should be pursued separately.

With the arrival at the Cape of the mercantile capitalism of seventeenth-century Europe and its imposition of property ownership by military means, capitalism began to insert itself into the web of life (to borrow Jason Moore's phrase),[9] changing the relation of Table Mountain's rock with the Cape Town water supply. This study forms chapter 1: a tale of property ownership made possible by the new Cartesian scientific cosmos of property measurement and mapping, and the closure of lands and water held in a commons.

Chapter 2 traces the Cape colonists' relationship with water over the next three hundred years, during which the dry inland Karoo region became a site of struggles over who got to access water and live, amid the rise of windmills in the late 1800s, accomplishing in South Africa almost exactly what was happening in the United States at the same time: the defeat of indigenous people whose knowledge of water points in dry lands had made the difference between life and death. The hopeless struggle of the !Xam against the commercial extraction of aquifer water for sheep farming that windmills made possible is paralleled now in the almost hopeless struggle of sheep farmers against the almost inevitable rise of the fracking derrick that will bleed gas from stone. The chapter traces the insertion of capitalism into the relation between water and life in the Karoo. As in chapter 1, chapter 2 argues for a continuity of colonial forms of relation into the neoliberal present: posing the challenge that if science is to be genuinely decolonial, it needs to reclaim the web of life from financialised relationships.

Part I, "Pasts Present," sets out the multiple natures (that is, multiple accounts of the world) that constitute contemporary political life in South Africa. *These are part of the conditions creating climate crisis in the wake of settler colonialism and modernity, globally.* I've aimed to work with the ways in which histories of water, land, and mountain continue into current struggles over water and land and fracking, asking what other ways there are to think about "the nature" of the relations among rock, water, and life in Cape Town and the Karoo. Both chapters evidence the larger argument that there is not one single version of "environment"—in these cases, Table Mountain or Karoo—rather, how something is understood to be "natural" or "environmental" is an effect of the ecopolitical relations of a historical moment.

Part II—"Present Futures"—explores current struggles over nature, in student decolonial activism and the possibility of rethinking the debate currently figured as indigenous knowledge versus science via studies of plant medicine in Namaqualand. Thinking about environmentalism in relation to land struggles, chapter 4 travels through settler-colonial history in the Eastern Cape and returns to the Cape Town region with attention to current struggles over land restitution and soils, with particular attention to the struggle between developers and farmers over the Philippi Horticultural Area east of the city.

Part III—"Futures Imperfect"—includes a critique of patriarchal primatology by way of a focus on the management of Cape baboons in chapter 5. In chapter 6, I offer an integrated exploration of the ecological regime shift in the Cape's kelp forests, from migrating lobsters to harmful algal blooms and the extinction risks faced by abalone, before focusing on fishers' struggles and fisheries management strategies, and the form of environmental science and activism developed by fishers and colleagues in the Abalobi project.

The coda serves as a conclusion, reviewing the studies to suggest approaches to integrated scholarship that can contest the kinds of captured environmental science that dominate current environmental governance in the Western Cape, the province I call home, and the region in which all of these studies are rooted.

————

In working toward an alternative way of understanding ecology, perhaps most important was that cycling around the peninsula on weekends, and then writing these chapters about different environmentalisms during the weeks, entailed a birth to presence:[10] the kind of presence that is neither purely subjective in a world of feelings and impressions, nor strictly objective in the sense that it pursues established concepts and categories. As I was seeking a line of flight from ideas that are dominant because they dominate every moment of ecological imagining, the approach that I found most useful was to follow connections that seemed tangential. By working around received ideas rather than through them, it became possible to trace not only the evident forms, but the gaps between them—as my friend and fine-art colleague Virginia MacKenny had taught me to do in drawing: sometimes it is helpful to work with the spaces between things, not just the thing itself.

"Thinking athwart" a situation, to borrow the words Elspeth Probyn uses to reimagine seafood sustainability,[11] makes it possible to think outside of the available vessels: concepts that are already named and their relationships prefigured. To think athwart an established concept may seem to some to choose a state of drift. Yet conceptual drift can be a lifeboat where inherited concepts themselves are sinking vessels. Where climate-smart agriculture, for example, brooks no criticism of its romance with commercial agriculture's genetically modified (GM) seeds, which are sold to policy-makers as drought resistant, they are not paying attention to a damaged vessel. Under proposed seed laws designed to suit GM-seed manufacturers, farmers who share seeds will be criminalised, as will those onto whose land GM plant pollen blows. The gift economy of seeds, so central to African rural society, would be outlawed. Corporate seeds will be stripped of their ability to reproduce, and seeds may render soil unable to be used by any non-GM seeds. Escape is needed from the conceptual apparatus that makes such an approach appear economically viable, scientifically objective, and technically efficient. What is "subjective" and "objective" in such an argument is not a form of world-making I can share: I cannot accept that these proposed climate interventions offered by seed companies are based on objective, neutral science. The claims by that form of climate-smart agriculture regarding what is true, just, and valid, and what matters, do not converge with the kind of world I wish to be part of making. To build a climate future that is just requires thinking athwart the concepts that are presented as objectively real. The need, therefore, is to question how and why something comes to be understood and accepted as "objective."

For many anticolonial thinkers, including Césaire and Édouard Glissant, the "objective reality" offered by colonial modernity was unpalatable. Concepts like "natives" and "tribes" created an ethnological version of Africa and its diasporic people that had only a distant, partial connection to the lived experience of Africans. The writings of both Glissant and Césaire decentralise the cataloguing of objects—natural resources and species—that was the focus of colonial natural science, and emphasise the poetry of knowledge and the poetics of relation as modes of knowing and making knowledge that simultaneously transform the possibilities of being.[12]

French writer Jean-Luc Nancy's work on presence-to similarly offers an alternative to the fixity of a world of received subjects and objects. His work responds to the crisis of the rise in ethnonationalisms in the early 1990s, and earlier in that century. In *The Birth to Presence*, he reaches for a philosophy of knowledge that is able to speak to more than pregiven identities and representations. Presence, to Nancy, is the encounter and re-encounter with that which has always appeared to us as "obvious." In that moment, "the obvious" is transformed—as is the observer, because self-assured identity and authenticity are tangled up with assumptions about "what is obviously real."[13] Presence is thus a way of encountering the world that involves a continuous re-forming of self—a process Nancy describes as continuous birth. The value of this approach to South Africans like myself who seek to question ways of making environmental knowledge, and to revise the inheritance of settler colonialism and the certainties and objectivities attached to "whiteliness," is immense.

Colonial and apartheid South Africa were among the more extreme and definitive experiments in modernity: the creation of the binary world of subjects (citizens) and objects (nature and natives). As a society, South Africa is built on the absence required for objectivity to exist as a mode of relation. Risking what Nancy calls presence—to a situation, its elements can touch us and affect us differently. It makes possible a profound transformation of not only what it is to know, but also what it is to be. Reaching for that presence, without the subjects and objects of modernity to generate regimes of absence and negation, is helped via ideas inherited from situations and relations other than the modernist staging of reality. Modes of presence-to landscapes articulated by Khoe and !Xam archives and Namaqualand healers, among others, offer forms of relation that are not predicated on subject and object. In and around places and landscapes where environmental questions are vexed, the coming chapters pursue their fingerprints amid the bootprints left by "authorised" knowledge.

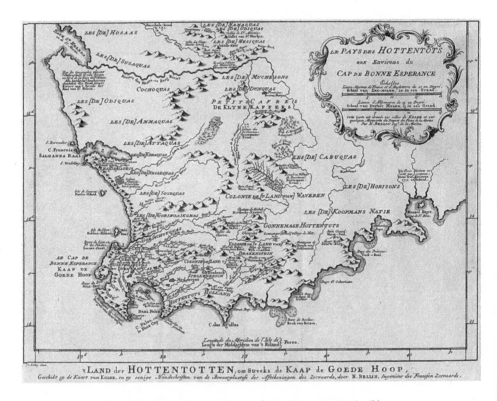

1.1 | Peter Kolbe's map of Khoena settlements in the Western Cape, Dutch edition, 1727.

PART I

PASTS
PRESENT

It is necessary to speak *of the* ghost,
indeed *to the* ghost and *with* it,
from the moment that no ethics, no politics,
whether revolutionary or not,
seems possible and thinkable and *just*
that does not recognize in its principle
the respect for those others who are no longer
or for those others who are not yet *there*, presently living,
whether they are already dead or not yet born.

—Jacques Derrida, *Specters of Marx*

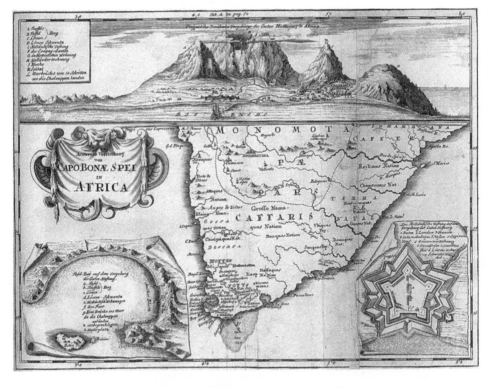

1.2 | Peter Kolbe's map of Khoena settlements, German edition, 1719.

CAPE TOWN'S NATURES

||Hu-!gais, Heerengracht, Hoerikwaggo™

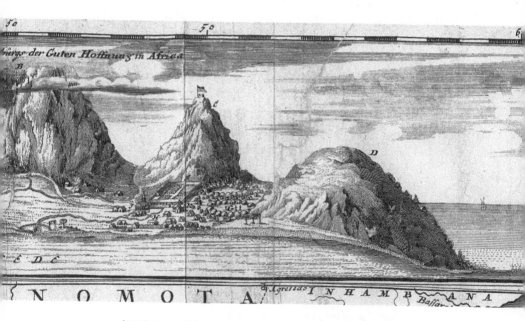

1.3 | Detail, Peter Kolbe's depiction of Khoena settlements in Cape Town, 1719.

"||*Hu-!gais*... is the name by which Cape Town is known wherever the Khoikhoi [Khoe, plural Khoena] tongue is spoken," wrote Theophilus Hahn in 1881, continuing:

> This name consists of two words, ||hū the root of a verb, meaning "to condense," hence ||hū-s, an old word for cloud. . . . !Gai is to bind, to surround, to tie, to envelop. ||Hu-!gais consequently means "veiled in clouds." And, indeed, every inhabitant of Cape Town will admit that this is a very significant name for "Table-mountain." We still say, if the clouds envelop the top of "Table-mountain," he has his "tablecloth" on.[1]

Now named Hoerikwaggo™ by Table Mountain National Park for its five-day trademarked hiking trail, ||Hu-!gais—"veiled in clouds"—was an observation of a generally prevailing weather; but also more than that.[2] ||Hu-!gais, Hahn noted, was also the name of the main Khoena kraal (circle of dwellings for people and cattle) on the slopes of Table Mountain and Lion's Head, above the castle that was built by Dutch colonist Jan van Riebeeck in the 1650s, and near to the streams and springs that flowed from the mountain rock.[3] So, ||Hu-!gais was both a weather pattern and a political name—and yet more than that. To understand the significance of Veiled-in-Clouds, you need to go a little deeper into Khoe thinking—itself not the easiest task, given that travellers and settlers who wrote about the Khoena tended to present either a romantic or a repulsive account of their Other.

A linguist, librarian, and traveller for over seven years with the Khoena, Hahn translated Tsūi||Goab, the name for what he calls the Khoe "Supreme Being."[4] Tsūi||Goab could also refer to "red dawn" or "red-daybreak," named for the red clouds in the dawn sky on a day of rain. The bringer of rain, Tsūi||Goab had other names too: !Khub (in Nama)[5] and |Nanub.[6] !Khub, argued Hahn, translated as "master" or "lord" or "rich man," though he makes the claim without noting the contradiction that the word *!khu*, to be laden with something, applied also to a pregnant woman in the word *!khi*.

Tsūi||Goab was also identical, writes Hahn, "with |Nanub, the thunder cloud [in which he dwells], and !Gurub, the thunder." |Nanub came from the root |*na*, "to filter" or "to stream." "It means especially that kind of streaming which a man can observe if he digs for water in the sand of a periodical river. That filtering and streaming together of the water . . . is |*na*. Therefore |Nanub is the filterer, the pourer, or to speak in South African Dutch, '*de Zuiverwater*,' an expression which well applies to the nature of the rainpouring cloud." In the same way as the Infinite was called the Red Sky, where he dwelt, Tsǔi‖Goab was also the Thunder, named for the thundercloud from which thunder came.[7] All three—Tsǔi‖Goab, |Nanub, and !Gurub—could be implored, "Let Rain!"[8]

Tusib was yet another name for Tsǔi-‖Goab or |Nanub. "*Tu* means to rain. Tusib, therefore, the Raingiver, or the one who looks like rain, who comes from the rain, that is, the one who spreads the green shining colour over the earth."[9]

Following Khoe dialects across Southern Africa, Hahn links Tsǔi‖Goab to the isiXhosa word "Thixo" (now commonly referring to the Christian god). ‖Hu-!gais was a place where the Raingiver could most often be found. In the *Dictionary of African Mythology*, Harold Scheub notes that Tsǔi‖Goab is also the figure of IKaggen in !Xam lore and was "the supreme being, the celestial god of the Khoi. He was omnipresent, wise, a notable warrior of great physical strength, a powerful magician. Although he died several times, he was reborn. He sent rain and caused the crops to grow and flourish. Tsǔi‖Goab was the first Khoi, from whom all the Khoi peoples originated. He was a creator, having made the rocks and stones from which the first Khoi came."[10]

With a linguistic link between Tsǔi‖Goab and clouds, ‖Hu-!gais—Veiled-in-Clouds—becomes a prevailing weather, a settlement, and a mountain where rock, cloud, the creator, rains, streams, and fresh growth are one in this rainy, wet, windy corner of an otherwise dry region.

Table Mountain took on a remarkably different description in Charles Darwin's writings in the 1830s. Focusing on rock alone, he observes, "Some of the beds of clay-slate are of a homogeneous texture, and obscurely striped with different zones of colour, whilst others are obscurely spotted. Within a hundred yards of the first vein of granite, the clay-slate consists of several varieties; some compact with a tinge of purple, others glistening with numerous minute scales of mica and imperfectly crystallized feldspar."[11]

Woven into Darwin's words is the measured excitement of the naturalist recognising a world-making fact: the volcanic granite that he had seen in Cape Town corresponded with that on the coast of Rio de Janeiro in Brazil—offering tantalising confirmation of the hypothesis that the continents of Africa and South America were once one. Nowhere in his account does he connect the mountain with weather or water, much less any divinity. That was not his concern, of course, for the nature of the history of the world was his interest, not the nature of a local ecology, much less a forgotten local knowledge that was generally represented as "belief" or "culture" not "knowledge of nature." But Darwin's approach also reflects the rise of a new kind of natural history: describing individual rocks, plants, and species in terms of the emerging science of natural history—a new universal science that overlooked its own situatedness in the larger political project of the time: building a science that expanded to the ends of the earth along with the empires of the time. Like the modernists who came before him and separated science from society and culture, Darwin in his science assumed that knowledge was based on observable things, devoid of connections to any society. Since his research questions came from a concern to produce an earth science that could link continents in evolutionary history, local cloud patterns and their relation to rocks were not among his interests. Moreover, as a natural historian, he had no way of situating the mountain and its waters in relation to coloniality, capital, power, and wealth. Framed in this way, for Darwin, Table Mountain is a collection of rocks—unrelated to water, or the rise of power and wealth and their relation to geological formations.

These obscurely striped and spotted sandstone formations with granite intrusions that catch mists and form rain, however, changed the history of global capital in the 1600s—and thus also the history of Europe and its colonies. For ships' companies and the Company's ships saw at ||Hu-!gais neither the empirical truths of geology nor the mystery-made-manifest of Tsũi||Goab. Rather, what defined the Cape for them was the possibility of securing an empire's trajectory to wealth if it could guarantee a fresh supply of water all year round, along with soils to be cultivated to restore sailors and ships to seaworthiness en route between Europe and the East, with its silks and spices and porcelain. The mountain's rocks and stones were not, in their view, intricately bound to water and life, but instead had utilitarian value: they could be put to use to build a dam around a spring, and a castle around a well, and a moat for the mountain's waters to defend the castle that defended the new political cosmos of ownership.

The gradual capture of ||Hu-!gais's land and waters by the Dutch East India Company (Verenigde Oost-Indische Compagnie, or VOC) made successive European settlements at the Cape possible, along with successful European merchant company expansions to the East. Enslaved people from the East were shipped through Cape Town as cargo from Batavia (now Jakarta), Bengal (between India and Bangladesh), Mozambique, and Madagascar to build corporate and national capital in Europe. Enslaved men and women and children from Guinea in West Africa were brought to build the town at the Cape, and the profits of their labour gilded cities like Amsterdam, London, and Lisbon. The relationship between rock and water at Cape Town facilitated new flows of global capital.

The dark rock walls of Van Riebeeck's Castle were part of a group of rocks thought by geologists to have formed about three billion years ago in what is now Sea Point and Robben Island. A little higher than sea level, the characteristic sandstone of the mountain's slopes was set down in the Precambrian era, about seven hundred million years ago (mya). In this sediment on an ancient continent, the only life-forms present were somewhere between micro-organisms and spindles and fronds.

About 540 mya, the granite that traverses the peninsula from Lion's Head to Simonstown, and that had so excited Darwin, intruded into the heavy sandstone formations from deep-earth volcanic magmas.[12]

About 280 mya, the grey sandstone pebbles and rocks that are now on the very top of the mountain (where cable-car travellers go picnicking) were probably sandbars in the ancient river channels of the Late Carboniferous era when the Karoo's vast ice formations melted under warmer skies. Karoo meltwaters flooded lower landscapes, taking with them silts, sands, and clays. A period of uplift and erosion and faulting in the Jurassic era, 180 mya, forced the mountain up and into the shape that Europe would come to describe as a table.

Table Mountain's geology changed the relations among Africa, Europe, and Asia not because of what was *in* its rocks, but because of what came *out* of them: the rainwater which they collected so abundantly and which, once it had filtered through a few hundred metres of sandstone, hit the older and more impermeable rock layers and came pouring out of the ground in cool, clear, sweet springs and streams that did not stop flowing, not even in the dry summers.[13]

In the 1600s, shipwreck survivors and travellers to the Cape had reported that the waters coming off ||Hu-!gais formed a river that, in time, the Dutch would name the Vaarsche (the Fresh River), as well as a spring that could be relied upon to fill the water barrels of ships anchored in Table Bay.

Popular Cape Town historian Patric Tariq Mellet notes that so many ships had passed Table Bay in the early 1600s that a group of Khoena had split off to form a group of traders:

> The Khoena group that the Europeans first interacted with were the maroon Khoena who called themselves the //Ammaqua or Watermans, who had moved away from tribal life and herding to almost permanently living by operating as traders and facilitators and servicemen to frequently passing ships. Other Khoena referred to them disparagingly as "Our kindred who had drifted away or expelled"—Goringhaicona. This new trader-community settled strategically at the river mouth of the Camissa as guardians of the water.[14]

In the VOC archives, a drawing from sometime in the early 1600s shows the plan to construct a dam. The soft wetland that filtered water was to be replaced by walls, apparently without thought that, in time, the absence of a natural filter would require laws and policing to protect the stream above it.

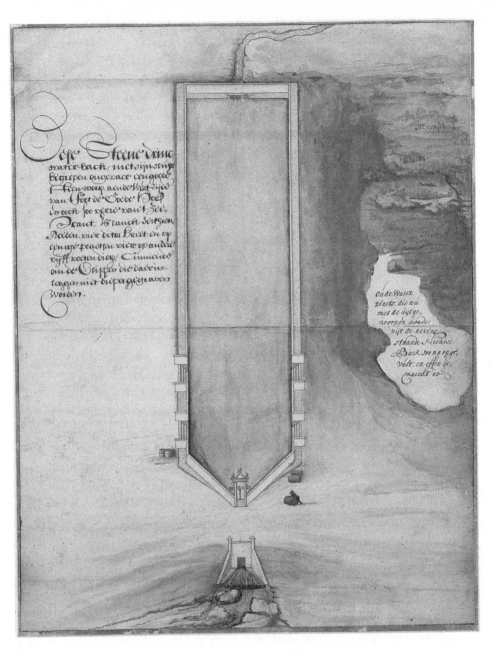

1.4 | Plan for a stone water dam at the proposed Fort de Goede Hoop
(Fort of Good Hope), Dutch East India Company, c. 1630.

An overflow pipe would lead to the sea, complete with an elaborate archway commemorating the VOC. No outflow is offered to where barrels could be conveniently filled. Instead, a black body cheerfully reclines over a barrel—presumably ready to roll them down and up a flight of rocky stairs to the stored water. Black labour was presumed, even then.

The bay below ‖Hu-!gais was, to the Khoena, //ammi i ssa, rendered now as Camissa, the "place of sweet waters."[15] It was even said by some that after a heavy storm, so much water poured off the mountain into the bay that sailors could drink from a pail lowered into the sea.

The Vaarsche's waters crossed what is now Strand ("Beach") Street in the city centre. They met the waves where the city's rail network now converges on Cape Town Station. A four-pointed mud-and-stone fort was built alongside the Vaarsche after the 1652 arrival of Dutch commander Jan van Riebeeck, designed around a well-point, with the purpose of taking over the water point from the Khoena Watermans. A few years later, the fort was superseded by a five-pointed stone castle. The waters of the Vaarsche were conscripted into its moat, in the hope that they would keep out the lions that would sometimes terrify the settlers at their doors, and the Khoena, against whom the Dutch went to war between 1659 and 1660.[16]

Khoena access to streams and springs had been eclipsed by the new regime of landownership of 1657 when the Dutch allocated property to farmers ("free burghers"), who forbade the Khoena from driving cattle over "their" lands and streams, and who recognised nothing of the importance of care for the springs and the water-beings that protected them.

The scale of Dutch hunting and fishing must have been, to the Khoena, incomprehensible. Van Riebeeck noted in his journal that 4,152 sealskins had been sent to Amsterdam in 1653: a huge labour given the few settlers at the time.[17] A 1652 diary entry notes that with a single cast of a net, "fully ten thousand fish were taken."

A diary entry from surgeon Johan Nieuhof at the Cape in February 1654 records, "We again went fishing, and in one draught caught as much as the shallop could hold: of them Heer Rietbeek [Van Riebeeck] took as much for himself as could be carried in fourteen wheelbarrows."[18]

The settlers' attitude to land and species reflected the spirit of the VOC, which, in the words of historian Nigel Worden, was "one of the most powerful, aggressive and acquisitive organizations in history. It was run by merchant capitalists for merchant capitalists, obsessed with making a profit—with buying cheap and selling dear. If what they wanted could be taken by force, and the operation of 'buying' dispensed with altogether, so much the better.

Monopoly, not free enterprise, was their ideal and such monopolies were created and protected by brute force."[19]

When the Dutch settlers arrived, Khoena nomads with the names Goringhaiqua, Gorachouqua, and Goringhaicona—known in historical archaeology as "the Peninsulars"—moved their flocks around the Cape Peninsula.[20] A large and prosperous Chainoqua settlement moved in the Hessequa region near what became Swellendam in the southern Cape ("the Hessequas"), and numerous settlements known as the Cochoqua ("the Saldanhars") lived near the Olifants River region near Saldanha Bay on the West Coast, with reports of very large flocks of sheep and cattle.[21] Some 200,000 Khoena are estimated to have lived along the coast from Mossel Bay to Walvis Bay at the time of European settlement.[22]

The diary of one Olfert Dapper records:

> In 1659 there arose between the Goringhaiquas or Capemen and our countrymen . . . a violent struggle over the occupation and ownership of the land about the Cape. The Capemen attempted to expel us, alleging that they had held this land from time immemorial. In this war the Gorachouquas (Tobacco Thieves) gave them continual support; and whenever an opportunity occurred of attacking our people they struck down many with assegais and killed them tragically. In addition, they plundered our countrymen of cattle lawfully bartered from them with merchandize, driving away the beasts so rapidly that they could not be fired upon, for they availed themselves mostly of opportunities when the weather was unsettled and rainy, knowing well that little could then be effected against them with firelocks.
>
> The knowledge of this trick they had obtained from one of their own people, known to them as Nommoä and to our countrymen as Doman. He had spent four or five years in Batavia . . . and had closely observed the habits of the Dutch. . . . Returning again to the Cape in the ships directed for Holland, he remained for some time in European dress amongst our countrymen, but finally went back to the Capemen, making them fully acquainted with the habits and customs of the Dutch and the nature of their weapons. He and another doughty warrior, called Carabinga . . . were always outstanding in battle. They took the lead in all the attacks, and owing to their remarkable agility could never be caught, not even by secret watches of men on horse or on foot. One morning in June 1659, after the war had already lasted three months, five Hottentots . . . were overtaken by five of our horsemen as they were running off with two cattle which they had stolen from a certain free burgher. A sharp skirmish

ensued. The Hottentots, seeing no possible means of flight nor desiring any mercy, defended themselves valiantly. They wounded two of the horsemen, one through the arm and under the lower ribs, and the other in the spine. But our countrymen repaid the debt by wounding three of them with the gun, and stabbing the other two dead with their own weapons. One of the three who were shot, a man named Eykamma, was taken to the Fort on a horse, with his neck pierced, his leg shattered, and a severe wound in the head; but Doman, with the other, escaped by jumping over a stream eight feet wide, after which flight proved their best weapon and salvation.

The wounded Eykamma, brought into the Fort, was asked why his people had made war against our countrymen, and tried to cause damage everywhere by killing, plundering and burning. Well-nigh overcome by the pain of his severe wounds, he replied by asking why the Dutch had ploughed over the land of the Hottentots, and sought to take the bread out of their mouth by sowing corn on the lands to which they had to drive their cattle for pasture; adding that they had never had other or better grazing grounds. The reason for all their attacks, he continued, was nothing else than to revenge themselves for the harm and injustice done to them: since they not only were commanded to keep away from certain of their grazing grounds, which they had always possessed undisturbed and only allowed us at first to use as a refreshment station, but they also saw their lands divided out amongst us without their knowledge by the heads of the settlement, and boundaries put up within which they might not pasture. He asked finally what we would have done had the same thing happened to us. Moreover, he added, they observed how we were strengthening ourselves daily with fortifications and bulwarks, which according to their way of thinking could have no other object than to bring them and all that was theirs under our authority and domination.[23]

The account of Nommoä (or Doman) giving back the Dutch clothes he had acquired and returning to the traditional loincloth became canonic in European thought about African Otherness. The story was mentioned by Peter Kolbe in his book about life at the Cape in the early 1700s, and it was borrowed by Jean-Jacques Rousseau, who made it the frontispiece for his book that established the figure of the Noble Savage in the European imagination of Africa. Siegfried Huigen notes:

Kolb's enlightened Hottentot [Khoena] gained its greatest prominence in Rousseau's *Discours sur l'origine et les fondaments de l'inégalité parmi les hommes*... (1775), in which Rousseau (1712–1778) attempted to ... [distinguish] three stages of human development: natural man living in solitude, man as part of a primitive hunting society and finally egoistic, civilised man. According to Rousseau man is happiest in the middle stage of primitive society—the "juste milieu"—and it would therefore have been best had he never left that stage. Rousseau supports his assertions . . . with borrowings from descriptions of primitive peoples, including the Hottentots. . . . Kolb's work, in the form of a summary in the *Histoire générale des voyages* (1748), was his main source of information about the Hottentots.[24]

Rousseau's reading of Nommoä's refusal to accept the Dutch expropriation of land fed a romantic account of Others as innately "primitive." That theory of tribalism would erase multiple struggles against injustice, including those in the apartheid era in which forced removals were justified by the creation of "homelands" for "cultures" or "tribes." "A man screaming," postcolonial founder Aimé Césaire would write two hundred years later, "is not a dancing bear."[25]

Securing rights of access to land and water required not only military might, but also maps and archival records based on a new knowledge: the science of mapping. Van Riebeeck and his successors brought to the Cape the techniques of cartography that were new in Europe at the time, together with ordinances (the "placcaat") and the capacity to compel compliance with cartography and its legislation in dungeons and a torture chamber. The latter included thumbscrews and a pulley with which those suspected of treason would be hoisted and dropped until bloodied and broken. The beginnings of science in South Africa went hand in hand with the development of law and the threat of the unmaking of a body via "legitimate" violence, whether through war on denizens or the torture of citizens.

"In order to give expression to the demands of Roman-Dutch Law which required the registration of title deeds and the boundaries of places," write Pascal Dubourg Glatigny, Estelle Alma Maré, and Russel Stafford Viljoen,

> land surveyors, a special category of skilled professionals who collaborated with architects, town planners and cartographers, became necessary. In the Netherlands their training included a thorough knowledge of geography and mathematics—especially of perspective [which] the *Duytsche Matematique* school [taught] to engineers without any command of the Latin language. . . . Since surveying was so important for the Dutch in establishing land partitions on which a permanent settlement at the Cape could be based, the first Commander, Jan van Riebeeck, requested trained land surveyors very soon after his arrival in 1652 to establish a settlement at the Cape of Good Hope.[26]

DISCOURS

SUR L'ORIGINE ET LES FONDEMENS
DE L'INEGALITE PARMI LES HOMMES.

Par JEAN JAQUES ROUSSEAU
CITOYEN DE GENÈVE.

Non in depravatis, fed in his quæ bene fecundum
naturam fe habent, confiderandum eft quid fit na-
turale. ARISTOT. Politic. L. 2.

A AMSTERDAM,
Chez MARC MICHEL REY.
MDCCLV,

Il retourne chez les Egaux.
Voyez la Note 13. p. 255.

1.5 | Frontispiece of Jean-Jacques Rousseau's *Discours sur l'origine et les fondemens de l'inegalité parmi les hommes* (1755), depicting the story of Doman, a Khoikhoi man once employed by Jan van Riebeeck, returning his European clothing to the Dutch settlers. The story became one of the foundations of Rousseau's concept of the Noble Savage.

Maps soon translated ||Hu-!gais into farmsteads and a town based on a new political cosmology of private property and water ownership. The mathematical science of cartography claimed legalised zones of exclusion as "facts." What it actually did was manufacture those facts. Within that claim to represent "only the facts" was what Isabelle Stengers calls "cosmopolitics": a word that speaks to *the capacity to name and render as "transcendent facts" the objects and authority that that approach had manufactured.*[27]

The Dutch surveyors worked with a baton and chains. Glatigny and ⸤ leagues note that as the European settlement in the Cape develope⸤ maps had to provide information on the success of the settlement for in Amsterdam. Unable to find French or Dutch maps that showed Khoena settlements at the Cape, Glatigny, Maré, and Viljoen focus instead on the way maps provided evidence of the progress of the fort, the gardens, the new farms, and the expanding settlement, observing throughout their visual rhetoric: empty lands are there for the taking. As a depiction of permanence, Glatigny, Maré, and Viljoen tell us, the maps set out both spatial claims *and* temporal claims—"in absolute contradiction with the way of life of the Khoikhoi, a nomadic people with seasonal migration patterns." They write:

> The Peninsular Khoikhoi were thus the first of several Khoikhoi groupings who were systematically ousted from the land. In 1657, Rijkloff van Goens and Van Riebeeck allocated land to a group of freeburghers in the vicinity of the Liesbeeck River and Table Bay.... Van Goens in 1657 ordered that land must be demarcated and distributed according to the Roman system. He further decreed that the plots would be granted according to the traditional Dutch method which outlined that "the plots [could be] as long and as broad as they wished." According to these instructions [the surveyor] Pieter Potter mapped farms for agricultural purposes, which, in effect, granted legal ownership to their prospective owners. *Van Goens further reckoned that disputes would be avoided since the demarcation of land followed a strict mathematical design.* In practice, of course, it was totally different, especially, as the "presence" of Khoikhoi kraals and access to resources was not taken into account.[28]

While immersed in Glatigny and colleagues' search for Khoena settlements on old maps for the writing of this chapter, I was invited to breakfast at an art gallery on the peninsula. To my astonishment, I found for sale a copy of a map of early Cape Town showing Khoena settlements on the slopes of Lion's Head, where Tamboerskloof and the Bo-Kaap are today. The map had German notes on it, but no source. Two years later, I learned that the map was published by Peter Kolbe, a German with a reputation for anti-establishment sentiment, whose illustrated account of the Khoena had been published as a German volume titled *Capvt Bonae Spei Hodiernvm* (The current state of the Cape of Good Hope) in 1719[29] and republished many times. Kolbe opposed stereotypes of the "Hottentots," which he had set out to correct, implicitly also questioning the turning of common land into Dutch private property.[30] Crucially, while the Latin edition's copperplate etchings of the Khoena were redone for the Dutch edition published eight years later, Kolbe's depiction of a large circle of round Khoena houses in Cape Town was excluded from the map and rendered as a separate drawing.[31] The political effects of this move were significant.

HOE DE HOTTENTOTTEN HAAR HUYSEN BOUWEN.

HOE ZY DE VISSCHEN VANGEN.

1.6 & 1.7 | Peter Kolbe's depiction of Khoena houses and fishing, 1727.

Glatigny, Maré, and Viljoen's study of the absence of Khoena settlements in early Cape maps is an important one in the history of science in South Africa, for it points to the way in which the first sciences in the region imposed a new mathematical logic on land, creating in the same moment a regime of "objectivity" without acknowledging that those very techniques of objectivity, measurement, paper, and ink, were expressions of a new political cosmology: that of private property, in place of a commons. And it also points to the ways in which a science that is dependent on its patrons is not independent, for it is subject to censorship and to cherry-picking. The idea that the Khoena could have objected to the maps was unlikely to have been palatable to patrons, publishers, or book buyers in Europe.

In maps of the time, the European cosmology of private landownership is presented as a neutral, independent, and socially transcendent fact. It is useful to draw on the thinking of Argentinian philosopher of decoloniality Enrique Dussel, and distinguish between seeing the problem as *science* and seeing it as *scientism*—the latter being an unreasonable faith in the claim that science is independent and neutral. The former is investigative scholarship for which any claim is open to question and reasonable answer; the latter is dogma about truth because it was generated by a scientist. "Scientism," Dussel writes, is "dangerous inasmuch as it fabricates the instruments necessary for the power of the center to be exercised over the periphery. At the proper moment, we shall have to question the naivety (with respect to the system as a totality) of scientists."[32]

Distinguishing between science and scientism—the former as a reliable way of producing knowledge through experimental processes; the latter as the belief that all scientists produce neutral, transcendent, natural knowledge that is permanently and totally separate from society, regardless of context—opens the way to recognising scientific achievements in producing knowledge of the world, while simultaneously opening up space for thinking critically about the irrational, transcendent political power that science accrues when it denies its locatedness in society.

Bruno Latour speaks of the three divinities of reason in the contemporary knowledge economy: technical efficiency, economic profitability, scientific objectivity.[33] I would trace these back to a time much earlier in the Cape, where "the gods of reason" in the 1650s authorised new claims to knowledge production—without accountability to any ethical precepts other than those of the voc.

Through the claim that the transcendent, independent, mathematical truth was mapped in cartography, and via the erasure of the Khoena presence in maps, ||Hu-!gais became subject to the Heeren XVII, the seventeen lords of the voc, upon whose continued patronage the settlers depended. The Dutch named one of the main canals (the "grachts") after these Lords, the Heeren, and the name Heerengracht remains the name for a major street in the Cape Town central business district.

It is difficult to imagine that the phonic similarity of the word *Heerengracht* to ||*Hu-!gais* (Hoerikwaggo™) had no potency; no in-your-face political translation of the mountain to those Khoena who were fighting to regain access to streams and kraals. Bear in mind that the concept of translation—at least in medieval English—was about not only a change of language, but a change of political allegiance from one feudal lord and his castle to another. Not only did one translate a language; one *was* translated when crossing a boundary and changing allegiance. Under a different banner, in a different political cosmos, as the subject of a new cosmopolitical order—one

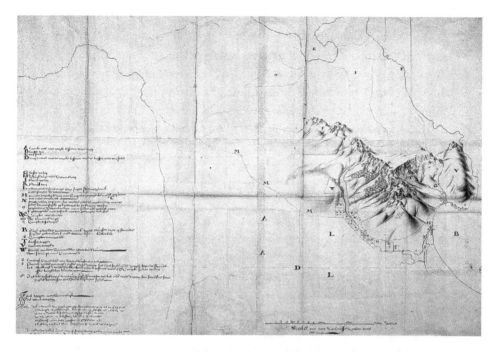

1.8 | Farm allocations to Dutch free burghers at Cape Town, c. 1660, from Table Bay to Newlands.

was translated into something different, as in *A Midsummer Night's Dream*, where Bottom changes into a donkey and Quince tells us, "Thou art translated!," to which Bottom memorably responds, "I see their knavery: this is to make an ass of me, to fright me if they could. But I will not stir from this place, do what they can."[34]

Under the Flag of Orange, the Dutch ensign, ||Hu-!gais had new masters in another world: the Heeren for whom land and water did not exist in an intimate relation, but were separate and extractable objects. Naming the settlement's main stream after the Heeren was not only a sign of a new political and cultural and economic mastery: it was the imposition of a different way of being human in the world—or, more specifically, outside it. Modernist thought—humans separate from nature, and knowledge separate from any viewpoint—was emerging in the cartographic technique of showing place via a view from everywhere and nowhere. It was a god's-eye view.

About fifteen years before the Dutch came to settle at the Cape, René Descartes had published his techniques for producing scientific knowledge. His strategies for knowing were to look for what is evident ("present to the eye's gaze"); reduce it into as many parts as possible; order and enumerate those parts; and then put them back together again as a long chain of inference.[35] By focusing on the observable, he argued, scholars were not interfering with the work of the Church, because the Church focused on matters of spirit. His cosmology, therefore, retained the dualism taught by the Church at the time.

That approach helped to keep his head on his shoulders (literally); even so, when he first published his *Discourse on the Method of Rightly Conducting One's Reason and Seeking Truth in the Sciences*,[36] he published it unsigned. In the end, his argument did not help; shortly after his death, his writings were banned by the Catholic Church.

Descartes's foundational split between belief and knowledge is the founding doctrine of modernist thought. That split made it possible for "science" to emerge out of the shadow of a Church that was objecting to new knowledges, like that of Copernicus and Galileo, which contradicted Church dogma.

Descartes's work on geometry and the measurement of space made it possible to create maps. His approach greatly advanced emerging techniques of mapping, rendering space and objects in it with geometrical accuracy. Much as emerging scientists sought to stay alive by differentiating their "pure knowledge" from the beliefs of theology, the new technology of mapping presented a visual approach to space as if it were neutral: a view from nowhere that was "pure knowledge"—devoid of theology or belief or values or society. On the contrary! At the Cape, representing Table Mountain's lands and water as owned property and resources for the profit of the Heeren XVII was as much a political cosmology as was thinking of ||Hu-!gais as a gatherer of mists, clouds, and rain. One version claims the authority and guardianship of king and company and a "neutral," "value-free" scientific method. The other is a claim of living in reciprocity with rains and cattle and beings who are guardians of the waters and lands themselves.

The earliest Dutch maps of property allocations and water resources at the Cape were hints of a relationship between science and violence that was to come. They also mark the Cape rollout of the paradigm of the Age of Expansion: of a world split into myriad separate things (recall Descartes's method) that were held together in an illusion of space as an empty map, that offered a visual rhetoric devoid of relationships, politics, or any social influences. This was the beginning of what Césaire would later describe as the "thingification" that came with colonisation. For the ideas of modernist thought which undergirded coloniality were of a world made of things connected only by their presence in space, from which they were extractible to whatever extent was humanly possible. Life and ecological relations were incidental and optional extras. That mind-set remains at the heart of the paradigm of extraction that has yielded an earth-changing, climate-changing geology now called the Anthropocene. And that mind-set remains taught in foundational introductions to the natural sciences. In a scalable model of knowledge from the scale of quarks and nanoparticles to the extent of the known universe, nowhere are relationships observable; much less are the relations of the observer to the observed. What exists in the scalable model of the universe are simply things at different scales. Yet worlds and knowledges are made day by day, in real time, in relationships, in economies, because of ecologies, and as new technologies make different objects evident and measurable and countable.

That approach to knowledge production is so universalising that the result, within the contemporary global university system, is that the only space where the knowledge produced by "Others" is studied has been in "ethnology" (including "ethnobotany"), in studies of "belief" and "superstition," and studies of myth and literature—the social sciences, not the natural sciences.

At the Cape, however, at least one early translator of Khoena thought did not find it quite to easy to sever knowledge from belief. Theophilus Hahn, when writing his science of Khoena religion, considered carefully the widespread Khoena idea that a water snake dwelt in each spring, and was wary of accepting it as either belief or superstition, or as a being that could be mapped in space. Making the case that the words for "snake" and "spring" are so similar that over time they merged in meaning, he wrote:

> The name for snake is |au-b . . . and for fountain |au-s. . . . Blood is also |au-b; bloody, full of blood, saturated with blood, |au-$_\chi$o. . . . But |au originally means to flow. And |au-b, the snake, or |au-s, the fountain, is nothing else than saying, *he flows* or *she flows*. The snake, however, is the one who flows over the ground. . . . The streaming and flowing of the cloud—that is, the rain—is also derived from the root |au. |Awib . . . is the rain; to rain, is |au-i or |avi—that is, to be streaming, to be flowing. |Au signifies also to bleed. . . . The colour red, |ava, also takes its origin from |au, to bleed. . . . To return now once more to |aub, snake, and |aus, fountain, we see how both words were predicative expressions, saying *he* or *she flows*. . . . Then, when the original meaning of *to flow* and *to stream* was forgotten, mythology got hold of |aub and |aus, and made sure *that in every fountain lived a snake*.[37]

Hahn's linguistic analysis points to the centrality of forms of movement—not spaces of fixity—in Khoena ways of noticing and knowing the world. Elsewhere in the world in some societies based on tracking, knowing the way something moves is a matter of survival, so the connection of snake, stream, blood, and rain here is significant: it suggests a form of knowledge among the Khoena and !Xam that pays particular attention to flow and movement and ethics. The figuring of "stream" and "snake" as entangled realities generated a relation with water sources that, crucially, ensured their protection and care in ways that are far more respectful than that of the contemporary City of Cape Town, which overwhelmingly treats its rivers as sewers. The Khoena respect for snakes in water sources, by contrast, speaks to a mode of knowledge in which worlds emerge through relationships and movement. Taking care of those relations and flows is taking care of the world.

The new mapmakers' object-based knowledges (think Darwin and the rocks of Cape Town) paid attention to the shape and form and place of a thing, with not much attention given at all to the processes through which things came to be named and known. Taking care of a world thus defined means taking care of the things in it—for reasons that vary from "because environmental science says we ought to" to "because we want to use them." The approach begs questions like "For whom?" and "Who is the 'we' here?" *because both are based on mastery.*

A knowledge based on relationships cannot elide one's own relationships, responsibilities, and reciprocities, which accrue to living in a world. In knowledges based on movement, a collection of dead specimens might be interesting but not that useful for teaching us how to live or survive alongside them. But in knowledges based on knowing things in space, a collection of dead specimens is (a) useful and (b) independent of the act of killing them to know them. Contemporary anthropologists approach this branch of knowledge studies as ontology. It points to the ways in which *the reason for making knowledge tends to become a central means of organising that knowledge.*[38]

Hahn's observations of Khoena linguistic attention to movement and flow helps to understand the Khoena appreciation of ‖Hu-!gais as a place where rocks gather clouds and rains and greenery. Hahn recalls a conversation with |Haigu$_x$ab, then the chief of the ‖Habobes, about their origin:

> He told me: "That very thing, the ‖Habobe, has been made by Tsūi-‖Goab in this country, and !Khub has made us, and given us this country. He gives to us the rain, and he makes the grass grow; and if we ask payment for our grass and water, we do the same what you white people do in asking payment for your lead and powder."[39]

|Haigu$_x$ab's insight into the forces that made the Dutch cosmos of that time was spot-on. The relationships that matter most in your oikos (your ecology-economy-ecumene), he seems to be saying, are those linking lead and powder and payment; our interests, by contrast, are rain and grass and gift.

The Dutch surgeon Johan Nieuhof, who arrived at the Cape in 1654, wrote:

> They [the Khoena] say, that there is One, whom they call Hunuma, who can give rain and drought, although they do not pray to him.... They listen very reluctantly when one speaks of God, also they punish those [among themselves] who do this, saying that such are too thoughtless, and speak of the Godhead with insufficient respect. If they are asked concerning the evil spirit they point with their fingers to the ground, *and point them also at you.*[40]

Who is the Evil One? You are! must have been a shocking encounter for Nieuhof, who for his part is a curious observer of the Khoena, whom he regards as barbarous—but he is unable to reflect on the barbarity of the ideas that the Dutch were imposing.

Many early writers had described the Khoena practice of leaving a gift after taking water from a spring. Imagine the Khoena outrage at seeing enough water being taken from a source to provision a ship's company for several months at sea, by barrels and wagon—without reciprocal gifting. Nieuhof describes an early battle over water in 1654:

> The Hottentots...muddied the drinking-water that we intended to bring aboard, which we resisted, seizing some of them; but they threw stones with such force that some of us were knocked down, and because they were in far greater numbers we were compelled to take to flight. At that time I had gone ashore to shoot some game, but before I could reach them our folk were already in flight. Next day I went with them, taking several armed men, with the intention of shooting at them if they again came to hinder our work; but so soon as they saw us coming armed they took to flight inland with wives and children and all they had.[41]

In 1713, after surviving two wars, the loss of access to water points, the loss of the right to move cattle through the farmlands of the growing settlement, and a growing number of cattle raids by settlers, there was an outbreak of smallpox. The Khoena fled, with so many dying at the sides of the roads that some described it as like a massacre. Written accounts suggest that Khoena settlements were decimated, with some groups reporting that only one in ten of their number had survived. "The Hottentots...died in their hundreds. They lay everywhere on the roads.... Cursing at the Dutchmen, who they said had bewitched them, they fled inland with their kraals, huts and cattle in hopes there to be freed from the malign disease."[42]

The Castle gave a skin of stone and iron to the bodies in its network—including its bodies of water. With its walls, the Castle rendered the bodies of those within its polis more impervious, less vulnerable to the risks of skin. At the same time, the bodies of those within and without it who had threatened its political cosmos were at constant risk of rupture: from soldiers or torture; from the filth of its dungeons; from battles; from scurvy; from the violence of slavery; or from slave revolt.

In establishing its well-point, the Castle had privatised the mountain's water flow. Streams and springs, mastered in maps, stones, and laws, and protected with gunpowder, became the property of the VOC.

Between 1655 and 1658, a number of important changes occurred. First, the water became foul, and after a visiting ship's crew took violently ill in 1655, the Placcaat 12 of 1655 was issued as the first environmental law in South Africa: "Niet boven de stroom van de spruitjie daer de schepen haer water halen te wassen en deselve troubel te maken"—"Do not wash or disturb the water above the stream of the spring where the ships fetch water."[43]

Second, the need to line the watercourses of the streams through the city was unavoidable, and it led to the laying of the grachts—open canals. This required labour, which also required expanding the settlement, which in turn required more food. The Khoena were unwilling to supply the settlement with all the meat and milk that it demanded for its expansion, and several settlers were allocated land as free burghers.

By 1657 the Dutch and Khoena were at war. Further fortifications of the Castle were determined necessary, which required even more labour to make bricks; gather seashells to make mortar; cut stone; and feed the expanding labour force whose tasks so outstripped their capacity that there was a near-mutiny by soldiers who were forced to hew timber by day and stand guard by night. They did not, they said, sign up as soldiers to be treated "worse than slaves," and many requested a transfer to Batavia.[44]

In consequence, the VOC at the Cape requested a shipment of enslaved people, who were captured and forcibly removed from the West African region of Guinea in 1658.[45] After their arrival, Van Riebeeck noted in his diary entry of 30 September 1658:

> Calm, sunshine. Death of 5 slaves and 1 female do., very likely because the men are mostly old and cannot bear the cold climate as well as the younger ones, who have, according to orders, been sent to Batavia. At present, including those recaptured, the entire number of slaves is 83, viz, 34

men and 49 women. The greater number are old and weak; and many half mad and cripples, who can do no work, and are merely a burden. They will, however, soon die off.[46]

Later began the Cape slave raids to Madagascar, as well as the shipment of enslaved people from the VOC-conquered Batavia. Recently published archival sources speak of the excessive violence of the new form of slavery at the Cape in which people were owned by the VOC as private property.[47] Their lives were short. The bones of many were rediscovered in 2003 on a site in the city centre, near the Waterfront (a recreational, tourist, and shopping mall built on the ocean's edge), in a street called Prestwich. Amid fierce public controversy, they were reinterred at a memorial site nearby.[48] As "cultural heritage," their mausoleum is now branded by an artisanal coffee house called Truth™. Its name (and trademark) adorns the outside of the coffee shop in body-length avant-garde-style chrome letters, as it does the bags of beans available for the coffee sippers' communion at home, and a high-end SUV owned by the company. Even Tokolos Stencils—the local equivalent of graffiti artist Banksy—would have a hard time designing a more revealing comment on South African history à la neoliberalism. To reinter a slave and make her bones pay for her memory through the cappuccinos of the wealthy does not change the relation to those bones: they remain objects with which their custodian has a proprietary relation. Shortly after Truth™ opened, local novelist Jade Gibson staged a protest, taking a skeleton made of tree twigs to the shop for a coffee.[49]

––––––––––

The central well in the Castle is lined with neat clay bricks, made from red river clay and crushed seashells, baked in the sun, and then laid in ascending circles by slaves who themselves would soon be interred nearby. Over time, the city fathers gave up the battle to regulate the grachts and banished them underground instead. Closed in behind metal grates and sealed in clay-brick sewers, the waters of the city were forced into the underworld and out to sea. They exploded in revolt: in 1905 a spark ignited a build-up of gas in a pipe under Adderley Street, shattering pavements and buildings and bodies, with the water spraying up as high as the third floor of adjoining buildings. The British, who were then in power, responded with a banning order even more stringent than those of the Dutch regime: the pipes carrying the fouled water were to be closed off altogether. Under apartheid, their outfalls were extended when the foreshore was built up. Their forgetting was sealed in 1994 when they were struck from the city's asset register, in the year in which democracy came to the parliamentary buildings next to the Company Gardens, where these same waters had once brought to life vegetables tended by the enslaved to supply soldiers, sailors, and settlers. But like the body hidden on the farm in Nadine Gordimer's novel *The Conservationist*, those buried without their attachments to the present will rise to haunt the living.[50]

Though the rock walls and well-point and water moat of the five-pointed Castle had long since ceased to guarantee food and water and therefore allegiance to whoever had military command of its walls, its symbolic value held long.

The orange flag of the Dutch had fluttered above it and its signal post on Signal Hill, from the mid-1600s until the British took the Cape in 1806. Britain's prisoners at the Castle included the Zulu king Cetswayo and his wives between 1879 and 1881; and in the 1890s and early 1900s, some Afrikaner prisoners of the Anglo-Boer or South African War. The flag of the Union of South Africa hung from it from 1910, and in 1961, when the prime minister known as the architect of apartheid, H. F. Verwoerd, declared the country a republic, the shape of the Castle came to adorn the chest of every South African Defence Force officer. Under Nelson Mandela's presidency, beginning in 1994, the new flag of South Africa flew on the Castle's buttresses, overlooking a fleet of matt-brown armoured cars. Mandela's South African National Defence Force (SANDF) used the Castle as its emblem until 2003.

I took my children there for a visit in 2013. It was an experiment: Could I tell them the stories of this place in a way that might enable the connection between water and history to matter to them? We explored the remnants of the well and the moat, as I tried to share what I knew about water and soldiers and Khoena and slaves and the beginnings of this city. It was hard to miss the change between then and now. The Castle, built to protect "nature," was now itself "just culture," the flotsam of a past, a "heritage" cost to be covered by tourism. The litter in its moat gave it the feel of a used ashtray. It was difficult to get the children to see past the mess and imagine it having a more vital presence in the city. As we left, we were dive-bombed by shrieking lapwing plovers determined to protect their offspring, who scuttled on the lawn alongside the moat. Laughing about the birds as the kamikaze of the Castle, we followed them around the cannons toward a range of stone buildings that would long ago have been part of the armoury. And at that point I saw the sign on one of the windows: "Protected by ADT."

That stopped me in my tracks. The place that the Dutch had built to protect fresh water for the bodies of those who established the beginnings of globalisation; the military installation that had victualled the planners of successive waves of colonisation across Southern Africa; the castle that housed the armies of the Dutch and (briefly) the French, the British, the apartheid regime, and the governments of Mandela and Thabo Mbeki; the Castle whose shape adorned every SANDF vehicle, installation, and breast until 2003 is now protected by ADT: the American District Telegraph Company.

Sometimes history is a stand-up comic.

How did a stone castle that protected the streams that protected the castle that enabled a Dutch multinational company to pursue its global seafaring in the 1650s turn into an icon protected by an American electronics multinational? How did it happen that political power based on the nature of stone waterways had turned into infrared detectors, magnetic pads, cameras, radio frequencies, and satellite connections to protect cultural heritage? What did it mean for my children's futures that our knowledge economy was paying so little attention to water that these streams, which had changed planetary history were now simply the city's colons? Why had I, as a social scientist, never yet thought about the change in the dependence of political power on rock and water, to a new materiality of electron flows?

When some benighted soul got the job of glueing the yellow-and-blue bit of ADT-signed plastic to the Castle, they were "only doing their job" in a vast and constantly changing network: a political ecology of capital that has mutated into new forms through times of conquest, coloniality, neoliberalism, ensigns, trademarks, and the Anthropocene. That bit of plastic stood as testament to the changing relationship between power and ecology—and a reminder that as that happens, a different idea of nature emerges, along with a different idea of what it is to be human, and what it is to produce (apparently) neutral knowledge.

The VOC approach to political power attended to the water flows necessary for a Heerengracht ecology, based on property, command, and control. Under the neoliberal regime of trademarks and electron flows, water flowed in pipes to the extent that it was matched by flows of money in electronic networks. Water current then had been a case of gravity plus stone plus policing. Water current now, in the age of water meters and ratepayer bills, depended on currency and electronic current.

The story of changes in the political ecology around the Castle tells a story: that Table Mountain's water does not have one permanent "nature." It has had many natures, from Tsūi ||Goab as the provider of rains and greenery at ||Hu-!gais, to the VOC as the protector of the Heerengracht, to the power of finance over electronic circuits governing piped flows.

It would be several years before I could explain that to my children. But I began to explain this to graduate students as the four natures of Table Mountain.

Nature 1 could be called ||Hu-!gais. For the Khoena, the words for blood and water were similar, both referring to flow. The Khoena called that flow a water snake, whose body also flowed. When you took water, you respected the snake: the flow. If you did not respect its flow, it could bite like a snake. You would get sick; be struck by what archaic English called "the flux"—diarrhoea. So the key idea was that the movements of life and water required respect: reciprocal co-operation ensured the continuation of flow.

Nature 2 could be called Heerengracht: the era of lines, hedges, maps, laws, and guns, which together generate a relation of mastery. A master is, however, almost always mastered. If you offend the people in charge of the Castle, you go to prison, where your body will discover the power of the wall. Your body will be thirsty, and will probably be made to bleed; you may be banished to the far side of many walls, and have to try to find your own water.

Nature 3: the world of hydrogeology and engineering. The structures of the city government evolve directly in relation to the need to increase the water supply to the growing city.[51] As the city expands, its sources of water get farther and farther away from the city itself, and its own water sources become progressively more fouled. Enclosed in pipes, foul water erupts in revolt. Engineering skills separate it from citizens, who must in turn be protected from dams, pipes, and canals. Water becomes megalitres and kilolitres; its flow is disrupted by a new political ecology of human-made rock—concrete—such as in the apartheid state's massive foreshore development in Cape Town, in which the shoreline is pushed far out to make a deepwater harbour, while the city's rivers are enclosed. Bodies of slaves and streams are both forgotten underground: neither remain in the city's memory.

The hard-surfacing of the city creates flooding in the Cape Flats, where most of the descendants of the Khoena now live, and lowers the groundwater table since water cannot sink easily through tar and concrete. The city's bulk water supply comes from the Berg and the Breede River watersheds far outside of the city, along with a number of dams in the wider region. Cape Town's springs disappear from the city's asset register. On a global scale, so many megacities are developing at the same time that the relationship of mastery over the earth increases to a planetary scale, ushering in the Anthropocene.

Nature 4: Hoerikwaggo™ marks the dawn of ownership and control through bank accounts. The rise of a new kind of digital feudalism: those who live online and with credit can purchase any amount of water, any kind of protection, any kind of object, including access to the Hoerikwaggo™ Trail on Table Mountain when it's open. But those who live offline and without credit, even on the edges of Hoerikwaggo™ (like the inhabitants of Hangberg, an impoverished fishing settlement flanking the more luxurious suburb of Hout Bay), are fought off the reserve by conservation officials when they try to build houses on it.[52] By contrast, the civil engineering company that rebuilt the Chapman's Peak toll road plaza facing Hangberg was allowed by SANParks to build a corporate castle on the site. The rock-water-plant ecology of Table Mountain, like the Castle, is heritage: Natural Heritage; Cultural Heritage. Neither is linked to the everyday lives of citizens.

These four points suggest four things about the relationship of politics and the natures it generates (the scholarly description is "political ecology"):

1 | Political power is not only about the control of social relationships; it is also the control of materialities that provide citizens with what their bodies need. Politics *is* ecological. It depends on controlling the supply of food and water.

2 | Part of ideology is its naturalisation—and in naturalising itself, it also naturalises its version of nature. The four natures of Table Mountain, in the history of Cape Town, demonstrate that what is understood to be nature is constantly changing as new technologies make possible different networks of power, based on the flows of different materials. So it is not possible to claim that there is only "one" version of nature.

3 | Under capital, streams and mountains have no legal standing, but companies are legal persons. So while in ||Hu-!gais nature a stream may have a double existence as stream and snake, in Hoerikwaggo companies, or the Pty Ltd., they have a double existence as companies and as legal persons, while streams and mountains remain classified as "just things." This is a political cosmology. Where a scientific research project takes the view that a stream or a mountain is just a thing, a resource to be harvested, it is not operating in a neutral, society-free zone. Yet the cosmopolitics of the time enables natural scientists to believe that what they know about companies is pure knowledge. They can name the one as "indigenous animism" and call it superstition, but they cannot call the other "capitalist animism," nor recognise it as the mythology of an era.

4 | Science that claims it is above power and outside society and providing its transcendent truths is better described as scientism. Scientism is a practice that is the opposite of science, which is based on permanent

questioning. Scientism proceeds with the certainty of its authority because of its identity as Science. And scientism can be as unethical as any other kind of fundamentalism or identity politics. Modernity has never been the opposite of belief. To quote Bruno Latour, Modernity has never been modern. It retains cosmopolitics at its core.[53]

How might it be possible to reclaim a critical, useful, life-giving science that is capable of unmaking the legacies of coloniality and its successors?

To answer these questions, we need to understand more about how life sciences and management sciences—including heritage management and conservation—are related to the material politics of our own time. Let us return, then, to Table Mountain's nature as Hoerikwaggo™.

III | HOERIKWAGGO™: NATURE PAYS

In the knowledge economy, political control subsists in flows of information that can be open when they are a closed circuit, and closed when the circuit is open—much like burglar alarms.

Water supply to a household or sometimes even a town may also depend on the flow of electronic data about water. As long as I pay the water bill, I can get water (unless, of course, the city installs a water management device on my property). But out in the shacklands, where people can rarely pay, one tap and one toilet can serve dozens of households.

In *Invisible Cities*, Italo Calvino crafted a novel that argued that a city is not only its buildings, but the terms its citizens accept as defining their being together:

> The inferno of the living is not something that will be; if there is one, it is what is already here, the inferno where we live every day, that we form by being together. There are two ways to escape suffering it. The first is easy for many: accept the inferno and become such a part of it that you can no longer see it. The second is risky and demands constant vigilance and apprehension: seek and learn to recognize who and what, in the midst of inferno, are not inferno, then make them endure, give them space.[54]

The terms that Capetonians have contemporarily accepted to govern water supply, however, are very much part of the inferno that characterises the era of trademarks, or TM. The "Values" section of the 2013 mission statement of the City of Cape Town's Water and Sanitation division says, "Our customers are the reason for our existence. The environment is our silent customer who shall receive an equal share of our services."[55]

When did citizens and nature become customers? In this mentality, when Nature pays, it is a customer. What about the parts of nature that don't "pay"? Can they be protected too? When did municipal water supply to citizens become a business with customers? Words matter, because they frame what actions will be considered rational and reasonable. If a customer does not pay, they do not have rights. What of a citizen? In what sense is it rational, reasonable, or logical for Cape Town's Water and Sanitation division to define nature as a paying customer? Why is that acceptable, but when ecologists and environmental lawyers call for rivers to have rights, they are regarded as esoteric and naive? The moneyed citizen receives piped water, pays and is paid; therefore, she is a customer with rights. But when a community like Sandvlei along the lower Kuils River in southeast Cape Town relies on their own boreholes, not a municipality supply, they are not "customers"— but their protests against having to live alongside the sewage effluent of the city's customers, discharged into the river from the overwhelmed Zandvliet Waste Water Treatment Works, fell on deaf ears.[56]

Definitions of relationships matter. The customer model, in government, generates a brutal mentality that is blind to the effects of its calculative thinking on finance-poor citizens, who, given South African history, will mostly be black.[57] While the city made much of the fact that it offered six thousand litres of water free to every connected household, in the years prior to the drought, the city is also filled with unconnected shacks in rambling settlements where a single tap is shared by many, none of whom is a "customer." In addition, households that cannot pay for water at municipal rates have their supply cut to a trickle. When you don't pay, it is like having a snake blocking your pipes . . .

In 2015 a revised version of the Water and Sanitation Department's mission statement appeared in the form of a "Consumer Service Charter for the Supply of Potable Water and Sanitation Services."[58] The words "environment," "ecology," "rainfall," and "drought" did not appear in it.

In the Environmental Management Department, the focus was on biodiversity. Its Integrated Environmental Management Plan (approved in 2001 and reviewed in 2008) made no mention of rainwater, the city's rivers, the mountain's springs, or the management of the Cape Flats aquifer.[59] A bulletpoint bucket list, it reflects PowerPoint presentation thinking, based on technical measurables and quantifiables with no hint of how these work together in an urban ecology. The document opens with a paragraph that could have been assembled by a word-picking algorithm: environmental beauty . . . oceans and mountains . . . floristic kingdom . . . culturally diverse . . . rich history . . . economic hub . . . unique environment. The words "rainfall" and "drought" do not appear here either. There is no mention of how the city's three separate water divisions—Water and Sanitation, Roads and Stormwater, Environmental Management—might be co-ordinated. Since Table Mountain National Park's twenty-five thousand hectares are under a law unto themselves, there is no mention of its springs, or its capacity to collect and direct water for the city's use, or to be part of the urban ecology. Table Mountain National Park's attention was on the Hoerikwaggo[TM] trail. Its focus is on making nature pay, using its customers (whose numbers are almost double those of the Kruger National Park annually) to fund nature conservation on the mountain and in parks elsewhere in the country.

The frustration of city planners is occasionally apparent. An in-house paper by city planner Candice Haskins, for example, offers an implicit critique of the city's approach to urban ecology:

> The Vaarsche River and many other small streams in the centre of Cape Town now exist almost entirely as underground piped systems. The 1940s to 60s was characterized by a period of intensive stormwater engineering, including realignment of the Liesbeek and Salt Rivers, construction of numerous canals (e.g. Liesbeek, Elsieskraal, Big and Little Lotus canals) and detention ponds, and upgrading of drainage systems in an attempt to alleviate and mitigate the effects of flooding in the rapidly growing city. Concrete canals, infilling, draining and dredging of wetlands to make way for development, and expansion of impervious areas (catchment hardening) have dramatically altered the natural hydrograph so that Cape Town experiences urban stormwater runoff patterns typical of most developed cities worldwide *viz.* reduced infiltration, increased runoff rates, volumes and incidents of flooding. . . . Typically catchment hardening results in up to 90% of precipitation being rapidly transported via urban stormwater systems to rivers as opposed to 10% runoff that would be expected under natural conditions where vegetation cover and therefore infiltration rates are higher.[60]

Streams in closed pipes, canalised rivers, and hard-surfaced roads cast the rains that fell on the mountain and its streets into the roles of problems to be managed, in the form of sewage and stormwater. With the rains sent into the sea, not the soils and aquifers, the city's drinking waters were piped from dams and rivers far away. The city's rivers, once home to hippos, are reputed to be among the dirtiest rivers in Africa.

Sometimes, from our campus classroom overlooking the city as far as the Hottentots-Holland mountains (which still bear that name), I would ask a class to identify the farmland and the dams, and think about how much water and food the city needed each day. But of course, they could not spot the major dams because there aren't any within the city limits, and as for farmland, there was only the Philippi Horticultural Area, an agricultural lung which itself was in a battle with the city officials bent on selling it off to developers. I would ask my students to think about the distance that food and water had to travel to the city's residences, every minute of every day, in order to feed everyone. Could a city designed like this, without thought to protecting farmland or nearby water sources, be sure it would never be a city of hunger and thirst? And if its design could not offer that certainty, did we not have a duty of citizenship to redesign it? I would ask them to debate whether, in the era of financialised democracy, the truth of the urban citizen has become "I pay, therefore I am."

After three years of low rainfall in 2015, 2016, and 2017, the city reached a drought crisis. Staring at the possibility of "Day Zero" in 2018, the point at which the distant reservoirs in the Cape hinterland would be empty, plans were put in place for residents to queue for a daily twenty-five-litre water ration under the eye of the military (that is, potentially in its gunsights). At that point, Water and Sanitation brought back into political life one of the springs on which the city was founded: the same spring around which the Dutch had planned the first dam (figure 1.4).[61] The recovery of the spring came after a decade-long battle by independent activist and urban planner Caron von Zeil to bring to city officials' attention the fact that Table Mountain generates four rivers and thirty-six springs, yielding millions of litres of water a day, all of which were going out to sea in forgotten pipes, and all of which were in a city that was, in the language of planners, "water-stressed."

Von Zeil's historical research was echoed in the work of Cape Town geographical information systems lecturer Siddique Motala, who showed, through a history of map overlays, how Van Riebeeck's wild almond security hedge from his Newlands farm on the Liesbeek River (adjoining the Kirstenbosch Botanical Garden) went all the way to its confluence with the Black River, and out to Paarden Eiland in Table Bay. That social and economic divide on the land, he demonstrates with overlaid maps, survives in the city's spatial patterns to this day.[62]

Von Zeil found accounts of the springs in the archives, and mapped the location of thirty-four through her project named Reclaim Camissa, after the Khoena name for the sweet waters below the mountain. Of those springs, according to the city council, eighteen were under its control, and of those eighteen, only one—the Albion Spring in Newlands—was deemed usable

|rinkable water before 2017. The Albion Spring alone supplied (according to council figures) approximately 3,180,000 litres per day (at its lowest measured flow in 2015), some of which was piped into the city's bulk water supply.[63] The remaining council-controlled springs ran away into stormwater drains. Notably, the Newlands spring that supplies the brewery owned by South African Breweries (SAB), at no charge to one of South Africa's largest multinational companies, is not included in the figures, as it is under corporate, not municipal, control. This nature does not pay, even though it is sold to SAB customers globally. The legality of this is questionable. The logic being applied is that the water exits the ground on private land, therefore it is privately owned. But in terms of South African post-apartheid water law (see chapter 2), water is not privately owned, whether or not it flows through private property.

In 2009 Cape Town had embarked on a massive new project—building a new castle on its seafront: not in the style of a five-pointed fort for political defence, this time, but a circle in the form of an African woven hat, designed to host matches for the 2010 FIFA™ World Cup. With the slogan "Green Goal: Hosting a responsible 2010 FIFA™ World Cup," the city embarked on a R25 million project to use the Oranjezicht spring water for the soccer field and surrounding Green Point park.[64] Von Zeil found her Reclaim Camissa ideas appropriated for purposes of winning FIFA™ green points. The inconvenient ecofeminism of her project, its memory of Khoena expulsions, and its design logics of water justice were erased. Angered that her work had been appropriated and translated into something with a set of values attached to private capital rather than the commons, Von Zeil closed her online commons of Camissa water knowledge.

In the same period, the national retail chain Woolworths drew on the water in the old Heerengracht pipe to service the toilets, air conditioners, and car wash at its corporate head office alongside the city hall, which, according to its financial report, would amount to "an estimated 27 375 000 litres of water a year . . . conserved by Woolworths."[65] In contrast to the SAB situation, the Woolworths plan to use the forgotten, piped stream was delayed as the city wrangled over charging for it.

Once approved, ||Hu-!gais now saves a corporate headquarters on the Heerengracht seventy-five thousand litres per day in car washes, chilled air, and executive bathroom flushes. The city that in Xhosa bears the name eNtshona Koloni, or the Colony, is the city that uses its veins for colons.

Like other modernist cities around the world, contemporary Cape Town has been designed with the idea of nature as something outside of itself, even though the twenty-five thousand hectares of Table Mountain National Park are at its centre, not its periphery. The park provides a last refuge for ||Hu-!gais's "nature," here defined as the biodiversity of the Cape floral kingdom. Though the park has appropriated that Khoena name so that "nature pays," it is not accessible to the many Capetonians for whom an annual access card is financially out of reach. The fences of the park are defended by armed police against the Khoena descendants in Hangberg, whose crowded settlement is too small, and who want to reclaim their own connection to the mountain.[66] Brutal street battles have followed.

Hoerikwaggo[TM]'s fence problem will never be sustainably solved by policing. This is because it is not a fence problem: it is an idea problem that requires a different conceptual ecology to resolve the larger problem of the exclusions that are inevitable in a city that lives outside its own ecology.

Thinking of the streams and their resurrection as corporate property, I am struck by a parallel from the early 1800s, when the British formalised the registration of slaves. Until 1816, enslaved people had no registration, therefore no visibility to the colony, and therefore no rules to govern their treatment.[67] The Registrar and Guardian of the Slaves kept records of deeds of sale and ownership, which in time became today's Deeds Office containing records of the ownership of property. The seal of the Registrar of Slaves and Deeds, and its seamless transformation over time to today's Deeds Office, tells us of the "thingifications" at work at that time: slaves and land were property. Becoming property, the enslaved took a step along the trajectory of settler "possessive logics":[68] from "becoming propertyless," and therefore losing one's legal personhood, to "becoming property" in someone else's possession.[69]

Cape Town's streams and springs had no political existence when they were struck off the city's asset register in the 1990s. Their reappearance in city life via sundry business proposals speaks to the forms of thingification in modernist settler-colonial thought. Their dramatic reappearance in the city's public consciousness during the multi-year drought that began in 2015 testifies to the devastations that have resulted in a world comprising only humans and things (subjects and objects) in modernist thought, and it underscores the necessity for rethinking the ways in which those designated "things" are excluded from political life. An ontology of persons and things, has devastated people and nature alike. In the words of Césaire that preface this book,

1.9 | British wax seal from Registrar at the Slave Office, set up to record slave ownership and punishments. The Slave Office later became part of the Cape Town Deeds Office, recording property ownership in the city. Image: Iziko Museum, Cape Town.

with modern European thought was borne a new process . . . a process of reification, that is, the thingification of the world. . . . The consequence you know, it is the appearance of the mechanized world, the world of efficiency but also the world in which people themselves become things. In short we are facing a gradual devaluation of the world, which leads quite naturally to an inhuman world on whose trajectory lie contempt, war, exploitation of humans by humans.[70]

The absurdity of regarding as progressive politics the upgrade of the enslaved to the status of owned property is clear to any thinking person. Should the same bringing-into-political-existence of streams and springs via corporate trademarks, in the neoliberal era, not provoke similar astonishment among our descendants who will have grappled with the effects on people and planet of thingification?

Amid the water crisis of 2017, the City of Cape Town proposed drilling another water source: the Cape Flats aquifer, while at the same time entertaining developers' proposals to buy up its zone of recharge (the Philippi Horticultural Area, or PHA), hard-surface it, build a commercially run prison there, and supply middle-class housing (for those who have the money to live in the ™ network). Nazeer Sonday and Susanna Coleman of the PHA campaign had long been arguing for food, land, and water justice in the PHA, demanding that the city enforce its laws on dumping and prevent the PHA from being carelessly stripped of its sand (which filters rainwater going into the aquifer) for glass and cement. Sonday points out that the PHA provides contemporary Cape Town with much of its food, a great deal of which goes to its most food-insecure areas, along with jobs and aquifer recharge. Located on the Cape Flats aquifer, the PHA is one of the very few water-secure agricultural areas in South Africa. Yet the city council, apparently imagining the city as if it was a spaceship unmoored from earthly ecologies, blocked and insulted Sonday and Coleman more times than the two care to recall. After extensive contestation, an official appointed a social science consultancy to collect different perceptions of the PHA for a vast fee—as if perceptions of reality were the issue; as if this is how climate change issues are to be addressed. It was a shocking example of the structural corruption of a system that should be called what it is: Social Science for Hire. The PHA case while heard in the Cape High Court while this book was in press.

At the same time, the city commissioned a private company to drill for water on the Cape Flats aquifer, with terrifyingly little thought to aquifer recharge, since without carefully managed recharge of a coastal aquifer, seawater will be drawn into it, as has already happened to similar aquifers elsewhere in the Western Cape. Farming with salt water is not an option. Neither is drinking it. And while the wealthy put down boreholes on their property as fast as the proliferating borehole companies could do the work, there was no management of their water extraction. One borehole driller with whom I was in contact told me that while salty water on the Atlantic coast used to go inland only as far as the seaside suburb of Table View, he was now finding salty water as far as Durbanville, some twenty miles from the coast.

Even as "Day Zero" approached, the city's relationship to the Newlands spring system bowed down to another castle: Castle Lager™, produced from the South African Breweries spring in Newlands. The city's strategic assessment of the future of its springs, in 2014–15, produced by the hydrogeological consultancy Geohydrological and Spatial Solutions International, had excluded the main Newlands springs from its commissioned report on the contradictory grounds that "the Newlands spring has no existing infrastructure that can be utilised. All infrastructure is the property of SAB Miller."[71]

That exclusion circumvented the necessity to consider whether it was strategic or in the city's interests to continue to allow one of the world's largest and wealthiest beverage companies to draw millions of litres from the commons for free during a drought—despite decades-long warnings from water specialists that the growing city was at risk of running out of water, not only because of its increasing population, but also because climate forecasts predicted a lower annual rainfall.

In a replay of the city's early history, Capetonians lined up at Newlands springs daily to collect limited quantities of water under guard, while a multinational took it for free. That multinational, SAB, is now amalgamated with Coca-Cola Shanduka Beverages, which had been part of the fortune made by current South African president Cyril Ramaphosa, founder of Shanduka Investments and one-time owner of McDonalds South Africa. Though Ramaphosa sold off his interests in Shanduka in 2014 when he returned to political life, the assets owned by his family trust have not been disclosed. Cape Town is witness to a new age of global water barons, built around a new kind of castle: this time one authorised by a ™.

––––––––

Post-apartheid South Africa's great knowledge crisis was AIDS, which taught us about mingled bodies. The flow of the disease mapped the entanglements of bodies across race, gender, sexuality, class, and property lines. Skin was no guard; neither was identity. The disease laid waste to the belief that social and political boundaries can stop the flow of materials. A virus threatening one person was a virus that could affect all. The lesson of HIV-AIDS is a reminder of the Khoena linkage of water and blood via flow: that waters and substances circulate through bodies rich and poor, black and white. The extraction of antiretrovirals from the water supply has become one of the new challenges for South Africa's water managers, nationally—not only in wastewater treatment works, but also where seawater near sewer outfalls was planned to be desalinated to provide drinking water to get through the drought (see chapter 6).[72]

Water passes through the bodies of all. The waters of Cape Town circulate through geological history: aquifers, rock, springs, and streams, and will continue to circulate to future generations. Via sewage and urban runoff, waters from Capetonian bodies enter the ocean, where pharmaceuticals, plastic particles, and household chemicals circulate through the bodies of seafloor creatures like sea urchins and barnacles, slowly ascending the food chain through lobsters and into seals, sharks, and large fish and bioaccumulating at each level. No fences and no laws halt flows of rock, water, and life. There is no division of "city" from "nature." There is no pipe that goes nowhere. The city is has made its own Anthropocene. To exit the current geological era without adding to the predicted Sixth Great Extinction or adding to the planetary load of expulsions, a paradigm shift in Cape Town water management is vital.

FRACKING THE KAROO

/kəˈruː/ kə-ROO; from a Khoikhoi Word,
Possibly Garo—"Desert"

2.1 | Tip of the *SubTerraFuge*, an installation in the Karoo contra fracking. Designed by
Nathan Honey and Isa Marques; built at AfrikaBurn in 2014. Photo: Lesley Green.

The dry semi-desert that is South Africa's Karoo began on the supercontinent Pangea that formed when Euramerica and Gondwana joined into one large landmass. Planetary temperatures dropped at that time, as the evolution of land plants elsewhere on the planet took up so much carbon dioxide that the earth lost its greenhouse roof.[1] At what was then the South Pole, the Karoo ice cap was kilometres deep, peaking between 359 and 299 million years ago (mya).

It was, many palaeontologists think, an early form of termites that ended the Karoo Ice Age (more recently renamed the "Late Paleozoic"). They gorged on the forests that had sprung up elsewhere around the planet, and their bellies converted the carbon that the trees had taken out of the atmosphere into methane, a potent carbon-based greenhouse gas.[2] Termite intestines produced such vast quantities of methane that the earth warmed. I like to think of this as the Antropocene.

Over the next hundred million years, glaciers scored the region's rocks as ice melted and thickened and melted again. Another hundred million years later, after Pangea had split in two between 175 mya and 55 mya,[3] the Karoo had become home and then graveyard to Jurassic dinosaurs and therapsids that marked the transition from reptiles to mammalian forms. Under ice, the Karoo basin was pressed down; as the ice thinned, that pressure was released and the basin was ruptured by volcanoes that threw skyward the purest form of carbon: diamonds.[4] Some of its rivers washed them into Brazil, whose continent was then still coiled into Africa like some ancient yin-yang symbol.

Fast-forward through the next eighty-nine million years of geological deposits, planetary wobbles, and mass extinction events[5] to about 1.7 mya: *Homo erectus* appeared in Africa in the Pleistocene.[6]

Slow-forward another 1.58 million years: *Homo sapiens* fossils appeared in the coastal region of South Africa, and inland on the South Africa–Swaziland border.[7] In the complex precolonial history of the area, the !Xam and the Khoena peoples came to dwell in the now-dry Karoo and the Kalahari Karoo Basin, and over time their neighbours came to be Nama, Tswana, Sotho, Swazi, Zulu, Xhosa, Shona, and others across Southern Africa.

Pause at the moment when colonisation commenced at the Cape about four hundred years ago, in the 1600s, followed by the struggles between the Dutch and the British that led to the British takeover of the southern tip of Africa in the 1790s. The discovery of river diamonds in the region of the Orange or Senqu River (to use the name it is given where it rises) led to an alluvial diamond rush controlled by local chiefs until early 1870, when they were overwhelmed by incoming settlers.[8] The discovery of diamond-bearing rock in the northern Karoo in 1869 swung the British Empire into inventing the local technosphere, in which metal mining structures, wooden beams, steam engines, long guns, and the muscles, bones, and guts of migrant labourers would combine to connect the volcanic residues of the Late Cretaceous era with the economic and political landscape in South Africa and Britain. At that time, 90 percent of the world's industrial diamond market came from the region. Diamonds gave mastery over geological matter to those who made machines to use them, for diamonds were the hardest known substance on earth. With them, their industrial owners could cut any other material.

Profits from the sale of Late Cretaceous–era diamonds from 91 mya fed the formation of cities, corporations, and institutions in Britain and her Cape colony. With money from the Rothschilds, the entrepreneur Cecil John Rhodes amassed a personal fortune from the diamond rush, taking control by means fair and foul of claims around the Big Hole of Kimberley, where the largest kimberlite volcanic pipe extruded. Appointed prime minister of the Cape in 1890, Rhodes set about establishing a legal infrastructure that favoured mining, and a social infrastructure that established race-based disenfranchisement, creating a class of Black labourers who would serve the emerging white-owned mining houses. A fight with Afrikaners over British authority would lead to two Anglo-Boer wars (also referred to as the South African wars) and sow the bitter seeds of Afrikaner nationalism and apartheid.

Determined to try to regain the wealth of the diamond industry, the Boers laid siege to Kimberley in 1899. The British won—making a cannon called Long Cecil in the workshops on the diamond mine—and wrested land from the Afrikaners, inventing crimes of war that Adolf Hitler would later use against civilians: concentration camps for women and children, destruction of domestic property, and economic plunder. Afrikaner farms were scorched, and livestock slaughtered. Black South Africans' land rights were stripped in 1913; black economic activity became largely confined to physical labour, much of which took place on the mines.

In the 1900s, the Carboniferous era from around 300 mya entered South African politics via South Africa's coal-fired power stations. In the 1960s, the newly independent Republic of South Africa, under the political leadership of Afrikaner nationalists, sought energy autonomy in order to pursue formal policies of race-based segregation. Nationalist technocrats commissioned geological surveys for coal, oil, and uranium. Coal-fired power stations are still being built in this country, despite South Africa being one of the highest per capita carbon emitters in the world.

A little over fifty years later, different molecules in the 300-million-year-old subterra are the subject of parliamentary debates. The target is Karoo methane, trapped in shale: a power source at the far end of recoverable petroleates.

On these gas molecules the governing African National Congress pinned its hopes for an economic revolution strong enough to stave off rumblings of political revolution, choosing as its allies the fracking arms of the oil companies Shell, Chevron, Falcon, and Bundu.

And, indeed, the engineering consultants who were advising on such extraction in South Africa seemed to believe that the cement they proposed to use to line and plug the thousands upon thousands of fracking wells and waste-water sites will be able, perpetually and to a depth of up to five kilometres, to withstand the pressures of gases and liquids, the movement of tectonic plates, and the forces of contending reason.

Cement is a magical substance in the cosmology of modernity, through which engineering has come up with ways of keeping fracking water or nuclear waste out of hydrological cycles and tectonic processes. Spellbound, engineers, lawyers, politicians, and corporate executives appear to believe that when cement is combined with environmental regulations, earthly forces and geological processes are stupefied and will magic themselves away. Cement is perhaps *the* artefact of modern humans' belief in our mastery over nature, and the conceptualisation of its role in hydraulic fracturing speaks to a confusion of states of matter with matters of state.

The idea that cement can perpetually keep apart the geospheres throws into comic relief the claim that the modern state's regulatory framework is adequate to govern the era of extraction. Indeed, it is not an exotification to say: *look at fracking landscapes and observe: cement is believed, by those followers of the gods of industry, to offer an immunity to geological time, and to the physics of flows, and to the forces of human history.* Such an irrational belief in the power of cement confers upon modern-minded humans the power to enact upon the earth the transformation of liquid to solid; the division between economics and ecology; the separation of human activity from ecological and planetary systems; and the transmutation of belief to knowledge. Occupying a special status outside of space and time, in this cosmology, cement appears to be assumed immune to tectonics, and impervious to osmosis. It is this belief that allows the illusion that discharging the frackers' 750 waste chemicals into a cement-lined piece of the subterra—among them heavy metals, carcinogens, endocrine disruptors, neurotoxins, and respiratory distressors—will confine them to a particular space, on a specific property, and in a legal territory, in perpetuity, even though the planet's history is that of flows between states of matter, and modern history is that of warfare on enemies in all six planetary spheres: techno-, atmo-, hydro-, bio-, cryo-, litho-. In cement, the geophilosophy of human exceptionalism is made concrete: the self-image of mod is denatured, dematerialised, separated from the planet itself.

After the Khoena had fled the Cape around 1710, their remnants formed alliances with !Xam (San or Bushmen) in the Cedarberg, Namaqualand, and Karoo, and the amaXhosa in the Eastern Cape. Without those alliances, they found themselves indentured servants, enslaved, or precariously vulnerable to being shot. By the mid-1700s, a series of wars in the Karoo crushed the !Xam resistance to newly proclaimed farmlands.[9] But dependence on erratic rains and small springs made farming a risky enterprise. It was the arrival of the windmill a hundred years later that made it easier to transform ecological relations in the Karoo so that they could become national economic assets.

The multibladed steel windmill designed for the dry American plains in the 1880s had proliferated in the Karoo by the time of the official 1905 census.[10] The windmill was a miracle. Its capacity to work in low winds and lift tens of thousands of litres of water per hour from the subterra enabled farmers to increase their sheep and ostrich flocks, and join dry land to new emerging global markets—the ostrich-feather boom, the military industry during the First World War that needed wool for soldiers' uniforms. It shifted poor farmers from a dry periphery, at least for a time, into the centre of the great new financial flows in globalising colonial markets.

The settling of the Karoo was indeed a matter of guns, germs, and steel windmills—with them, air and aquifer entered into a new relationship with the landscape. There was less need to wait for rain. Wind could make water. An entire industry of water diviners sprang up, whose wobbling sticks promised farmers blue gold.[11]

Thus began phase 2 of settler hydropolitics in South Africa, this time not in Cape Town, but in the Karoo, where farmers and labourers could stock arid land with vast numbers of sheep and ostriches. More farms meant more fences, and more people meant more hunting; the great migrating springbok herds that had taken days to pass Karoo towns like Graaff-Reinet, Cradock, and Beaufort West gradually disappeared. Confined by barbed wire, ostriches and sheep consumed plants in great squares and rectangles until the tastiest plants too were among the disappeared, unable to hold soil in place, generating deserts, badlands, and soil erosion,[12] leaving poisonous or inedible species[13] in a landscape that needs more than twenty years to re-establish common Karoo shrubs, including:[14]

Brownanthus ciliatus (Haarslaai): 17–19 years
Ruschia spinosa (Doringvygie): 18–22 years
Galenia fruticosa (Vanwyksbrak): 23–29 years

Tripteris sinuata, also called *Osteospermum sinuatum* (Bietou or Skaapbos): about 21 years

Pteronia pallens (Scholtzbos): about 85–100 years.[15]

Generations later, ostrich farms have been through boom and bust along with the global ostrich-feather market, rescued mostly by the promotion of ostrich meat in elite supermarkets. By the end of the twentieth century, sheep farmers were operating at a fraction of their grandparents' productivity, at a time when wool prices had fallen in real terms and bank interest rates had spiralled.

By the last quarter of the twentieth century, the Karoo required ecological restoration. But at the same time, agriculture was increasingly tied in to global commodity trading, which required expanding and intensifying even as the apartheid state's protective measures and benefits that had built up white agriculture were falling away.[16] Not a few white farmers feared that their land might be expropriated, or ownership turned over to farmworkers, many of whose families had lived on the farms for generations, and whose children had suffered poor education. Nelson Mandela's new democracy looked favourably on environmentalism, however. The light-bulb moment for many farmers was this: turn farms into nature reserves. Then capitalise on tourism; let the land recover; build hunting lodges; ally with conservation; buy up neighbouring farms if you can; enforce pre-emptive evictions; and rely on casual labour.[17] A growing number of city-based and international buyers bought up land for hunting, taking advantage of the weakened local currency, building up the Karoo into an area of "lifestyle game farming," with many speculating that game ranching would, in time, become a lucrative commercial industry that could help turn the tide on extinctions of blesbok, bontebok, Cape mountain zebra, black rhino, gemsbok, eland, black wildebeest, and even leopard, lion, and cheetah.[18]

But at the same time as paddock fences came down, the newer, higher external fences of game farms spoke of a new trend in the Karoo: the expulsion of low-skilled labourers who had lived all their lives on farms and been poorly educated at farm schools. The result: retrenchments swelled informal settlements.[19] While wildlife ranching was addressing extinctions, it was also contributing to expulsions from farms to shanty towns, and creating a bioeconomy of killable wildlife with high-value hides and horns.[20] The market-led gentrification of the Karoo private nature reserves meant creating around animals both pockets of wealth and shanty towns of poverty. The same fences that stopped the herds from escaping and prevented poaching also put their long-term sustainability at risk, because fencing off extreme human poverty from even modest environmental wealth renders impossible the building of environmental alliances among residents. The good fences that make good neighbours are those that are tended by both.

This is important, because in the twenty-first century, the material base of political economy has shifted globally. Nationally significant wealth potential no longer inheres in water (for the moment), but in the energy that can power up electrical current: in petrocapital, in uranium, and in the secretive shell companies favoured by the resource extractivists who have corrupted democratic institutions through sophisticated tax-avoidance strategies. The global order has shifted from hydropolitics to what anthropologist Dominic Boyer calls "energopolitics": political power based on energy supply, amid the rise of mega-wealthy multinational corporations whose executives are so cosy with economic policymakers and other government officials, to the detriment of environmental governance.[21]

For the few small bands of !Xam left in the Karoo in the 1800s, the clanking windmill ushered in a hydropolitics that marked the end of their world. For many farmers of the Karoo in the contemporary era, the spectre of the fracking derrick that would usher in the era of petropolitics suggests the end of theirs.

II | A REPARATIVE JUSTICE VIA PETROPOLITICS?

"Fracking" was a word first uttered in the Karoo around 2007, when Shell South Africa announced an interest in the tiny gas bubbles trapped in the Karoo's shales. A court challenge led to a two-year moratorium on fracking. On its expiry in 2014, then-president Jacob Zuma announced that fracking would go ahead. A public outcry followed, brought to a head by the announcement from Cancer South Africa that carcinogens were rife in the chemicals used in fracking, along with neurotoxins and endocrine disruptors. In response, the government appointed a Strategic Environmental Assessment team, led by ecologist Bob Scholes.

In their first public consultancy meeting in the Karoo in November 2015, however, the assessment team took a shot across the bow "when Cape Town-based Graaff-Reinet public relations and public management entrepreneur Stewart Giyose stood up and objected to the racial composition of the meeting, at that point evenly divided between white farmers and black mining entrepreneur groups. Giyose demanded that the 'majority of the people [i.e., South Africans of colour] should be here to have their say.'" He and two others claimed that the meeting had not been properly advertised to the local community, and in the end the meeting was postponed.[22]

It was an early indication that debate on Karoo fracking was breaking along the fault line of race: between black entrepreneurs and government officials who argued that a new wave of petrocapitalism would address the impoverishment left behind by the hydropolitics of the colonial and apartheid eras on lands still dominated by white farmers and, on the other side, white environmentalists claiming that the Karoo was a near-pristine treasure needing protection.

In the United States, opposition to fracking had been rendered mute by a framing of the argument as a matter of "rational economics" over "irrational environmentalists."[23] Translated into the South African context, "petropolitics versus greenies" became black developmentalists versus white environmentalists. That fault line split into a chasm when the major South African body opposing fracking, the Treasure the Karoo Action Group, forged an alliance with Afriforum, a conservative body representing the interests of white farmers, avoiding questions of land justice and historical ecologies.

At that point, I felt I had no option but to walk away to try to think about what kind of Karoo environmentalism I could support. For the alliance brought into the sharpest relief the difficulty of forming an environmental public in South Africa that was not defined by a romantic idea of nature, without thought to the ongoing structural violence of grinding poverty in the new shanty towns that were springing up across the Karoo as farmers evicted workers, so often in the name of nature and ecology.

Yet: how could I walk away—surely it was obvious that petropolitics had constituted some of the major decolonial struggles in Africa, the sagas of which continue to unravel now in the Panama Papers, and stories of secret deals between European economies and African oil? Nigeria's petrodelta struggles with Shell,[24] which led to the execution of the novelist Ken Saro-Wiwa, have much in common with Ecuador's recent struggles with Chevron over toxic spills in the Hoarani rainforest.[25]

Echoing those relations of deception was the artwork *SubTerraFuge*—installed in the Karoo, in the proverbial middle of nowhere, on a farm next to SANParks' Tankwa National Park—it was a multi-tower installation about the relations on which fracking depends. Its name is a conglomerate of **Sub-:** *under or below a particular level or thing;* **Terra:** *land or territory;* **Fuge:** *expelling or dispelling either a specified thing or in a specified way,* and a play on **Subterfuge:** *deceit used in order to achieve one's goal.*[26] Constructed as part of the annual "AfrikaBurn" festival in 2014, it connected earth and sky, living and dead, the permanence of industrial infrastructure and the temporality of fire, and referenced the tall infrastructural installations that would be required to turn this part of the Karoo into an extractive petropolitical economy, amid the windmills, church steeples, and conical cedars planted in old graveyards across the Karoo. The *SubTerraFuge* was a comment on the *subterfuge* through which the language of economic growth measured in capital is assumed to trickle down through layers of stratified humans to the benefit of all. Such subterfuges continue to divide the necropolitical—a politics of death—from an ecology of life across all spheres, including the university, whether in macroeconomics or mining engineering.[27] (The *SubTerraFuge* was designed by Nathan Honey, one of the set designers for *Mad Max Fury Road*, starring South African actress Charlize Theron.)

So: in the face of such a powerful statement, and amid the intense struggle mounted by the sole Karoo defenders under the banner of the Treasure the Karoo Action Group, how could any justice-focused environmentalist take a step back from the fracking struggle?

What unnerved me was the broader context in which racial legacies were woven into the battle. An uncritical alliance with AfriForum, and the claim that the Karoo was all but pristine, were positions I could not support, and would almost inevitably lose given the historical need to address race-based economic injustice. The questions I was asking were different. How might activists reconstitute an ecopolitics based on justice, in the Karoo? Was petropolitics, with its historically untrustworthy promise of trickle-down and its dreadful ecological record, the only guarantee of post-apartheid reparations in the Karoo, or was there another way to join the dots to bring about a reparative land-based ecological justice? Would game farms achieve that, with their track record of expulsions? Was a different hydropolitics needed? What was I not seeing?

Nelson Mandela's former minister of water affairs and forestry, the late Kader Asmal, published his last-ever opinion piece expressing concern about fracking in the Karoo in 2011. Fracking, he argued, posed a significant risk of water contamination and scarcity. But, he noted, "We're lucky. We have—as part of our constitution—a Bill of Rights that contains . . . a right to 'an environment that is not harmful to (our) health or well-being.'"[28]

Asmal was surely the first-ever professor of human rights law to have become a water minister, and surely the first-ever African minister of water whose obituary was in *National Geographic*. Asmal's genius was to apply his expertise in human rights law to water as he sought to undo the laws, infrastructures, and management rules of settler-colonial hydropolitics.

For Asmal, the fracking argument would be won by assessing evidence in relation to a human rights framework that included the right to a healthy environment. This was his hydropolitics at work—Hydropolitics II, or a hydropolitics of justice. I could imagine many well-meaning justice activists and academics making that argument, and indeed had heard several scientists propose that a strong justice and regulatory framework would prevail: "We just need to make sure we gather the evidence, and governance structures will ensure the right to a healthy environment."

However, as I studied scientific evidence and court cases against fracking in the United States, the ground of contest appeared not to be about the right to a healthy environment at all. Instead, what was fought over was *the nature of the evidence itself.* After a major court case in the state of Pennsylvania, in which four of five justices found that fracking was harmful to the public, the fiscus, and future generations, a special audit of environmental regulation failures found multiple failures of governance, almost all of which related to the hiding of evidence. In Pennsylvania, what petropolitics had presented as the incontrovertible, objective science of nature that would lead to great wealth was a science that had been fractured by the power of industrial energy combined with state collusion.

The more I studied the situation in different American states, the more it seemed to me that a combination of lawfare and fragmented science was corrupting the conventions of evidence. By requiring expensive and often logistically unattainable standards of proof; settling out of court; discouraging local government officials from investigating; requiring non-disclosure agreements in exchange for compensation, and even hiding behind intellectual property law, the energy industry's version of reality was sealed off from empirical disproval. The corporate version of truth was an "occult" view, to borrow astronomers' language for when a larger object like Venus disappears behind a closer, and smaller, object like the moon. Missing in scientific assessments of "the evidence of harm from fracking" was an analysis of the effects of industry practices on the available evidence. In that context, questioning their science (always "The Science") left critical scientists open to accusations of being unreasonable, irrational, illogical, anti-science, or unpatriotic: in a word, "biased," which in the research world is as magical a word as "Stupefy!" was to Harry Potter, because both words call actions to a halt.[29]

In addition, it seemed that I was looking at a petroleum industry that *promised* to base itself on a science of fracking, but *in practice* was fracking science. I've published elsewhere a detailed study that critiques the major review of fracking published in the journal *Environmental Science and Technology* (*EST*).[30] The journal is authorised by the American Chemical Society and is one of the most-cited journals in these sectors, so an industry review paper in *EST* is important. The 2014 *EST* paper "A Critical Review of the

Risks to Water Resources from Unconventional Shale Gas Development and Hydraulic Fracturing in the United States" constituted a definitive expert intervention in a polemical matter.[31]

That a team of distinguished scientists concurred, in that paper, that there is evidence of significant risk to water from the hydraulic fracturing industry is important, although their certainty that for every problem there is a mitigating engineering solution is not one that I share. Their conclusion that "the direct contamination of shallow groundwater from hydraulic fracturing fluids and deep formation waters *by hydraulic fracturing itself*, however, remains controversial"[32] draws directly on the rhetoric of the oil industry, which claims that underground activities are separable from surface activities, and that when reduced to what happens below ground at the point of the fracking drill, hydraulic fracturing had not been conclusively demonstrated to be damaging to vital sources of groundwater and aquifers.

That conceptual disappearing trick of the whole to a tiny part of its total activities is at the heart of the fracking matter. In the documentary *Unearthed*, director Jolynn Minnaar interviewed Mark Boling, the president of development solutions for Southwestern Energy, one of the largest gas drillers in the United States, who set it out neatly:

> What the companies are thinking of in their mind when they say "hydraulic fracturing hasn't caused anything" is they mean the actual activity of the completion down at 4,000 feet or whatever. I think what a lot of the public is thinking is hydraulic fracturing is the whole thing. It's the drilling of the well, it's the casing of the well, it's the completing of the well, it's all of that. If that's your context, then I can understand why they would say, "You're not telling me the truth."[33]

According to this scenario, and the EST article, the industry as a whole is seen as a neutral one with bad component parts, and perhaps occasional incompetence, that simply needs to be improved via research and regulation. The publication of the paper in a respected scientific journal enables scientists at universities to continue to sign off on research projects aiming to improve hydraulic fracturing techniques, safe in the certainty that their work can attend to human errors and oversights, and that these can be remediated and regulated. It also enables regulators and governments to argue that any polluting events are not the result of hydraulic fracturing itself.

But the idea that hydraulic fracturing itself can be conceptualised as independent of the accidents and damages with which it has been associated is a move that occults (hides) seismic effects; loss of health and well-being relating to toxicities in air, water, and soil; illness; death; livestock losses; radiation increases; light pollution; and noise pollution.

The combination of lawfare via non-disclosure agreements and intellectual property defence, added to the fractioning of the concept of "what fracking is," constitutes a corruption of the conventions of evidence.

These observations throw into question the efficacy of Asmal's proposals. In the fossil-fuel era, where a single oil multinational can be wealthier than several countries, shale gas companies and allied environmental regulators have been too slick to allow the argument to move into the terrain where they could not win—the human right to a clean environment. That argument would of course be incontestable. So it had, at all costs, to be kept out of "society" and within "nature," where a patriarchal Poppa Knows Best version of science could keep people spellbound. The situation exemplified the problems that inhere where contemporary parliaments deal with matters of society, and leave matters of nature to "The Science," where there is no discussion or accountability over how knowledge is produced.[34] Curious about parliamentary oversight of the fracking debate, I attended a Parliamentary Portfolio Committee meeting in 2014, where parliamentarians would be in discussion with representatives of the Department of Mineral Resources (DMR). Over tea, I commented to the DMR's science advisor that he must be pretty busy, serving the department that oversees mining in a country where minerals are an economic mainstay. He laughed at my remark. "I don't only serve this ministry," he said. "I'm the science advisor for four ministries!"

Built into South Africa's parliamentary organogram, then, is the assumption that science needs no discussion. This organogram reflects a cosmology: that science is Science, and no questions are expected from anywhere outside of the institution of science. In this kind of political life, nature is a "Naturestan" known only by the scientists who claim to understand its depths. Absurd results follow: in the name of science, the DMR, at that meeting of the Parliamentary Portfolio Committee, proposed American Petroleum Academy guidelines for safe fracking in South Africa—but US guidelines for pipe thickness are for depths of up to two kilometres, whereas in the different geological situation in the Karoo, wells would be drilled to a depth of five to six kilometres underground, involving substantially different pressures.

Without sufficient independent science, petropolitics would be able, from its extra-parliamentary seat in Naturestan, to claim the authority of science and the moral high ground of economic growth; manage the law of evidence; claim to be non-political, and evade parliamentary oversight over what questions were asked and how they were answered. Indeed, one of the

...rliamentary Portfolio Committee members' questions, in that oversight committee, was for the DMR to please explain the difference between shale gas and fracking—a bit like asking the Ministry of Agriculture to explain the difference between potatoes and farming. So much for rigorous governance!

Addressing the petropolitical Naturestan requires more than adding ownership equity to its practices. Petropolitics promises liberation by the barrel: that by sacrificing ecologies, historical human injustices will be restored. But petropolitics involves a profound and total transformation of the social networks, logistical networks, and biogeochemistry of life and activity. It is a rearrangement of the material relations of a regional ecology and economy, and a justification of regional "zones of sacrifice" with the claim that economy is separate from ecology. But when biogeochemical zones of sacrifice are created, zones of ecological death may follow for generations, engendering what Achille Mbembe calls "necropolitics"—a politics of death, not a politics of life,[35] and what Conor Joseph Cavanagh and David Himmelfarb describe as "necropolitical ecologies."[36]

The push to frack the Karoo brings into sharp relief the necessity for reconstituting a political conversation about ecology—not the ecology of the Naturestan as somewhere that only scientists may go, but an ecopolitics that recognises that human bodies exist in the ecologies, the geologies, that we make and leave behind—and which flow. Climate change points to the fact that there is no pristine nature that can be kept apart from damaged zones. Moreover, regulating damage to the environment in South Africa solely with reference to the possibility for harm to human health, or the potential benefit to the economy, is inadequate. South Africa's environmental clause, in its Constitution, needs a revision.

The need to constitute a new ecopolitics is the most important political challenge of our times. As long as pristine nature is the leitmotif of the environmentalists—that the environments that are worth protecting are those that are pristine, like the Karoo—it will keep scoring an own-goal. Aha! the petropolitical or economic development lobbies will respond: you are wrong! Karoo nature has been damaged by sheep farming; therefore, in terms of your own argument, fracking may go ahead, and we hereby claim the moral high ground in the name of addressing the very settler-colonial history you deny.

In order to generate an environmental science and law capable of addressing the unhelpful inherited split of nature and society, ecologies need a better presence in Parliament than one scientific advisor for four ministries. Parliamentarians should be equipped to ask advising scientists not only about their results and methods, but about whether their questions and concepts were adequate.

Activists need to be asking legislators, Why do we have corporate legal persons in our economic and financial systems, but no legal standing for ecological actors such as aquifers or soils that are crucial for economics and inseparable from survival? Kenya, Ecuador, Peru, and Bolivia have begun to work into their social contract–based constitutions the beginnings of a natural contract.[37] Their constitutional and legal revisions suggest ways in which ecologies can be protected not only as an extension of human rights, but with respect to the well-being of soils, water, and landforms themselves.

The fracking question, and the concern to frame a new ecopolitics, also raises questions about the time frames of regulation. Fracking puts at risk the Karoo's aquifers, which were formed and filled on a fossil timescale. Harming them, without knowing either their extent or their interconnections, would potentially cause effects at fossil timescales. How could environmental regulation address not only the rights of future generations a hundred years hence, but the rights of future forms of life? By what right, reason, or logic can fossil fuel extraction be governed in time frames that suit financial and electoral cycles? Regulators need to consider what it means to draft environmental regulations for activities that will permanently displace highly toxic molecular compounds, using seepage-preventing materials such as cement which retain little to any effectiveness after fifty years, and which are resistant neither to the seismic effects that the industry itself will cause, nor to the movement of tectonic plates, as will occur in the future.

Then there's the question of whether the framing concern, in ecopolitics, should be the commons or private property. Fracking sets in motion molecular flows that cross state, property, and bodily lines. Fracking cannot be managed by property law. What would need to happen in law in order to address its biogeochemical flows? Can they be managed? If these flows cannot be managed in law or on land, in what sense are they legal?

Land redistribution is an urgent task in South Africa. Land redistribution with poisoned water and soils would leave to future generations a new, cruel historical legacy of apartheid: this time in toxic reparations. It is correct that contemporary economic power lies not only in land and water, but also in energy. The key is to ensure that energy is not a new subterfuge for an old necropolitics—that it is, rather, part of a politics of life and well-being.

III | KAROO ECO-POLITICS: THE CASE FOR RENEWABLES

Dominic Boyer, founding director of the Center for Energy and Environmental Humanities in Houston, Texas, comments that the fossil economy is fossilised. It's a particularly apt point in South Africa, which is reliant on a single centralised energy provider that is dependent on coal and nuclear power, and characterised by long, inefficient supply chains.[38] Shifting the relationships with bulk utility companies globally has been key to the success of countries' efforts to shift to wind and solar energy, something that is essential if the production and distribution of electricity are to be democratised.[39]

Resolving this would be key to the Karoo and its long-term future. Consider that South Africa's Koeberg nuclear power station produces 13,668 GWh (gigawatt-hours) per annum on average via two reactors, while the Karoo solar energy offers an average performance of 3–6 kWh (kilowatt hours) per square metre per day. Assume an actual output of only 4 kWh, and multiply that by 365 days per year, and every square metre of the Karoo has the potential to produce 1,460 kWh per square metre per annum. Of course, some days would be cloudy—but we can take this into account by averaging it out. If I take the Koeberg figure and convert it to kilowatt hours, it is 13,668,000,000 kWh. Divide that by 4 kWh per square metre, and the number of square metres needed is 3,417,000,000. The square root of that is 58,455 metres, or an area 58.5 by 58.5 kilometres of the Karoo under solar panels to make energy at the equivalent rate of Koeberg. If one allows for gaps between the panels, and roadways among them, the final area would be marginally bigger.

Giving over farmland to solar farms is not in itself ecologically friendly: for the land itself, solar panels may be quite destructive. Are there alternatives? Imagine if new housing for displaced Karoo farmworkers was roofed with solar tiles, and tied in to the grids, with outputs that both supplied each house and sold excess to the grid, as is possible elsewhere in the world. That would be an empowering and decentred approach to energy production, and a viable alternative to the fossilised fossil grids whose protocols of ownership, illogically, are structuring solar farms and wind farms.

"Mining the sky" is the logo of one of the solar companies working to set up large solar energy plants in South Africa—working with today's energy, today, instead of mining the solar energy stored in the earth from the years before the dinosaurs. Current solar farms and renewable energy projects could be strengthened by changing policies to remove current caps on the amount of renewable energy that may be produced, and by investing in an energy infrastructure that could take feeds into the national grid from many places. Most important of all is the need to think about what relations will offer a permanent basis for a reparative ecology in the Karoo.

Transitioning to renewables, in South Africa as elsewhere, has seen big utility companies seeking to control the transition in ways that preserve their monopoly. A post-fossil future requires a different political ecology of energy:[40] short supply chains; many power production centres; the pluralisation of energy futures with interlinked energy producers;[41] the democratisation of energy debates.[42] But whatever energy networks are in the future of the Karoo, none will exist outside ecopolitical relations.

The technologies of the windmill and the fracking derrick and solar arrays each create different political, economic, and ecological relationships among rock, water, and air; each yields different social formations and establishes different ecopolitical relationships. What resources are there to draw on, to imagine an ecopolitical future in this landscape that has seen genocide and ecocide, and which is again at risk in the current regime of extractions, extinctions, and expulsions?

In !Xam, the Karoo was the "garo," the desert or dry land, in which the knowledge of water holes is only a part of what a person needed to know to survive: the larger whole is to protect the relationship that people, land, and living beings have with rain. Pippa Skotnes's index entry for "rain" in her curatorship of the Bleek and Lloyd Archive of !Xam stories runs to some twenty A4 pages when dropped into standard typescript. Glimpsed in them is a Rain that is not a thing, but a Being who transforms landscapes and plants, animals and people.

As a Being, Rain could caress gently or carry away houses. Rain could whisper or roar, or be angry and shoot with hailstones. Rain makes its approach known through animals: the lizard comes down the tree; the tortoises walk uphill; the puff adders move; the springbok come. Rain affects caterpillars and ants. Rain comes with the wind. Rain is linked to falling stars, finding food, flowers, swallows, and hunting. "When the Rain is angry with anyone, people may be carried off in the whirlwind."[43] The old people say that when the rain is angry, it wants to kill people.[44] Rain can be requested; Rain is to be respected. Rainmakers' work is to lead the Water Bull across the landscape. Rain is the legs of a large cloud that is the body of the Water Bull. Rain has a scent and a wind. Rain's darkness resembles the night. Rain brings cold; Rain becomes ponds. Rain brings food (uientjies, or little onion-like bulbs in the ground) and springbok. The hearts of the dead Rain-bringers fall into water pits; they become part of the fount of water.[45]

Asking readers to think with the presence of the Rain Being, in all its indicators and effects, is not to ask for belief in the Rain Being, but to allow the !Xam who lived in the Karoo before windmills to compel thinking and feeling about relations with water in a dry landscape as something other than kilolitres in a dam or the depth of a borehole. According to Isabelle Stengers, "To name is not to say what is true but to confer on what is named the power to make us feel and think in the mode that the name calls for." She goes on to write of the idea of "Gaia":

Naming Gaia is naming a question, but emphatically not defining the terms of the answer, as such a definition would give us, us again, always us, the first and last word. Learning to compose will need many names, not a global one, the voices of many peoples, knowledges, and earthly practices. It belongs to a process of multifold creation, the terrible difficulty of which it would be foolish and dangerous to underestimate *but which it would be suicidal to think of as impossible.* There will be no response other than the barbaric if we do not learn to couple together multiple, divergent struggles and engagements in this process of creation, as hesitant and stammering as it may be.[46]

The Rain Being of the !Xam may be a figure who has the power to remind us that while industrialists have devised Big Yellow Monster earthmoving equipment and trademarked them "EarthMaster" or "Caterpillar" or "Earth-Mover," corporate multinationals (whether petrochemical or renewable) are not in charge of the rains on which the Karoo ecology and economy depend. At issue is not the reality of the Rain Being, but the fiction of corporate mastery. The Rain Being stands as a figure reminding markets and technologies and legal regimes that they may extract, but they cannot live without the relations that compose life in this dry landscape.

The idea of a responsive earth is all but ungraspable to those who grew up with the illusion of industrial mastery over an unresponsive, mute, inert earth. Different kinds of figures, Stengers proposes, can compel different ways of thinking.

In the Karoo, the figure of the Rain Being is a reminder that the dry interior of South Africa is not a silently enduring vista awaiting extraction via windmills or fracking derricks or uranium miners or solar panels. With gifts of water the Rain Being calls to life the dust, plants, and creatures of the Karoo, inviting those who would live here to share in that convocation.

PART II

PRESENT FUTURES

In his play *A Tempest*, Aimé Césaire works with the relation of the Monster-Slave, Caliban, without whom the Magician-Duke, Prospero, could not exist because he would have died on the island on which he was shipwrecked. Yet Prospero believes that he has "saved" Caliban and that without him, Caliban could not exist. Caliban is the name Prospero has given the man who calls himself Uhuru, rather than a name from the Caribbean that is an anagram for "canibal."[1]

Césaire uses the pairing of Prospero and Caliban, Master and Slave, to explore the relationship central to modernity and coloniality alike: that of mastery. The Monster-Slave—like the poachers mentioned in the introduction—is defined via a moral violence that in turn justifies the violence of the salvation that the Magician-Duke brings to the world. What Césaire achieves in *A Tempest* is to show that the Magician-Duke is dependent on his Monster-Slave, but believes that he is the self-made man, and cannot see that he is trying to save his Monster-Slave from the monstrousness that the duke himself has scripted.

Césaire's exploration of the relations between subject and object, Master and Slave, is an argument that opposites produce one another. As such, they are more like yin and yang than opposites in two separate columns. Master and Slave are performances of ideas about the self; they are not essences, but relations. The idea was revolutionary in the 1940s and 1950s, when it first took on the logics of coloniality—as do the movements #BlackLivesMatter and #RhodesMustFall, almost three-quarters of a century later.

The idea of the self-made man was exemplified in popular accounts of Cecil John Rhodes, the colonial prime minister who dispossessed so many black South Africans of their land. The narrative of Rhodes reflects one of the gods of reason of our times: the belief that white privileges have been earned. The self-made man is one of the great gods of reason of the modern era; as a figure, it serves to persuade its adherents, in the face of contradictory evidence, that they have achieved or "done it"—whether personal success or knowledge-making—all by themselves. When Césaire renders white privilege in the figure of the Magician-Duke, he makes it possible for us to understand the shadow side of coloniality.

A change in this god of reason in our time demands a process of engaging with the figures of the Magician-Duke and the Monster-Slave as they have come to figure in the social sciences and the sciences; in the deepest recesses of the psyche; in the territories of necrotic landscapes; in the geographies of the monstrous that have been wrought on the earth; and in the white-man's-burden form of environmentalism and nature conservation. What the paired Magician-Duke and Monster-Slave enable us to see is that the power of the master is both political and discursive, and that the enslaved's self and the master's self are tangled up in one another. In this duo are found the figure of the subject and the object—what Nelson Maldonado-Torres describes as the war at the heart of modernist thought.[2]

Post-Holocaust, many Jewish philosophers pursued a similar route. Think of Jean-Luc Nancy's *The Birth to Presence*, which is part of his project to get away from the risk of nationalist political tyranny that arises when an ontology of Being is transposed into the social realm. Think of Emmanuel Levinas's description of the "face-to-face" encounter that forces a Nazi prison camp guard to engage his humanity, rather than his category.[3] Jacques Derrida's career began with a treatise on the abstractions of geometry,[4] reflecting on the ways in which the production of Cartesian grid space separates humans from the world; after his visit to South Africa around the time of the assassination of the anti-apartheid activist Chris Hani, Derrida rethinks the ontology of Being via hauntology;[5] a few days before Derrida's death, he concludes the manuscript of *On Touching*.[6]

Derrida's intellectual journey from his first book—a critique of geometry as a philosophy that separates one from the world—to his final work on Nancy's thinking about touch and presence as philosophy is paralleled by Leopold Senghor's account of Africa's "touch-reason," which he saw in opposition to abstract reason. Senghor distinguished between "Being Black" and what he called "Living Black." The former he spoke of as being part of a category subsumed by its relation to white mastery; the latter, as a way of relating to the world. And he distinguished what he called the European desire to "master nature by making it an instrument of their will to power" from the potential he saw for knowing the world via an African "touch-reason"—"not to oppose itself to nature, but, in a reciprocal embrace, to unite itself with it."[7]

What Senghor points to here, however, is a step further than whiteness. For him, white supremacist thought is central to the modernist imagination of the world, in which it is not only a theory of the person (in which unbridled individuality leads to greatness), but a theory of the categorisations of the things in the world itself, in which the world is there for the Great Man's inventory, and therefore also for the taking. Narratives of individual greatness are part of a theory of the world where things are just "there"—being *qua* being—rather than being brought into world-makings through specific relationships and practices. Caribbean thinker Édouard Glissant reaches for this in *Poetics of Relation*, as does Isabelle Stengers in her work with Alfred North Whitehead's approach to knowledge production.[8] Unmaking the effects of the similar subject-object pair has been the focus of contemporary European philosophers such as Stengers, Bruno Latour, and others in respect of the nature/society divide.

In their different ways, all of these thinkers claim the possibility, and the life-giving release, of a different way of relating—not only between people, but also with ecologies and creatures.

For Césaire, one of the earliest postcolonial thinkers, surrealism makes it possible to grasp that there is something missing in modernity–coloniality. Surrealism is very different from sarcastic irony. Irony has a place, but also very specific histories; it translates poorly; and it risks slipping into an anti-politics—because when using irony, we are saying the opposite of what we mean through tone or through staging, without doing the hard work of developing a new language. Irony is a language of opposites in which we assert what we are against, but not what we are for. Irony has a necessary political moment: it is useful when a bureaucracy begins to unravel, when the old is no longer working. But it is limited. The work of the humanities—poetry and art and film and literature—is storying that for which we cannot yet find words as we find ourselves living in a world that none of us grew up expecting. Poets like Césaire become leaders because they speak to the need to remake language. Through surrealism, they enable us to overthrow the gods of reason, and to reach for that which is unseen, unnoticed, in a world that is not a machine—but that is being taught, and treated, as one.

Surrealism is a necessary step to grasping a bigger picture—and to undoing the legacy of coloniality, and supporting globally the practices of living in the earth, not on it, in the coming very difficult years of catastrophic climate change.

Glissant, Césaire's Martiniquan colleague, worked toward undoing a world made in the image of Being–Non-Being via a poetics of relation. Surreal poetics makes it possible to destabilise words and grammars that are themselves cast in sentences of subjects and objects.

Poetry is the art of translation.[9] The work of researchers and scholars and the trainers of experts is to allow ourselves to be translated from the space of property and territory to a place in which we are asking of plants and streams, dolphins and deserts, the kinds of questions in which the answers do not land them in our checkboxes of either Beings or Non-Beings. Decolonising oneself and one's expertise in the Anthropocene means learning to live in relation to rock and water and air and life: reclaiming "the commons sense," to borrow from Gutwirth and Stengers—that sense of earth that beings, living and non-living, share.[10]

There are many languages and heritages of thought that have cultivated the arts of living in partnership with creatures and critters, and the responsiveness that inheres in materialities and their movements; hence the critical importance of decolonising thought, resisting the categories and stasis that are the dominant intellectual heritage of universities, the inheritance which confers on us the authority of expertise in a democracy. But what is that authority? Is it the authority to tell others what to do, or to pose more useful questions? That black experience must contest the assumptions of scholarship invested in white privilege was observed by Jean-Paul Sartre in his essay "Black Orpheus," as long ago as 1948:

> When you removed the gag that was keeping these black mouths shut, what were you hoping for? That they would sing your praises? Did you think that when they raised themselves up again, you would read adoration in the eyes of these heads that our fathers had forced to bend down to the very ground? Here are black men standing, looking at us, and I hope that you—like me—will feel the shock of being seen. For three thousand years, the white man has enjoyed the privilege of seeing without being seen; he was only a look—the light from his eyes drew each thing out of the shadow of its birth; the whiteness of his skin was another look, condensed light. The white man—white because he was man, white like daylight, white like truth, white like virtue—lighted up creation like a torch and unveiled the secret white essence of beings. Today, these black men are looking at us, and our gaze comes back to our own eyes; in their turn, black torches light up the world.[11]

#SCIENCEMUSTFALL AND AN ABC OF NAMAQUALAND PLANT MEDICINE

On Asking Cosmopolitical Questions

We have not abandoned hope that a dialogue with the Martians might lead to the reconquest of Eden. But in the meantime, earthlings that we are, we also have the right to reject a choice limited simply to the alternatives of hell or purgatory.
—Thomas Sankara, "Imperialism Is the Arsonist of Our Forests"

3.1 | Graduate student protestors atop the plinth at the University of Cape Town, from which had been removed the statue of Cecil John Rhodes, April 2015. Photo: Lesley Green.

After a meeting on decolonising science at the University of Cape Town (UCT), YouTube download figures record that a four-minute clip from it had been viewed almost 500,000 times within four days. It trended on Twitter under the hashtags #ScienceMustFall and #GravityMustFall—and on YouTube was second only to a video of a shark breaching a diver's cage with the diver trapped inside. A fortnight later, #ScienceMustFall views had passed one million.

South African universities shut down en masse in October 2016. The University of Limpopo closed for the remainder of the academic year; the University of KwaZulu-Natal suffered the burning of its law library and other buildings. The Cape Peninsula University of Technology and the University of the Western Cape also lost buildings to students protesting high tuition fees, the private security companies and police brought in to control protest actions, and the legacy of coloniality in the curriculum. Nelson Mandela University closed its doors after arson on its nature reserve amid which some students threatened to braai (barbecue) their zebras. The University of the Witwatersrand hired private security companies who ran battles off campus into the streets of Johannesburg, where the police could turn their full force on protesters. UCT closed for a few days, then a week, then more than a month.

Where #RhodesMustFall and #FeesMustFall had in 2015 managed to mobilise students across the board to march on the Union Buildings in Pretoria to reduce university fee increases, the 2016 student movement fractured into multiple groups with several agendas. At UCT, they varied from those who organised meetings on the transformation of curricula in different departments and faculties, to those who dumped sewage in computer labs, to those who threatened to burn the chemical engineering building and invaded an infectious diseases laboratory on the health sciences campus, threatening to trash it.[1] A year later, veteran journalist Jacques Pauw revealed that the embattled then-president Jacob Zuma—perhaps fearing an Arab Spring—had actively worked to subvert and fracture the movement.[2] Amid the turmoil, some who appeared in campus protests were not students at all. One of those was a leader of Black First Land First (a new movement focusing on land restitution that had splintered from the political party known as the EFF, or Economic Freedom Fighters, and later registered as a political party known as the BLF). A year later, Black First Land First was widely discredited for taking funds from agents of state capture in revelations of a national corruption scandal. Some of the prominent protesters turned out to have been on the payroll of security agencies.

While some agencies sought to steer the protests in divisive directions, however, and while rage, insults, and violent class disruption made simple ethical pronouncements impossible, South African student protests had begun something irreversible: a serious dialogue on the necessity of transforming the university. Among the many targets were the university's unaffordability to most black families; the canon in its disciplines that overwhelmingly taught the work of white authors; and university managers and academics who were experienced as insensible to black experience on campuses.

During the shutdown at UCT, far away from the headlines, several departments—notably economics, historical studies, and architecture—hosted extended and often incisive discussions on and off campus on the economics of education; the demographic of an overwhelmingly white professoriate; the decolonisation of the curriculum; and the possibility of free education. In the third week of the shutdown, a student-led group calling itself Science Faculty Engagements followed suit, initiating dialogue on the decolonisation of science. Of the dozen or so hours of discussion that took place in five long meetings, four minutes trended as #ScienceMustFall. The speaker was Mickey Moyo, an undergraduate politics student and a representative of #FeesMustFall 2016:

"If I personally were committed to enforcing decolonisation, [I would say that] science as a whole is a product of Western modernity, and the whole thing should be, like, scratched off. . . .

"We have to restart science from, I don't know, an African perspective, from our perspective, of how we have experienced science. For instance, I have a question for all the science people. There is a place in KwaZulu-Natal called Umhlab'uyalingana. They believe that through the magic, you call it black magic, they call it witchcraft, you are able to send lightning to strike someone. Can you explain that scientifically because it's something that happens?"

"It's not true!" (The interjection comes from the audience.)

"You see?" (The speaker holds her hands out in his direction, palms up, as if to say, "Here it is! Exhibit A!" The chair calls the interjector to order, requiring an apology and an agreement to abide by "the rules of the house." He apologises and agrees. The floor returns to the speaker.)

"So I will finish. See, that very response is the reason why I am not in the Science Faculty. I did science throughout my high school years and there was a lot of things that I just . . . (She trails off, flicking her hands palm up in frustration.) . . . Western modernity is the direct antagonistic factor to decolonisation, because Western knowledge is totalising. It is saying that it was

Newton and only Newton who knew or saw an apple falling and then out of nowhere decided that gravity existed and created an equation and that is it. Whether people knew Newton or not, or whether whatever happens in Western Africa, Northern Africa, the thing is that the only way to explain gravity is through Newton, who sat under a tree and saw an apple fall.

"So Western modernity is the problem that decolonisation directly deals with. To say that we are going to decolonise by having knowledge that is produced by us, that speaks to us, and that is able to accommodate knowledge from our perspective.

"So if you are saying that you disagree with our approach it means that you are vested in the Western and Eurocentric way of understanding, which means you yourself need to go back, internally, and decolonise your mind, [and then] come back and say, 'How can I relook at what I've been studying all these years?' Because Western knowledge is very pathetic, to say the least. . . . Decolonising the sciences would mean doing away with it entirely and starting all over again to deal with how we respond to the environment and how we understand it."[3]

Played back to back, the million views of these four minutes in the clip's first fortnight online would take about eight years—click by click, this became for many in public life *the* defining speech of the struggles over the curriculum of the university. No other video clip of Fallist struggles received as much attention. A crevasse opened in the support base of the student movement. Students and academics who were sympathetic to the call to engage with the legacy of the colonial in campus curricula found themselves cast in the untenable position of being "anti-science." That would be catastrophic for academic reputations in a country where being anti-science had a recent, painful, and tragic history during a national struggle for HIV patients to gain access to antiretroviral (ARV) drugs.

Moyo's speech also made public the contempt for "the top university in Africa" (UCT) that had come to preoccupy the Fallists, as Occupiers of "the occupiers," which, in the parlance of the time, meant whites occupying an African space. Of the many resonances in Moyo's speech, the most discordant note was that of the "indigenous knowledge–science war" that had in 2009 ousted the sitting South African president in a tussle over "African" versus "Western" knowledge in the treatment of AIDS and HIV.

Predictably, in online fora where "I scalp, therefore I am" seems the guiding principle, out came the most vitriolic of trolls. Not only had the speech touched a nerve; it had hammered the national funny bone. Anyone who had never before been a comic had laughs to score and political existence to assert on issues that had been central in South African life since coloniality: rejecting superstition; asserting the supremacy of reason; doing so in the cause of opposing anarchy; annihilating a young black woman's claims in public without censure; suggesting she test gravity by jumping off a tall building.

I am not going to defend the letter of what Moyo argued. Some of her words remind me of the kind of speeches I made as a teenager during a brief spell as a fundamentalist evangelical, in which truth and error were equally overstated. Of more interest to me was that she was reaching for an issue that was at the core of the colonial project—which justified itself as the emissary of truth in the face of what it believed to be darkness and superstition. Even more important: the difficulty she had in articulating a more subtle argument was a product of a university and a high school education system that split the social sciences from the natural sciences, and taught science as if it had no social grounding at all, in the process unmooring scientific expertise from this world and attributing to "experts" a political power and social status that was both denied and feted at the same time. That she was unable to articulate a careful and considered argument about science in society was an indictment of an education system in South Africa where science studies are not taught, of a scholarly community in which terms that yield the indigenous knowledge–science dualism have not been rethought, and an indictment of a scientific community that has allowed authoritarian decision-making—especially in matters of environmental governance—to claim the name of science and survive unquestioned.

Her comments and the rejoinders, however, deserve a different response to that of the trolls: the first, in respect of Isaac Newton; the second, on the binary between traditional knowledge and science.

II | NEWTON AND THE HISTORY OF SCIENCE

First, Moyo does not question the existence of gravity, as was implied in much of the #GravityMustFall commentary: she questions the assertion that in the whole of human history, only Newton had observed that things fall, and she questions a history of science that claims European preeminence of the knowledge that things fall. While clearly Newton's grasp of mass as a force that generates gravity was a first in the discipline of physics, it is equally historically true that science does not only come from Europe—Chinese and Arab and Hindi scholars made major contributions to mathematics and astronomy and medicine—and that many of Africa's great libraries and cities housing this knowledge were destroyed. She is correct to observe that where the history of science is taught as if all of it comes from Europe, it is untrue.

Second, Moyo evidently does not know the details of Newton's universal law of gravitation, which Newton expressed as $F = G\frac{m_1 m_2}{r^2}$, "where F is the gravitational force acting between two objects, m_1 and m_2 are the masses of the objects, r is the distance between the centers of their masses, and G is the gravitational constant."[4] What Moyo is contesting, however poorly she articulates this, is the popular claim that Newton's scholarship gives to those who operate in the name of science the right to claim theirs is an ultimate and universal truth that in turn may confer on its exponents mastery over nature and therefore unquestioned authority in matters of governance. There is in fact a great deal in science that contests the universality of Newton: Albert Einstein's work on relativity demonstrates that Newton's laws do not apply universally, and quantum mechanics is not compatible with Newtonian mechanics. As for Newton, he himself was critically aware that his work on gravity had described but not explained it, and he prevaricated on whether gravity was a material force or a divine force.[5] The irreconcilability of quantum mechanics and Newtonian mechanics provokes necessary questions about the universal conceptual paradigm of physics.[6] In other words, even Newton's law of gravitation, so axiomatic to the authority of contemporary physics, must be situated; it is not universal. Crucially, the political implication is that those who present themselves as the heirs of Newton are not inheritors of an ultimate truth that demonstrates universal mastery.

I do not support the kind of interpretive work that seizes on Einstein's theory of relativity to proclaim that all science is relative, and therefore cultural relativism rules. That argument has had a catastrophic history in South Africa because it was central to the ideas of former president Thabo Mbeki, who, in presiding over the unavailability of ARV medicines for people infected with HIV, incorrectly divided the issue along the binary of African versus Western science—a topic I've discussed elsewhere.[7] Through the later work of Bruno Latour and the work of philosophers of science, including Isabelle Stengers, Karen Barad, and Didier Debaise,[8] I have learned to insist on staying with an empiricism[9] that embraces also the conditions of observation and the cosmopolitical assumptions of its observers (as described in chapter 1).[10] This approach enables one to welcome the work of both Newton and Einstein, and to see both as bound to specific situations by the questions they were asking, and the tools and techniques available to them. Situated knowledge is a different approach to cultural relativism.

Moyo's question attends to the arrogance that has accrued to whiteness and coloniality via scientific claims to mastery over all nature, as if scientific knowledge is necessarily universal, and as if all of the history of science is

European. It is not: formal mathematics, biological taxonomy, astronomy, geometry, and medicine all have deep histories in China, Japan, the Middle East, the Amazon and the Andes, indigenous Scandinavia, and Africa—all places where historical disruptions have seen libraries lost along with the social structures that had enabled a class of scholars to flourish.

Newton himself expressed his frustration with the scientific fashions of his day in favour of the ways "the ancients" had known space and geometry. "Descartes achieved the result ... by an algebraic calculus which, when transposed into words ... would prove to be so tedious and entangled as to provoke nausea, nor might it be understood," he wrote. "But they (the ancients) accomplished it by certain simple proportions ... concealing the analysis by which they found their constructions."[11] In other words, the ancients—like Euclid's geometers—offered ways of knowing something in proportions: a mode of knowledge very different to the metrics offered by Descartes's algebra. At issue, for Newton, was the value placed on converting things to formulae. To be understood, Newton argued, curves needed to be traced by the geometry of motion and proportion, not defined by imaginary points in a Cartesian plane that satisfied an equation. Moyo's criticism of the value placed on abstract equations where an understanding of movement would be adequate is a point on which Newton is in agreement. As philosopher of science Michel Serres noted: "I understand the weaver to be a pre-mathematical technician more ancient than the surveyor."[12] The complex fractals in African weaving, exquisitely translated into the language of geometry by Mozambiquan scholar Paulus Gerdes, come to mind.[13]

Even more noteworthy is that Newton's work on alchemy, theology, and biblical prophecy has been all but eliminated in most accounts of his science. It is not possible to write a historically accurate biography of Newton that reduces him to a "cold mathematician who calculated the orbits of the planets, espousing a deterministic view of nature."[14] Indeed, the history told of Newton as one whose work was purely "hard science" derives from the division of his papers after his death, which were auctioned off separately in categories of scientific and esoteric.[15] In that tiny act of dividing his papers, a watershed was created. The side of Newton that was welcomed in the history of science has been a Newton who, to quote feminist science scholar and quantum physicist Karen Barad,

gave voice to ... a deterministic world: placing knowledge of the future and past at Man's feet. Prediction and retrodiction are Man's for the asking, the price is but a slim investment in what is happening in an instant, any instant. Each bit of matter, whether the size of a planet or an atom, traces out its designated trajectory specified at the beginning of time. Effects follow their causes end on end and each particle takes its preordained place with each tick of the clock. The world unfolds without a hitch. Strict determinism operates like a well-oiled machine. Nature is a clockwork, a windup toy the Omniscient One started up at time t = 0 and then even He lost interest in and abandoned. . . . The universe is a tidy affair indeed. The presumed radical disjuncture between continuity and discontinuity was the gateway to Man's stewardship, giving him full knowability and control over nature. Calculus is the escape hatch through which Man takes flight from his own finitude. Man's reward: a God's-eye view of the universe, the universal viewpoint, the escape from perspective, with all the rights and privileges accorded therein. Vision that goes right to the heart of matter, unmediated sight, knowledge without end, without responsibility. Individuals with inherent properties there for the knowing, there for the taking. Matter is discrete but time is continuous. Nature and culture are split by this continuity and objectivity is secured as externality. We know this story well, it's written into our bones, in many ways we inhabit it and it inhabits us.[16]

It is this rendering of Newton as the author and authorisor of ultimate knowledge, which Barad problematises, and which Moyo objects to. They are correct to distrust that authorisation, for so too did Newton. In a letter to a colleague, he wrote:

> You sometime speak of gravity as essential & inherent to matter: pray do not ascribe that notion to me, for the cause of gravity is what I do not pretend to know. . . . That gravity should be innate inherent & essential to matter so that one body may act upon another at a distance through vacuum without the mediation of any thing else by & from one to another is to me so great an absurdity that I believe no man who has in philosophical matters any competent faculty of thinking can ever fall into it. Gravity must be caused by an agent acting constantly according to certain laws, but whether this agent be material or immaterial is a question I have left to the consideration of my readers.[17]

Newton does not claim mastery of the knowledge of gravity. Why, then, do his self-proclaimed inheritors claim that he did?

III | "TRADITIONAL KNOWLEDGE | SCIENCE"

The work of rethinking the representation of science as forming a binary with indigenous knowledge could have been addressed in South Africa in the struggle from 2003 to 2009 over medications for treating HIV-AIDS, had it been framed by then-president Mbeki in terms other than African versus Western science. The result was that citizens and doctors allied behind the Treatment Action Campaign (TAC) to secure ARVs for those with HIV and AIDS against a president who was determined to pursue African solutions for African problems, and who rejected HIV science (even the idea that HIV was a virus) as an invention of the West. This was an absurd argument. The idea that scientific questions could be resolved by identity politics or political commitments is a mistake that transposes the category "knowledge" with the category "belief." A more useful question would be how those categories arose historically, and why they remain unhelpful. Nonetheless, the transposition has been encouraged in the publications of some international social science which situated itself in support of indigenous political claims. By contrast, while progressive South African social scientists who described themselves as postmodernists rejected the indigenous knowledge argument as "an invention of reality" (in the name of non-racialism), they were silent on the power that science was assuming in society simply "because it was Science"— which is as much a play of an identity card as anything else. In my own work on this issue from 2006 to 2012, when I tried to engage authoritarian expressions of Science (in the singular, with a capital S), I was denounced by a "pro-science" academic colleague as an "AIDS denialist," and even accused of "aiding and abetting genocide."[18] But when I spoke about trying to reappraise the rejection of claims to indigenous knowledges within the social sciences, I was accused by one of my seniors of cultural relativism, which, in South Africa at the time, amounted to being accused of being an apartheid apologist.

Many had thought that after the TAC had won the fight for ARVs to be dispensed through the state health system, belief had been banished from science faculties in South Africa. Scholars like Nicoli Nattrass, a health economist, wrote what many thought would mark the epitaph of the debate: her book *The AIDS Conspiracy: Science Fights Back* is a vindication of Science in the singular and with a capital S.[19] Case closed!

Or not.

A similar struggle soon reared up quietly, about the scientific base of another TAC: this time, the total allowable catch in fisheries. The TAC is a figure that is set annually by government fisheries managers to determine catch quotas for commercial fisheries. In a series of parliamentary debates, its scientific base was explicitly rejected by several members of Parliament who called for the scientists and fisheries managers to draw on indigenous knowledge in their calculation of catch quotas.[20]

And in KwaZulu-Natal in 2011, after several people in a rural homestead had died after lightning strikes, Nomsa Dube, the politician responsible for the provincial Ministry of Co-operative Governance and Traditional Affairs, asked for more research on why there were so many lightning strikes, and why they were affecting mostly black people. Given that the capacity to send lightning is widely associated, in Southern Africa, with dark magic, she was pilloried by the black and white press alike. "Does the Honourable representative not know," asked one pundit after another, "where lightning comes from?" The story went viral internationally: a South African government official had asked for scientific research into the cause of lightning.[21]

Hardly any thought to ask for the basis of her claim. The issue became (the pun is inescapable) a lightning rod for displays of the superiority of science and rationalism, and of smirking pundits themselves lost no opportunity to display their superior expertise.[22] If they had bothered to check with the South African Weather Service, they would have found that there were about seventy thousand lightning strikes in KwaZulu-Natal and Pondoland in 2014, an unusually high number.[23] It is possible that this rise was related to an increase of extreme weather events associated with climate disorder, since one of the country's major insurance companies had noted a 402 percent increase in storm-related claims for March 2014 as against March 2013.[24] There was a national conference in November 2015 on the increase of lightning, which in KwaZulu-Natal could kill an entire flock of sheep that might be sheltering under a tree, and has killed many families in the past few years. The question that conference asked—according to the opening speech by then–environment minister Edna Molewa—was whether the increasingly severe storms were related to climate change, and whether that increased incidence was also the proverbial lightning rod for social stressors and suspicions.[25]

The grammar of Ms. Dube's question in 2011, posed in her second language of English, was clear: she had asked for scientific investigation of what had caused "the lightning" that had killed a particular family. Her question was about a specific context, not a universal truth. But stripping the context was necessary to justify the preening pronouncements by the masters of universal truth and reason.

All of these situations—the struggle for ARVs, the fisheries management struggles, the question about increased lightning strikes—track the national fault line in South Africa: the relationship of science and traditional knowledge.

After Mbeki resigned the presidency, in large measure because of his unpopularity over his mishandling of HIV, discussion of the colonial legacy in knowledge and tertiary curricula was muffled on most South African university campuses by the likelihood of denunciation. But the generation known as the "Born-Frees"—those born after Mandela took over South Africa's presidency in 1994—were senior undergraduates by 2015, and had by then discovered that being a Born-Free was merely a brand name in a country whose liberation had been swallowed whole by neoliberalism and the ensuing corruption of the state. The result: democracy and a new constitution had not delivered a "trickle down" to the poor, but a "gush up" to the privileged.[26] Multigenerational dispossession of land, jobs, and education was still the reality.

When a generation finds their senses of self to have been forged in a lie, a revolution against adult hypocrisy is inevitable. Arundhati Roy's words ring true: when mass protest is nourished by the memory of generations of repression and humiliation, it yields "a kind of rage that, once it finds utterance, cannot easily be tamed, rebottled and sent back to where it came from."[27]

That revolution required a rejection of the kind of education that was standardised to address disciplinary agendas framed in the Northern Hemisphere that often offered little (if any) connection to the concerns the Born-Frees faced. Principal among these was that the university system required mastery of specific pieces of a largely Euro-American canon to graduate, making mastery of often irrelevant framings of knowledge to be a career gateway. African knowledge was still in exile on campuses: some to the intellectual Bantustans (apartheid's "cultural homelands") of ethnopharmacology, ethnobotany, and ethnomusicology; others within marginal academic divisions where critical race studies was taught in a way that lacked meaningful engagement with the sciences, not least because, in the wake of the HIV-AIDS knowledge crisis, South African social scientists were afraid of being cast as "anti-science." The need was for new conceptual tools, since postmodernist science studies often treated science as an expression of identity politics. And while identity politics certainly permeated much science, science cannot be reduced to identity politics (much as not all red trucks are fire engines). Scientific research methods cannot be evaluated on the basis of identity politics, although the questions that are posed and the concerns that are addressed may well be framed within identity-based concerns. Much as identity politics cannot be evaluated by scientific methods, scientific methods cannot be evaluated by identity politics, because they are not the same modalities of inquiry. To claim otherwise is to make the category mistake made by Mbeki. Yet the greatest stumbling block to persuading students not to "replace Science with identity politics" was that authoritative science itself was consistently defended by its own identity politics.

To unravel the mix-up of modalities requires more than defending categories of knowledge and belief, and shifting them across the categories black and white, Africa and West. It requires finding a starting place other than that defined by "bias" or "belief" or "identity."

Case studies are always a good way to begin when there is a need to find a different way to understand a problem. I want to begin with the struggle over ARV medications in South Africa.

It took a concerted national and international movement led by the TAC, during Mbeki's presidency, to secure a government commitment to rolling out ARVs. It was a political victory when a new minister of health was appointed who was a medical doctor and would take seriously the evidence of virologists—that HIV is a virus; that ARVs are effective in managing it—and the evidence of health economists who argued that the cost of rolling ARVs out at scale was less than the cost to the fiscus of the welfare interventions needed to cope with the impact of suffering and death in communities and households.

However, the cost of the debate over knowledge in South Africa was very high. The ARV debacle provided the perfect "infernal question," to use Stengers's description of a rhetorical question that aims to end one's will to think out of the box, and close down any discussion.[28]

Mbeki's efforts to find a science for treating HIV using African knowledges framed the research question in relation to area and identity—a catastrophic error that is so well known in the sciences that it has a name: the category mistake. Viruses are stopped in their tracks by neither the sign of Africa nor the sign of Europe.

Mbeki was taking issue with the cultural and political authority of a certain form of knowledge—the position argued by many postmodern and postcolonial critics. In other words, Mbeki correctly framed a question about the political authority of establishment science over the knowledge of nature, but incorrectly asserted that the solution was an identity politics of science. A battle is won or lost on the chosen battlefield; identity was not going to win in a struggle over science.

Mbeki's critics framed the options for scientists as cultural relativism versus science; "junk science" versus "real science."[29] Reclaiming a pure knowledge of nature independent of society, activists sought to re-establish a science outside of politics. That position was, in their view, critical in order to secure treatment for HIV-AIDS, but it was also a position beset by uncomfortable internal contradictions. The argument that traditional knowledge was the enemy put these progressive activists on the side of big pharmaceutical companies—an uncomfortable position that put activists in a position of opposing both science studies and efforts to bring indigenous knowledge into the universities. In humanities faculties, postmodernist scholars in the social sciences found themselves taking a hard-line position opposing the idea of indigenous knowledge studies, yet unable to engage the evident reality that their own theoretical approach—a postmodern cultural relativism—lay at the core of the very presidential thought that they so bitterly opposed.

Notwithstanding the successes of the TAC in securing affordable HIV treatment for millions, notwithstanding the importance of Mbeki's struggle to challenge the dominance of Euro-American intellectual heritage in sciences and universities, notwithstanding the common ground they shared in opposing the abuse of patenting by pharmaceutical companies, the struggle of TAC v. Mbeki became a drama of cosmic proportions, *because it was grounded in cosmology itself.* The TAC founded its argument on the unquestionable nature of scientific truth, as if science was independent of society. Mbeki founded his argument on culture and imperialism, arguing that science was part of a divided society that was black on one side and white on the other. Thus composed, their struggle became a tussle of science versus identity, nature versus culture—replaying the unresolvable fault line at the core of modernist thought. The debate was characterised by aggressive denunciations and fierce polemic, creating a climate on campuses in which it was almost impossible to think beyond the two positions.

Tragically, their impasse was resolved by Jacob Zuma, an anti-intellectual who used the stand-off to settle an old score with Mbeki, who had once fired him as his deputy. Mbeki lost the presidency to Zuma in 2009 with the maximum possible indignity permitted by party protocol. Aaron Motsoaledi, a medical doctor, was appointed as minister of health and rolled out a life-saving national HIV-treatment plan to national acclaim, scoring Zuma the political kudos he desperately needed. Zuma closed Mbeki's prized indigenous knowledge phytochemistry laboratory at the Medical Research Council facility in Delft in Cape Town, as well as Mbeki's project on the Timbuktu libraries in Mali, which contained priceless ancient texts of African astronomical knowledge written down in old Arabic scripts.

Through Motsoaledi's science-based interventions, many lives were saved. Many children who were at risk of being orphaned still have their parents. South Africans' life expectancy improved. But the struggle to question the relationship of coloniality and knowledge was decisively lost, setting back by many years the discussion about African knowledge, modernity, and coloniality. "Hard science" was repositioned as the ultimate authority over nature. But it was a brittle hardness, and one that would come to be sniped at in many fora, particularly via the other TAC in South African political life: the total allowable catch in fisheries management. And while research findings that demonstrated that the division of nature from society was impossible to sustain were at the forefront of sciences like neurology, epigenetics, and climate change, the majority of South African medical and environmental science debates have retreated into the same defensive position adopted by the Treatment Action Campaign: holding that science is independent of politics; that the only place for matters societal in science is in the gathering of social data for scientific models, or in debates over scholarly ethics.

The uneasy truce remains this, in much South African public debate over the place of science in democracy: nature itself must remain untouched. Politics must stay in its place.

African knowledge thus returned to the Bantustans of the universities: situated not in knowledge research, but in ethnoknowledge research programmes, or in divisions for the study of regional knowledge. African knowledge was from that point to survive in South African universities courtesy of epistemic charity; as superstition to be overcome, a regional knowledge specialisation, or a source for an authentic African voice with which to confirm the values, efficacy, and practices of environmentalism. Thus, Mbeki's attempts to promote postcolonial knowledge debates backfired—setting the debates he sought to initiate back by at least a decade, and setting in place one of the conditions that would foment violent student outcries against eurocentrism in South African university curricula in 2015. For that reason, the field of science studies, even though well established internationally, found barely any purchase in South Africa. It took #ScienceMustFall to show how much damage had been done.

IV | THE ABC PROJECT

Is it possible to think about different forms of knowledge without reducing one to the opposite of the other?

At the height of the battle over African plant medicine in 2009, a small collective of researchers in anthropology, botany, and chemistry coalesced around a question posed by UCT organic chemist David Gammon: how might natural-products chemists and bossiedokters—medicinal plant healers (literally, "bush doctors")—better understand one another? Over time, the in-house joke became that ours was the ABC project, not least because it felt as if we were beginning to learn to work together in new ways.

The group focused on the medicinal properties of plants in Namaqualand, about five hours' drive north of Cape Town, in the succulent Karoo biome where botanist Timm Hoffman had collected data on plant growth once a month for some fifteen years. In the process, he had come to know several of the region's healers. David was interested in running assays on the plants, and arrived at my office looking for an anthropologist interested in the question of how to approach "indigenous knowledge." I invited Joshua Cohen to focus his PhD in anthropology on the issue.[30] Amelia ("Millie") Hilgart developed her PhD in botany and chemistry on seasonal plant toxicities, and Nicola Wheat focused on chemical assays from the plants.[31] Over time, the team grew to include visiting philosopher of science Helen Verran (who had trained as a chemist) and university research transformation specialist Robert Morrell.

Thinking together slowly over five years was a rare privilege. Once the graduate students' field research was almost complete, we had the task of trying to synthesise our findings. It was much harder than any of us expected. Of course, there was an easier conversation between chemistry and botany—they both spoke the language of molecules—but finding a way to include anthropology in the conversation was harder. The first problem was finding words in English to describe practices that have come down to us through coloniality and modernity: the language of cures versus curses—from which descriptions of the latter would predictably flow into the category of "belief" and/or "superstition," and from there into "nonsense." All of us wanted to avoid that watershed, and to avoid slipping into epistemic charity and cultural relativism. None of us wanted to be associated with an argument that scientifically verifiable truth could change like a chameleon to match the scholar. There had to be a way to engage the work of the healers differently—but how?

With the thoughtful facilitation of Robert, the insights of Helen, and plenty of coffee and humour, the conversation stayed open. In the paper that we eventually published in the *South African Journal of Science*, David framed the core research question:

> Over the last decade or two, the discipline of natural products chemistry has been caught in a tension of introspection, on the one hand, and something of a renaissance on the other. Leading practitioners like Cordell and others have been calling for an urgent reappraisal of the importance of natural products research. . . . In issuing these challenges, they however do not acknowledge the role of indigenous people living within and from the biodiversity, in terms of knowledge production or dissemination. This issue is notably taken up by Etkin and Elisabetsky in their analysis of papers published in the *Journal of Ethnopharmacology* over a 25-year period since the inception of the journal, where, despite the stated intentions of the journal, they conclude that: "Much of what is reported as ethnopharmacological research is comprised by decontextualised catalogues of plants and lists of phytoconstituents and/or pharmacologic properties

[and] few researchers in ethnopharmacology seem to be interested in the people whose knowledge and identity are embodied in these plants. While some studies are based on plants drawn from indigenous pharmacopoeias, most of what is published as ethnopharmacology has a weak, if any, ethnographic component."[32]

Joshua Cohen's anthropology dissertation focused on the categories and practices of the bossiedokters in their work with plants and patients. His findings, as summarised in the *South African Journal of Science* article:

> Two key terms—*krag* and *wind*—emerged. The first of these terms, *krag* (power, vitality, strength), is a commonly used Afrikaans word that can be thought of as a kind of "body energy" that waxes and wanes with the ups and downs of everyday life. In order to aim their patients toward health, it is important for kruiedokters [herb doctors] to be able to "cultivate" their patients' krag. To this end, various tools might be used by the kruiedokter— jokes, guitar playing, food, and of course *bossiemedisyne* [plant medicine]. Many plants are directed at alleviating *krag*-sapping symptoms: high blood pressure, diabetes, colds and flu, swelling in the limbs. The *krag* in the plants themselves, that which enables them to do their healing work, is closely associated with the *krag van die natuur* [the power of nature], this in turn being closely associated with *die Here* [the Lord].
>
> Beyond all other things, *kruiedokters* are skilled at, and known for, their ability to deal with the various kinds of *wind* [wind] that cause so much trouble in people's body-person. There are, generally speaking, two kinds of wind: *natuurlike* [natural] and *toor* [magic], both of which retain the propensity for movement and change, the same as the wind that blows in the veld. The first kind of wind can be a "simple" build-up of gas in the intestines and stomach, but can also be "picked up" in the world and "lodge" itself in any part of the body—from the head to the muscles of the leg. Health and well-being are often linked internally, to things flowing properly around the body through organs which are not "stressed"; and externally, to people being able to move along the proper "path" of life as laid out for them by God. Therefore, in order to return their patients to these flows, kruiedokters use various techniques to unblock these winds, including plants [like] *kougoed* (*Sceletium tortuosum*) . . . and other substances to aid the expulsion of wind, and massage.
>
> Knowledge of the second kind of wind, *toorwind*, is an important area of *kruiedokters*' expertise. These are winds that are "sent" by malicious others, through a range of media including *gif* [poison] placed in someone's food, in dreams, or through magical "traps" placed along their

victim's path. Such winds can grab on very tightly to their sufferer in the form of a *bose gees* [evil spirit] that "sits" in the victim's body, growing, sapping their "*liggaam se krag*" [body energy]. *Kruiedokters* exercise what is seen as a God-given, evil-tackling talent, to cleanse their patients of such winds. They use "fighting roots," that is, the roots of the various "storm" plants in protective xaimpies [medicine bundles] placed in the home or carried around on the person. *Kougoed* encourages the movement of wind out of the body. Existing in what might be called an intersubjective space between a *kwaadaandoener* [evil sender] and their victim, the treatment of these kinds of winds involves the *kruiedokter* effectively placing themselves, as a kind of defender, between the attacker and the attacked.[33]

How to translate the bossiedokters' work into languages and practices that might make sense in the sciences?[34] After one or two moments where the collaboration risked collapse, the three simultaneous projects reached a startling point of convergence—that to have a transdisciplinary conversation about plants, we needed to start with a different question. A paper describing the process of thinking together was published in 2015. For me, the key moment in the writing process was the following paragraph, which, though readable in a few seconds, represented years of thinking about how to resituate the debate between indigenous and scientific knowledge in a different set of questions:

> The primary orienting, although often unstated, question in chemical studies of plant medicine concerns pharmacologically active ingredients for antibacterial, antifungal, antiviral and/or anti-inflammatory properties. Yet this orientation depends on equating health and illness with the eradication of a particular taxonomy of pathogens. If the orientation to health includes a wider array of toxins and taxonomies that contribute to the experience of having energy or vitality (*krag* or the different, although not entirely dissimilar [Chinese] concept of "*qi*"), then biochemical research need not necessarily begin with the particular pathway of seeking compounds related to pathogen elimination. Cohen's work suggests the examination of pharma and the flow of energy in the body. This suggestion raises questions about plants in an ethnopharmacology of Namaqualand, which would include cleansing, balance, attention to social harms and the [related] toxicities of stress.[35]

In searching for combat molecules—antibacterials, antifungals, antivirals— our research team had unwittingly drawn from a particular intellectual heritage in which disease is a matter of a physical pathogen, and, therefore, research on plant efficacies involved searching for molecules associated with fighting that kind of pathogen. Yet by limiting our understanding of plant efficacies to attacks on a pathogen, a very large part of the healers' work with plants could be neither seen nor recognised nor understood. If the healer was turning to the plant to restore vitality, strength, and well-being (krag), her or his "ask" of the plant was not limited to the supply of guerrilla molecules to fight illness. Namaqualand plants, of course, have plenty of those: Nicola Wheat's study found that 70 percent of these dryland plants carry the kind of combat molecules that can be identified through chemical assays. But the ask of the plant doctor was not limited to these, as health defined as krag is a concept that includes not only the absence of pathogens and the presence of a pathogen task force, but also the ease of energy flows in the body, whether via blood or neurons or adrenalin, and the protection against the very real inner stress caused by toxic relationships.

Our problem was that our concepts of what causes illness or wellness— the presence or absence of pathogens—made us look for pathogen fighters in the plants. Conceptualise health differently, and different things become significant. Pathogens; build-ups that cause blockages; social stresses—all of these have specific modes of existence where they are seen and named. But the specific intellectual inheritance of the biomedical health sciences makes it difficult to see them. It is much easier to name one mode of existence "truth" and another "belief." It is much more challenging to ask what it is that one is *not* seeing—but that is where the productive possibilities are, for a different kind of conversation within universities on knowledges and intellectual inheritance. The accomplishment of the project, for me, was that by the end we had learned to step outside of a particular way of inheriting the university in order to—in the words of Stengers and Despret—"construct a different kind of question, one that had the chance of receiving different kinds of answers."[36]

Posing questions about the assumed certainties of intellectual inheritance has the potential to open conversations across different ways of knowing. It makes it possible to consider how and why different things (like specific molecules) come to matter, and come to be named and known in specific ways. Resolving the distinction between belief and knowledge, African and scientific knowledges, has not been a matter of identity, or even a matter of identity-based epistemology. It has been a matter of ontology—what we have understood to be the things that exist, that really matter. Empire might be built for the most part on claims to be empirical, but the empirical is not reducible to the work of empire. Empire is, however, a means of translation: asserting what matters; how things are related. That recognition means that the empirical needs a wider frame. The issue is to recognise that what we name and define comes to be named and defined because it matters in a wider network of references. Escaping coloniality requires escaping the territory in which that wider network of references is mistaken for the universal truth; a singular Nature known by only one Science.

I trust that it is clear to the reader that I am not suggesting that there is no truth, no possibility of empiricism. On the contrary, this is a proposal for a radical empiricism that understands the ways in which social networks—constantly shifting and mutating and growing—generate specific concerns, and therefore come to name and know different matters, differently (as per chapter 1 on the multiple natures of Table Mountain's waters).[37] "Multiple natures" is a legitimate term only as long as it is not once again slipped into the old wineskin of "culture"—because then we have once again swapped the players on the chequerboard, but not changed the game. *The question "What am I seeing?" needs to be expanded to include "How and why am I seeing what I am seeing?"* Defending good science requires addressing with equal seriousness the question of what good science is. What are its concerns? What questions were posed? How and why did those questions and concerns come to matter? In what respects does a specific technology that measures a particular object affirm a prior world view about what matters? What other kinds of concerns and what other kinds of objects might matter to different networks? Addressing these questions requires an empiricism that embraces how things come to be named and known. German philosopher Peter Sloterdijk explains this empiricism as *a move from essence to event*.[38] If the study of science expands its enquiry to how and why specific "things" (like combat molecules) come to our attention, we can begin to do better science when we not only study the things that we know and are counting, but also ask how and why those specific things come to be named, and known, and why they come to matter.

Science measures what its instruments enable. What is perceivable through laboratory techniques and instruments—in our study, liquid crystal mass spectrometry—gives access to one particular minute aspect of nature; in this case, the light spectra emitted by specific molecules. But that technology should not be confused with "the whole true nature," because the use of the technology itself is interrelated with a set of prior ideas that should be open to question. Liquid crystal mass spectrometry, for example, will show the spectrum of visible light. What of other forms of light that are not perceptible by the human eye?

Another example: in obstetric medicine, a foetal heart rate monitor provides numerical evidence that a courtroom judge is more willing to accept than the evidence of an experienced midwife who reads a birthing belly with the clinical skill she has built into her hands, together with intermittent readings from a handheld foetal stethoscope. Why is a court more willing to accept one form of evidence as an indication of good practice than another—especially when that monitor requires a woman to lie still in the worst possible position for giving birth? Does the technology serve to better the practitioner's chances in a court of law, or improve births in a birthing room?

The midwifery example demonstrates that *ideas of what is significant and measurable are not independent of society*. What is of primary concern to one community of scientific practice may not be a priority for another. *Specific concerns frame specific questions and therefore also the answers*. The work needed in our plant medicine study was to understand the interplay of heritage, technology, language, and our research questions. Learning the importance of these allows the development of subtler, more nuanced, and more useful understandings of the relationship among science, society, and knowing.

When science resists engaging with philosophical questions about *what* it sees and *how* it knows, it slips into an authoritarianism that runs counter to the spirit of scientific enquiry. But science that accepts questioning of founding assumptions, specifically that of the division of nature from society, is a more robust science because it is able to respond to the issues that put its claims at risk. Without that, science is acutely vulnerable, especially in contexts like South Africa, to the accusation that its knowledge is simply a belief.

Where authoritarian science is unable or unwilling to recognise its social and political dependence on established agreements, it is unable to defend itself against the claim that it is invested in white privilege, or in capital. At that moment, the authority of authoritarian science is circular: it depends on the acceptance that it is authoritative. Such science risks becoming scientism—a dependence on the independence of science from society, rather than an expanded science that is capable of accounting for its entanglements with society and humanness. When scientism is allowed to flourish in government-hired consultancies, it is a danger to democracy because, at that point, science, law, and policing become inextricably entangled. The scientist's request to the social scientist to "help us get compliance with the science" is politically dangerous. It lends itself to authoritarian politics: the kind of politics that "seeks behaviour change"—via the imposition of policies assumed to be ultimate truths, backed up by fines or criminalisation of individuals. But a politics based on the shared desire for health and well-being, and collective action, will have a much more democratic outcome.

Scientism asserts the kind of authority familiar in critiques of patriarchy: do not question the pater. And like other forms of patriarchy, its circular defence of its authority lasts only as long as people are compliant. As such, it invites defiance. Decolonial scholarship, however, can and should be based on valid, verifiable sciences. The first move that is needed is to distinguish between science and scientism.

The assertion that scientific observation is separate from human society is not challenged only by decolonials arguing for the legitimacy of different ways of knowing the world; as mentioned above, it is also challenged at the forefront of science itself, in fields like epigenetics, neuroscience, and climate science. Epigenetics studies the life events that switch specific genetic markers on. Neuroscience explores the social situations in which pathways in the brain take form. Climate science explores the terraforming impact of humans on the planet (along with that of plants in the Carboniferous era or, in the earliest geological era, the ocean microfauna that put enough oxygen into the atmosphere to support complex life-forms). Cutting-edge science makes it impossible to sustain the philosophy that there is a division of nature from culture.

The surprise that came from our plant study was that once we attended to the concerns that framed knowledge, we could ask questions that no longer took what we were seeing, naming, and knowing as given fact. Instead, by attending to the reasons for knowing things, we could move from asserting "what is" to understanding the interaction of the knower with what is known. It was a shift from seeing existences as predetermined to understanding that the reason for asking a question changed the observer's conceptualisation of the problem. This is something Aimé Césaire had recognised more than half a century before, in his critiques of colonial science and scholarship about blackness (see the introductory note to part II, this volume).

How might these ideas facilitate a different decolonial conversation with science? I'd like to conclude in an imaginary dialogue with #ScienceMustFall.

CODA: #SCIENCEMUSTFALL

1 | If I personally were committed to enforcing decolonisation, [I would say that] science as a whole is a product of Western modernity, and the whole thing should be, like, scratched off.

I wonder whether we might be able to agree that the issue raised here is scientific authoritarianism, or scientism, rather than scientific research methods per se? In addition, I want to suggest to you that in questioning coloniality, one is also questioning the core structure of modernist thought. The split at the heart of modernist thought between subjects and objects, knowledge and belief, is something that philosopher of science Bruno Latour talks about a "product recall" of modernity to address that.[39] And feminist philosopher of science Isabelle Stengers speaks of the violence imposed on women, specifically those accused of being witches, that attended the birth of the split between knowledge and belief.[40] Val Plumwood explored the relationship between science, patriarchy, and the mastery of nature.[41] All of these resources might be useful as you reword your argument into something that is not "all or nothing"—which will take away the ammunition used against you, like that much-touted argument that you ought to throw away your cellphone.

2 | We have to restart science from, I don't know, an African perspective, from our perspective, of how we have experienced science.

Scientism is tied in with a specific set of ideas about knowledge. I think it is useful to understand the ways in which notions of what is nature, and what is reasonable, rational, and "neutral," have changed historically (see chapter 1). Most especially, you may find useful the notion of cosmopolitics, which includes the cosmological assumptions in terms of which scientific reliability is turned into an authorisor of political relations and actions.

When you speak about "how you experience science," I think I understand that you are speaking of a broader African experience of knowledge production that enforces compliance with its assumptions about what is real and how things are related. Ethnological studies of black life were the focus of postcolonial writers who rejected their assumptions almost in totality. Contemporarily, I see that same struggle playing out in the realm of environmental management sciences which often serve to explain away or even justify environmental injustice. Zephaniah Phiri-Maseko, for example, the internationally regarded Zimbabwean farmer renowned for his farming techniques in a dry area, spent six months in British leg irons for farming in a wetland area where farming was not permitted (see chapter 4).[42] De facto, uses of science like these serve to legitimate racialised authority. When *Rational Standard* retorts to you that "facts don't care about your feelings," it might be useful to ask editor Tim Crowe, a former conservation biologist, to explain this with reference to the case of Mr. Phiri-Maseko.[43] I also think that it will be useful for you to read Stengers's work on cosmopolitics, for she sees in the scholarly work of identifying the political cosmology of scientism the means to counter its toxicity.

3 | For instance, I have a question for all the science people. There is a place in KwaZulu-Natal called Umhlab'uyalingana. They believe that through the magic, you call it black magic, they call it witchcraft, you are able to send lightning to

strike someone. Can you explain that scientifically because it's something that happens?

I think this is a terrible example because in setting it out, you fall into the trap created by modernist thought itself: pitting magic against science. If the terms of the analysis prefigure the answer, it is not a useful question. I am interested that you say it is something that happens; it suggests you may be committed to observation and empiricism, even though you are questioning the limits of that empiricism. The interesting question that your example poses is whether we are trapped into explaining things only in terms of belief or science, culture or nature, or whether there are different ways to formulate the problem. I am convinced that there are alternatives and that the most useful route to go is situating knowledge within the concerns and questions that frame it—as we did in the ABC project above.[44] In sum: in order to do transformative work, you need to ask questions that transform the terms in which the answers have been posed.

4 | Western modernity is the direct antagonistic factor to decolonisation, because Western knowledge is totalising.

Yes: coloniality and modernist thought have been two sides of the same coin. The problem comes in where you don't distinguish between "modernity" and "Western knowledge"—this is a problem for two reasons.

First, the geographical argument is a disempowering one. The Western Hemisphere includes the whole of Latin America, all First Nations of North America, the still colonised people of the Arctic, and all those many scholars who oppose modernist binaries in that "West." Where your argument slips into area-based identity as the determinative factor of what makes science, you are a partner in rendering invisible exactly the knowledges that scientism renders invisible.

A more useful alternative may be to work instead with the capture of science by the agendas of capitalism, patriarchy, and the dogma of economic growth. If you frame a critique of the alliance of contemporary science with corporate interests, you are not making invisible your allies, nor are you making an enemy of all your potential allies.

Second, *"indigenous knowledge studies" and science are not opposites.* Using them to judge or validate one another is a category mistake. If they are defined as opposites (as if Canadian Cree and Kalahari !Xam and Scandinavian Lapp knowledges exist together as a single meaningful category), the ensuing inherent contrast with science is resolvable only via acts of epistemic charity, "the curse of tolerance" (who wants to be tolerated?),[45] or the creation of a Bantustan within the university in which "ethnoknowledge" can be celebrated. Those solutions are satisfactory for neither the scientist nor the decolonial nor the indigenous knowledge specialist. The problem is that the categories "indigenous knowledge" and "science" derive from the primary dualism at the core of modernist thought: belief v. knowledge; value v. fact; society v. nature; subjectivity v. objectivity. The only way out of this cul-de-sac is to understand that knowledge is produced in relation to concerns that frame specific questions, and that those concerns may change depending on the life experience of the person asking the question.

5 | It is saying that it was Newton and only Newton who knew or saw an apple falling and then out of nowhere decided that gravity existed and created an equation and that is it. Whether people knew Newton or not, or whether whatever happens in Western Africa, Northern Africa, the thing is that the only way to explain gravity is through Newton, who sat under a tree and saw an apple fall.

I think you would find in Niccolo Guicciardini's biography of Newton a sense of the chasm between the popular version of Newton, which you object to and which uses his scholarship to claim universal authority, and the questions Newton himself posed about his own work.[46]

One example that you might find useful to think with is pi: while π was celebrated as a Greek discovery (in other words, a Euro-American discovery, even though ancient Greece had vital ties in both Africa (Egypt) and Europe), ancient Japanese temple geometers had worked out the proportion by different means.[47] You could also talk about how the history of mathematics depended on the Hindu invention of a numeral for zero, or you could speak to the work of Arabic scholars in medicine, mathematics, and astronomy. Basket weavers and beadworkers across Africa have exceptionally complex understandings of the increases and decreases necessary to create a form, as Gerdes has demonstrated in his analyses of their geometries.[48] As I understand it, you are arguing that the veneration of specific European founding fathers of science is at the expense of all others' intellectual accomplishments. And you're arguing that the use of that to legitimate authority over political life is an expression of white supremacist thinking. I couldn't agree more.

6 | So Western modernity is the problem that decolonisation directly deals with. To say that we are going to decolonise by having knowledge that is produced by us, that speaks to us, and that is able to accommodate knowledge from our

perspective. . . . Decolonising the sciences would mean doing away with it entirely and starting all over again to deal with how we respond to the environment and how we understand it.

Challenging modernity really got the howls going, didn't it? Frame your argument in terms other than the binaries proposed by the very modernist thought that you reject.

7 | So if you are saying that you disagree with our approach, it means that you are vested in the Western and Eurocentric way of understanding, which means you yourself need to go back, internally, and decolonise your mind, [and then] come back and say, "How can I relook at what I've been studying all these years?" Because Western knowledge is very pathetic, to say the least.

You are saying that the emperor is naked. That takes courage. Your more important point got lost in the need to needle the professoriate. That was a costly loss: a small victory that lost a greater battle. And while I recognise your desire to smash supremacism and patriarchy, I think that to win an argument for the transformation of science, it is useful to distinguish science from scientism—the latter being the unquestioned authority of experts without any understanding of the relationships between truth claims and the authorisation of patriarchy, coloniality, capitalism, white supremacism.

"RESISTANCE IS FERTILE!"

On Being Sons and Daughters of Soil

The individual, the community, the land are inextricable in the
process of creating history. Landscape is a character in this process.

—Édouard Glissant, *Caribbean Discourse*

"From time to time I have sought the Bible for understanding and perhaps I
can direct you to Ezekiel 25.17," wrote Mark Caruso, an executive of an Australian mining company called Mineral Sands Resources (MSR) in a letter to
residents in two South African villages who were opposing his mine. "And
I will strike down upon thee with great vengeance and furious anger, those
who attempt to poison and destroy my brothers. And you will know my
name is the Lord when I lay my vengeance upon thee. . . . I am enlivened by
[the] opportunity to grind all resistance to my presence and the presence of
MSR into the animals [*sic*] of history as a failed campaign."[1]

Caruso's letter was reported in a *Sunday Times* feature. Directed at activists opposing titanium-rich mineral sand mining in the villages of Xolobeni
and Lutzville, his words are quotations from the biblical book of Ezekiel and
the book of Tarantino known as *Pulp Fiction*.

After processing, titanium dioxide or TiO_2 becomes entangled in global
"green" commodities networks including lighter, fuel-sparing airplane parts;
the "green" whiteness that is in white paint instead of lead; sunscreens that
use titanium instead of the zinc that damages coral reefs; as well as white
paper, cosmetics, and food colouring.

Xolobeni is on the shores of the Indian Ocean; Lutzville near the shores
of the Atlantic. One village speaks the amamPondo dialect of isiXhosa; the
other speaks Afrikaans. Both settlements aim to live off the land: one by

4.1 | Placard from the meeting room at the Philippi Horticultural Area campaign, Cape Town, 2017. Photo: Lesley Green.

maintaining traditional rights; the other by reclaiming the lost arts of food production.

Caruso's letter was sent in October 2015. It was the assassination of the local anti-mining organiser Bazooka Radebe in Xolobeni in April 2016 that prompted our research cluster, Environmental Humanities South (based at the University of Cape Town), to ask whether it was useful to describe the era of geological climate change as the "Anthropocene," based on the idea that the human species ("Anthropos") is damaging the planet, or whether the term "Capitalocene" was a more useful diagnostic.[2]

The issue is not simply a question of semantics. A wrong diagnosis is an inaccurate definition of the problem, and leads to the wrong research questions, which in turn lead to policy interventions that will try to fix what ain't broke.

The term "Capitalocene" draws attention to the two dozen or so corporates that are responsible for the bulk of carbon dioxide pollution.[3] It underscores that commodity trade–based extractivism in mining and foods contributes disproportionately to climate disorder. And the term underscores that *not* "all humans are complicit," as if there is a war of all humans against the earth. For Deborah Danowski and Eduardo Viveiros de Castro, the problem is not humans versus earth, but humans versus terrans—terms that contrast earth mastery with earth partnership.[4] The situation at Xolobeni crystallises that analysis.

Bazooka Radebe was one of the activists opposing the mine in Xolobeni, which would provide short-term gains in exchange for long-term land loss. Shocked by his death, my colleague Michelle Pressend offered to arrange a screening of a film about Xolobeni struggles called *The Shore Break*, co-produced by environmental lawyer Odette Geldenhuys with director Ryley Grunenwald.[5] The documentary focuses on the evidence that the state and the mining company had bought off local traditional leaders to get access to land for extractive mining, pitting traditional leaders, together with mining interests who enjoyed active support from the national government, against Xolobeni residents wanting to live off the land.

At the discussion after the screening, the situation at Lutzville was raised, where the same company had caused a cliff to collapse. Both local struggles were in difficulty because the national government was taking the view that promoting economic development and protecting the environment exist in an either/or relationship. Other speakers linked the situation to the greater Cape Town area, where the campaign regarding the Philippi Horticultural Area (PHA) was about to head to the Cape High Court to fight the same battle against the city council, which was planning to sell off the last remaining productive farmland within the city limits to developers and sand miners, putting at risk both the city's future food supply and its water supply as the area recharged the Cape Flats aquifer. All three situations involved sand, soil, and an argument for development, and questions of what it would mean to address climate change and land rights. All three situations involved citizens of colour defending their right to live in the land's ecology. In all three situations, government officials framed the problem to be addressed as $D = (ZS/BEE) \Leftrightarrow hr^{td}$. In English: the development (D) of zones of sacrifice (ZS) divided by Black Economic Empowerment (BEE) groups leads to historical reparations (hr) by the power of trickle-down (td), and vice versa. For our new environmental humanities programme, the question was: how did the unfolding situations at Lutzville, the PHA, and Xolobeni challenge and change environmental research methods and analysis?

What did a struggle for land rights mean if land and those who seek to live off the land could be sacrificed to a mining business? How does one explore an apparent continuity between forms of land mastery under settler colonialism and neoliberal extractivism? How was the mastery that had been woven into the fabric of colonial and apartheid racism now perpetuated in South Africa's "post-apartheid" governance?

What was at issue was not just land acreage and corporate gains in the Anthropocene, but what it meant to be an environmentalist in South Africa in the Capitalocene. What kind of ecological politics could reconnect ecology with economy?

Even the most cursory glance at South African environmentalism is a confrontation with the unbearable whiteness of green, not only because most environmentalism in South Africa is overwhelmingly white, but because it is necessary to confront the ways in which environmentalism has become a space in which a quotient of South African whiteness, and white supremacy, has sought to reconstitute itself within the field of "nature," where there is as little political accountability as is possible to a parliament that limits itself to matters of "economy" and "society." A first step to resituating environmentalism in contemporary democracy is to confront the curious entanglement of the refuge of "nature" and "scientific authority," with the last political space in which dogmas that enable white supremacism have a chance of passing unchallenged. Unravelling that "mashup" is critical if black environmental struggles, and green environmentalism, are to join forces to challenge the split between "development" and "environment."

Nature reserves are the definitive face of South African environmentalism. Since only "experts" and paying reserve customers can participate in a nature that only exists behind a fence, nature reserves situate environmentalism in South African public life as an "elite activity," and therefore inevitably a mostly white one. In turn, a moral high ground places ecology over economy, which in South Africa associates with the categories white and black. Add the deeper layers of the settler-colonial psyche, in which whiteness saves Africa from itself, and add the overwhelmingly white voice of professional science with its claims to objective, unvarnished truth, and the situation offers an all-too-easy resting space for the sort of white supremacism that is evident in "the rhino wars," where white waBenzi (Mercedes Benz drivers) drive around with stickers like "Save the rhino, hunt a poacher." The one shown in the introduction bears the name of Palala. An online search shows that Palala owns a lion-breeding farm that specialises in supplying white lions for its hunting lodge.[6]

Perhaps the most egregious example of "whites saving nature from Africans" is an international "humanitarian" organisation based in New York that sends US military veterans to Africa to "save wildlife." VETPAW, as mentioned in the introduction, uses clickbait images of white girls holding big guns and draped in bullets, or white guys in camo with big guns saving cheetahs, or white guys and white girls with big guns surrounded by grinning black children, and endless African sunsets. VETPAW's South African Facebook page supporters call for "poachers" to be hunted, castrated, shot, or tortured. In the name of "green," such posts seem to be immune to prosecution for hate speech.

VETPAW is not necessarily the extreme end of the spectrum. The "Save the rhino, hunt a poacher" sticker was on a car at the petrol station near my home. On Facebook, a local mixed-race hiking group tolerated a dog lover's call for a poacher's castration under a picture of a poacher arrest. A local gun shop offered a cut-price special: fire fifteen bullets from an AK-47 at a target labelled "Rhino Poacher." You don't need to be a linguist to know "green" code-speak for "black." The question is why white supremacy in South Africa has been so wildly successful in recrafting racist speech under the banner of green, and making it look dignified, moral, and socially palatable.

Achille Mbembe's *Critique of Black Reason* offers ways to think through the problem when he describes "race as being both beside and beyond being. It is an operation of the imagination, the site of an encounter with the shadows and hidden zones of the unconscious."[7] "The African politics of our world cannot be a *politics of the similar*. It can only be a politics of difference—the politics of the Good Samaritan, nourished by a sense of guilt, resentment, or pity, but never by an obligation to justice or responsibility."[8] His account of whiteness as a claim to historical time speaks to the ways in which a significant proportion of white environmentalists frame their role in addressing current extinction risks, arguing that in their view: "Only the White race possessed a will and a capacity to construct life within history. The Black race in particular had neither life, nor will, nor energy of its own. . . . It was nothing but inert matter, waiting to be molded [by] a superior race."[9]

Mbembe's rendering of whiteness is as a "certain mode of Western presence in the world" (for example, in respect of the defence of African wildlife). Notwithstanding the dependence of consumers and tourists on an extractive economy grounded in excessive ecological and structural violence, and notwithstanding the history of settler-colonial land grabs, it is, in the unbearable whiteness of many greens, the poacher who becomes the guilty party for destroying the animal kingdom and the humanity of Humanity, both at the same time. Hence, the poacher justifies a new wave of Whiteness (capitalised here because it speaks to a racialised political identity) that justifies cruelty without measure.

For white readers who struggle with this account of Whiteness, let me be clear. neither I nor Mbembe conflate Whiteness with all who are born light-skinned; I speak of a set of dispositions associated with everyday Whiteness-based supremacies: being saviour, judge, expert, and martyr (after Marilyn Frye).[10]

As I see it, the dogma of recrafted Whiteness legitimates itself by framing itself as saviour: planetary stewards (against black poachers, for example). Moreover, green whiteness readily leverages the expertise of science—scientism—to take advantage of the unquestionable assertion of the authority of natural science in matters political. Reversing the apartheid positioning of blackness on the side of nature, the wild, and animals, green whiteness becomes nature, and the judge of what is good and what must be saved. (Black, of course, remains animalised via the figure of the poacher.) Green whiteness capitalises on the extra-parliamentary privilege of scientific authority. Blackness, in that view, becomes the source of ecological damage, without recognition that settler colonialism has an environmental history of ecocide ("That was bad people only!") that is real and ongoing as the commodities market fosters almost limitless extractivism as the basis of the economy. Green whiteness offers no understanding of the history of parks, the creation of the category "wild," or African "postcolonial" debts that are still being paid to colonial nations.[11] Green whiteness pays no heed to the conundrum that South

Africa's "wild nature" is sold to the Heathrow classes as "real Africa" when the majority of South Africans cannot afford to enter the parks for even a day trip. Green whiteness cannot contemplate the contradiction of protecting its "real Africa" from Africans. Untangling green versions of Whiteness from environmentalism requires undoing the tangle of negation upon fantasy upon haunting upon forgetting, amid another tangle of reclamation and reclaiming, with categories half-made and half-unmade, strangely remade, and still sold for profit.

With "real nature" fetishised in profitable nature parks, green whiteness seldom considers the loss of the commons, for its nature exists behind fences, and everything else is private property. "The land question"—that source of black rage and white fright—is reduced to property deeds, not a relation of mastery, so green whiteness remains comforted by the certainty that nature has a place, and they can go home and spray their lawns with herbicides to get rid of the weeds, and don't need to oppose agro-industrial farming, whose pesticides have killed insects and bees on an ecocidal scale.

Green whiteness cannot succeed in mobilising an environmental public, or in building the kinds of alliances that will counter neoliberal approaches to extractivism.

Because I am "Green" by name, the term both tickles me and, I hope, may in some way underscore for readers that the challenge to find a line of flight from green whiteness is not a matter of "othering" well-intentioned environmentalists, but of perhaps offering a line of flight from the ways in being "Green" risks slipping into Whiteness 2.0. In questioning these slippages, I have had to confront my own history, and question my own environmental assumptions and practices and their entanglements with my own inheritance of South African racism. For this reason, I've chosen to offer an account of race and environment in the next section in a story that is interwoven with my own genealogy.

II | METABOLIC RIFT: THE NATURE OF SETTLER COLONIALISM (FROM "THEM" TO "ME")

In the 1840s a young journalist working in Britain started to follow the story of the closing of forests to peasants. When they were no longer able to collect fuel, or gather wild foods, in common lands, their household ecologies could no longer function, and most had no option but to move to towns in search of paid labour.

The journalist's first major story was "On the Law of the Theft of Fallen Wood," published in 1842. In it, he compares the laws that had closed the commons in different countries of Europe. As factories and cities increased in size, so too did the need for commercial agriculture on a scale that could feed many more than those who lived off it. As more farming estates were driving more people off the land, the journalist observed, soils were becoming depleted.[12]

The journalist's name was Karl Marx. His concern with the alienation of workers and the depletion of soils threads throughout the three volumes that developed out of his story on wood: *Capital*.[13] Marx keenly observed the depletion of soils when large-scale agriculture stripped lands of their nutrient cycle, and argued that new ecologies of life and monoculture farming under capital had broken the metabolic cycles that fed soils.[14]

Metabolic rift expanded globally across oceans into the colonised lands with the ever-growing demands of the industrial economy. The European food crisis, David Montgomery argues, was solved by exporting people and importing food from the colonies.[15] Without resettling mouths and outsourcing land and muscle, Europe's industries would never have been built.

In 1819 some ninety thousand Britons applied for the "plentiful and fertile" lands advertised in South Africa's Eastern Cape. Four thousand were accepted.

For settler descendants like this writer, "colonisation" is something "others" did only for as long as you don't research your own history.

When I did start tracing my ancestry, I found that one of my paternal great-great-great-grandfathers, James Cawood, was one of the six sons of David Cawood, who came from Yorkshire and settled alongside the Fish River in 1820, on the edge of the "neutral zone" negotiated by the British with the amaXhosa.[16] By 1848 his twenty-one-year-old grandson, James Fordred, had enough wealth to sail to London to buy a printing press.[17] His capital came from ivory and hides; there is a record of the Cawoods shooting a hundred elephants in a single day.[18]

In 1828 Colonel Charles Somerset's troops, along with the Cawoods, allied with Chief Hintsa of the amaGcaleka Xhosa to defeat the Ngwane fleeing Shaka.[19] Six years later, at least one of the Cawood brothers accompanied Colonel Harry Smith's expedition in which Hintsa was slain.[20] His assassination and humiliation in death is one of the most terrible moments in South African history.[21]

On at least one side of my ancestry, settler colonialism involved the elimination of human and animal contenders for the land. Expulsions and extinctions have the same roots.

On the other side of the family, my maternal great-great-great-grandfather, Nathaniel John Morgan, published in 1833 his account of the amaXhosa

4.2 | Thomas Baines's depiction of the British settlers' ivory market, Grahamstown, South Africa, 1866.

in the first South African scientific journal, the *South African Quarterly Journal*.[22] In it, he gingerly opposes the argument that the Xhosa and the Khoikhoi had to be subjugated and displaced for the English settlement to succeed.[23] The publication includes an account of amaXhosa history and politics, basic isiXhosa vocabulary, and what he had learned in Xhosa villages about farming local land:

> Within sight of their kraals, generally on the opposite side of the ravine or kloof, are situated their corn fields or gardens. [They] inclose an extensive piece of ground [and] bring into cultivation the moist and fertile parts only. In the fields, the cultivation of which is often the labour of several families, are erected temporary huts to afford a shade for the children who, when the corn is sown, are stationed there to prevent the entrance of cattle, and as the corn ripens to keep off the birds. . . . In these gardens they cultivate . . . corn, melons, pumpkins, beans, and a little tobacco. . . . In preparing and cultivating the land, they first clear the ground of weeds, then they throw the seed on the surface, and cover it lightly with the soil, using small wooden spades; and when it appears a little high above ground they again carefully destroy the weeds, thinning the corn and throwing a little earth about the stem. When it is ripe enough to be gathered, they cut off the heads, and either hang them up in their huts or

place them on a frame raised some height from the ground. . . . After it has been kept some time in this manner, they beat the grain out, and put it into small holes prepared for that purpose in the centre of their cattle kraal; each hole is capable of containing about two sacksful. On the top they throw a quantity of the stalks to absorb any moisture that may happen to penetrate through the earth and the manure that is placed over the stone covering the entrance. These granaries are opened only at particular times. Corn so secured will keep sweet and good for a great length of time, though, if the season has been wet and it is stowed away a little damp, it sweats and becomes sour. . . . Their attention is however chiefly engaged by their cattle: these they herd, protecting them with great care by night and day.[24]

Morgan led a party that settled outside Grahamstown, and he served as a staff surgeon for the British Army; his writing gives some insight into how much Xhosa villagers taught him about farming, history, language, and customs. His picture of the partnership of plants, soils, cattle, and people made a case for continued amaXhosa stewardship of land. A friend and ally of the British philanthropist Dr. John Phillips, he politely presented evidence that opposed the position of both the Dutch, who had driven the Khoena from the Western Cape to the Eastern Cape, and the British, who had brought in settlers and the military to hold a moving line between empire and natives.

Morgan was not a successful farmer. He lost his farm and died shortly afterward. Settler conservatives like those from whom I received my paternal DNA won the political argument both in the Eastern Cape and elsewhere in South Africa, supporting sentiments like, "I contend that we are the finest race in the world and that the more of the world we inhabit the better it is for the human race."[25] Those words were penned by Cecil John Rhodes in 1877, whose name was given to my paternal grandfather, Cecil Fordred, in 1898.

Later the British prime minister at the Cape, Rhodes connived to strip Tswana chiefs of diamond lands in the Kimberley region in the 1880s. Thus it was that with wealth largely derived from Tswana-held diamond lands, Rhodes bequeathed to the nation the forestland that would become Kirstenbosch Botanical Garden; a zoo in which captive lions would roar adjacent to the prime ministerial residence; grounds for a university in Cape Town; and the Rhodes Scholarship programme, which in time would educate presidents and prime ministers, including Bill Clinton.

By 1885, through the occupation of land by thousands of families like mine, the "Sketch Map of South Africa Showing British Possessions" notes

4.3. | Front page of Nathaniel John Morgan's article, 1833. Source: Bolus Herbarium Library, University of Cape Town.

the occupation of Tembuland (*sic*), Amampondo, Transkei, Namaqualand, Great Bushmanland, Zululand, Swaziland, Amatonga, Basotholand, Griqualand East and Griqualand West, and Matabeland, with Damara Land and Great Namaqualand registered as German territory.[26] Some stamps in my childhood stamp collection are from Rhodes's BSAC—the British South Africa Company—his name for the country that came to be known as Rhodesia and is now Zimbabwe.

My grandfather was born on a farm in the eastern Orange Free State in 1898—almost certainly land taken by the British from Afrikaners during the Anglo-Boer or South African Wars.

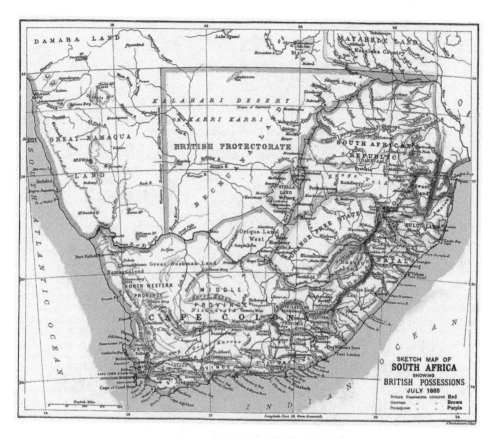

4.4 | "Sketch Map of South Africa Showing British Possessions, July 1885,"
as published in the *Scottish Geographical Magazine*, 1885.

On farmland nearby, when Grandpa Cecil was a six-year-old learning his ABCs and horseriding, a Zulu family by the name of Mazibuko was among the many sharecroppers in the district. Their son was born in 1904, and his English name was Robert. When he was buried in 1994, he would be remembered as the Tree Man who had taught the arts of making "black gold" (described in a passage below) for trench farming.[27]

At the age of fifteen, Grandpa Cecil was digging British trenches, having signed up to march in a Scottish kilt and fight for king and country in World War I. His winter uniform would have come from Karoo sheep watered by fossil water drawn by the new windmills. Scarred by chlorine gas and haunted by trench memories, he came home to dig for British gold in South African mines. Later he became a municipal health inspector, responsible for pest extermination in the fast-industrialising coastal town of Port Elizabeth,

4.5 | *The Rhodes Colossus Striding from Cape Town to Cairo, Punch,*
10 December 1892. Artist: Edward Linley Sambourne (1844–1910).

which was growing along with the vehicle-manufacturing sector. Cecil died in 1978, a year after Steve "Frank Talk" Biko had been interrogated to death near my grandparents' home in 1977, two years after the student riots of 1976.

———

When Mazibuko was nine and my grandfather fifteen, the Native Land Act went from bill to law in four days. It was June 1913. The law was rushed through Parliament by the British at a time when many feared a Boer uprising while England focused on the coming war.[28]

What Marx had noted in Europe—the slow and steady expulsion of peasants from the land into a capitalist economy—was completed in a period of weeks in South Africa. Black lives were separated from forests and cattle plains; they were forced off farms, off riverbanks and seashores, out of villages, and into mines, farms, and factories. As of the passing of the act, no black person was allowed to buy land. Forced into native locations, they required written permission to leave them.[29] Writer Sol Plaatje cycled around South Africa that August, documenting what he saw:

> It was cold that afternoon as we cycled into the "Free" State from Transvaal, and towards evening the southern winds rose. A cutting blizzard raged during the night, and native mothers evicted from their homes shivered with their babies by their sides. . . . Kgobadi's goats had been to kid when he trekked from his farm; but the kids, which in halcyon times represented the interest on his capital, were now one by one dying as fast as they were born and left by the roadside for the jackals and vultures to feast upon.
>
> This visitation was not confined to Kgobadi's stock, Mrs. Kgobadi carried a sick baby when the eviction took place. . . . Two days out the little one began to sink as the result of privation and exposure on the road, and the night before we met them its little soul was released from its earthly bonds. The death of the child added a fresh perplexity to the stricken parents. . . . The deceased child had to be buried, but where, when, and how?
>
> This young wandering family decided to dig a grave under cover of the darkness of that night . . . in a stolen grave, lest the proprietor of the spot, or any of his servants, should surprise them in the act.[30]

The same parliamentarians who had passed the Native Land Act a few weeks before then enacted a bill establishing Cape Town's Kirstenbosch Botanical Garden on 1 July 1913. That the centenary celebrations of Kirstenbosch Botanical Garden took place in 2013 amid separate memorialisations of the centenary of the Native Land Act at the University of Cape Town affirmed that South Africans—including myself, at the time—had yet to recognise the continuity of the nature/society divisions in modernity, coloniality, apartheid, and nature reserves.

Industrial agriculture, cultural nationalism, native reserves, and nature reserves are knotted together around the central problem of private property and the rise of a global food system to feed workers forced into towns by the closure of the commons. What had been separated from ecology was economy. In South Africa, fantasies of whiteness and blackness slow-danced through history from one side to the other. Native reserves—later Bantustans—were labour reserves (economy) justified as cultural ecology (nature). Nature reserves were justified as natural ecology, separate from the economy's agriculture and from cities and mines. In the current era, where economic imperatives are framed as black, whiteness has reconfigured itself around ecology. As South Africa is finally reckoning with land restitution, there is a question that needs to be asked: in decolonising settler-colonial environmentalisms, what will it take to relink ecology and economy, and at the same time escape their entanglement with racialised fantasies? Can modernity's "metabolic rift" be addressed?

III | "SONS AND DAUGHTERS OF SOIL"?

While many black South Africans fought in World War I, Robert Mazibuko's memory of trenches, gold, and public health was quite different to my grandfather's. In 1994, after the first democratic election that made Nelson Mandela president of South Africa, Joanne Bloch interviewed him:

> I was born in 1904 near Ladysmith, at a place called Spioenkop.... At that time, agriculture, the soil and animals were so important to the African people. They were our banks and mines. We lived because there were animals, pastures and water....
>
> In the early 1940s I worked as the principal of Hlophenkulu High School in Nongoma. I taught all the boys and girls to plant their own crops on the Mission lands. We planted lots of fruit trees and vegetables and the community was very pleased....
>
> Poisons and fertilisers were not for me. Although they make plants grow very fast, they also die very fast. Also chemicals cannot work without much rain and moisture, so if it doesn't rain, they don't work.... Another bad thing is that chemicals kill all the insects in the soil, even the good ones like earthworms....
>
> We need to learn some lessons from our ancestors. They were part of the land and at one with it. We need to look after our environment and use it as well as they did.
>
> The African people knew so much about conservation because they moved around with their cattle. In this way, they learned which grasses, plants, trees and animals were useful, and where to find clean water.
>
> Today we don't move about from place to place as our ancestors did. More people are born every day, and our cities are growing and growing.

Millions of trees have been cut down, which has hurt our earth. Big farms where they grow crops like sugar cane for money have made the soil terribly weak. It has become too loose.

When there is wind, soil is blown left and right, and its richness and goodness is blown away. Also when it rains heavily, the rich soil is washed into the rivers. It is now more important than ever to look after our earth. In African culture, the people always respected the land, the trees, the plants, the animals, birds and insects. For example, there were many different birds that no-one was allowed to kill—like owls, nkombose, ngete, eagles, vultures, storks and tickbirds.

These birds were too useful to kill. You could not kill a secretary bird either. It was important because it killed snakes. No one could kill an elephant or a hippopotamus for any reason at all. Antbears were very important because they ate white ants, which destroy trees and homes. So nobody killed them. Some animals, like buck, could only be killed to make useful things out of their skins, like karosses and little slings for mothers to carry their babies in.

In the old days, certain trees were very important too. Most people were not allowed to touch the inkayi tree. Its wood was very hardy, so only warriors could break its branches to make fighting sticks. Many other trees and plants were important as herbal medicine was made from them. For example, no-one was allowed to destroy the tree called mhlehampethu, because it was used to treat sores on animals. No fly would come near an animal once its sores were covered with this muthi. . . .

Our ancestors understood that in nature everything is linked up or interconnected. The animals, insects and birds need the trees and plants. The trees and plants need the animals, insects and birds. People need all of them. When last did you see an owl? These days owls are very rare, because there are so few grasses left in the veld. Without these grasses, there is no place for the rats to live in the veld. Without the rats, the owls have no food, so they die.

Many other South African wild animals like antbears and secretary birds are rare now. When I was young I remember there were flocks of birds flying in the sky, eating insects like locusts and moths. Today it is very rare to see those birds. Yet if we looked after our environment, these birds would come back in their hundreds and so would the other rare animals. What can we do now? The very first thing to do to improve our environment is to plant trees. They should be indigenous trees—trees that come from South Africa and grow well here. They should also be trees that have fruit that is eaten by birds, animals and people. When there are many of these trees, monkeys and owls and other birds will come to live in them. Their droppings, and their bodies when they die, will make the soil rich again.

A donga or gulley in the veld is a sign that there is a problem in that place. All the soil is washing away when it rains. If we find a donga, we must try to grow some trees there. The trees' roots will hold the soil in place. The best seeds to plant are acacia seeds, because these trees are so strong. . . .

We can plant aloes or trees like willows, so that animals do not tramp about there and dirty the water. Wetlands are like sponges which must not be broken or crushed. If you care for them by planting indigenous trees there, this means that there will be more water. This water will be fresh and clean, so people will not get sick from drinking it. Also, the land around the wet place is more fertile if there is more water. . . .

Conservation is vitally important for the future of our country. If we plant trees and grasses and look after our wetlands and wild birds and animals, we will be doing a lot. But there is something even more important we need to know.

How do we practice conservation in our own gardens? How can we use our own little piece of land to help ourselves and help our earth too? We have to learn how to enrich our soil, because without the soil, there is nothing.

. . . You must put back into the soil what you take out of it because nothing is useless to the soil. It feeds on all our rubbish. That's why the first thing to do in your garden is to start a compost heap—to make what I call black gold. Black gold is compost—made from old leaves, peels and sweepings from the kitchen, mixed with grass and animal manure. Once this breaks down, it is dark in colour, and I call it black gold because it will last much longer than any chemicals. Chemicals make plants grow very fast, but black gold gives plants the chance to grow naturally, taking many elements from the soil which are good for human bodies.

There are many plants out in the veld which have deep roots and which put goodness into the soil. These are the plants and trees I help people to find and grow because they make the soil healthier. People can also grow the vegetables that grew long ago, like wild tomato which will grow anywhere. Birds eat it and, because they cannot digest its seed, wherever the birds make droppings, more plants grow.[31]

In the farming practices of award-winning Zimbabwean farmer Zephaniah Phiri-Maseko, "the marriage of water and soil" in dry lands is a practical philosophy in which food growing, soil and water, personhood, multiple species, and the community of the living and the dead meet in household farmland. To tend the soil is to tend the household. Christopher Mabeza's study of

Phiri-Maseko's land practices reflects his gentle humour—with which he survived many an encounter with local agricultural extension officers, gaining a limp he took to his grave after six months in British leg irons as a young man.[32] A Malawian by birth, Phiri-Maseko was allowed to farm only in a dry area of Zimbabwe with a tiny spring. Via partnerships with soil and animals and members of his household, he found a way to "rhyme with nature" to make sure every drop stayed on his land.

Land has a powerful presence in African literature, which anthropologist Francis B. Nyamnjoh argues is the ethnography of Africa from within.[33] Some of the most successful African environmental movements have focused on reconnecting people and soil through tree planting. Thomas Sankara's tree-planting project in Burkina Faso was one of the bases of his work to revolutionise that country's economy.[34] More recently, in post-genocidal Rwanda tree planting has become part of the massive public campaign that is part of the national healing, in which people work together for a day, once a fortnight, to plant saplings.[35]

Kenya's Wangari Maathai, founder of the Green Belt Movement and winner of the Nobel Peace Prize, saw in soil erosion and its associated drought the legacy of trees lost to misguided logging and tea plantations. Regenerating Kenyan political life tree by tree, the Green Belt Movement reclaimed household ecologies along with water systems and soil metabolisms, notwithstanding the opposition of occasional agricultural extension officers, trained as they were in the sciences of industrial-scale farming. The retranslation of scientific expertise into the practices of the household is evident in Maathai's book *Unbowed*:

> We organized meetings where foresters talked to the women about how to run their own nurseries. But these were difficult encounters. The foresters didn't understand why I was trying to teach rural women how to plant trees. "You need a professional," they told me. "You need a diploma to plant trees." They then told the women about the gradient of the land and the entry point of the sun's rays, the depth of the seedbed, the content of the gravel, the type of soil, and all the specialized tools and inputs needed to run a successful tree nursery. Naturally, this was more than the women, nearly all of whom were poor and illiterate, could handle.
>
> What the foresters were saying didn't seem right to me.... So I advised the women to look at the seedlings in a different way. "I don't think you need a diploma to plant a tree," I told them. "Use your woman sense. These tree seedlings are very much like the seeds you deal with, beans and maize and millet, every day. Put them in the soil. If they're good, they'll germinate. If they're not, they won't. Simple."

And this is what they did. . . . Soon the women started showing one another, and before we knew it, tree nurseries were springing up on farms and public land around the country. These women were our "foresters without diplomas."[36]

For Maathai, the preservation of soils through tree planting was a gendered project that responded to the patriarchal authority that local agricultural science had become. Local knowledge and concerns were framed as "woman sense" because, in rural Kenya, women were managing households where soils had degraded, economic times had become tough, and men had migrated to the cities in search of jobs. With trees to create shade, stabilise the soils, and stop desertification, women were able to grow the food needed for households. Where Phiri-Maseko had spoken of "rhyming with the soil," Maathai worked to foster an attitude of "gifting with the world":

Kikuyus used a gourd, in which they carried porridge or beer, as an offering or gift. Whoever received the gourd would polish it with oil before returning it. Over time, the gourd would become beautifully varnished by this repeated polishing. The deeper the color of the gourd, the more generous you had been—and the more connected you remained to the world around you. . . .

These gestures of giving capture both the spiritual and the practical elements of gratitude and respect for resources. Our connections to the planet and each other are reinforced simultaneously. The spirit of not wasting, because we assign value to something, is found in many traditions, but not often expressed. We could benefit from spending more time polishing our gourds for each other.[37]

Maathai's thinking was criticised by some on the grounds that she promoted in the place of colonial science an anticolonial African pastoralism linked to a romantic version of Négritude. Her critics have argued that she inverts the colonial narrative and is therefore not freed of its conceptual structure—that hers is a narrative tied up in idylls and essences. But, citing research on Great Zimbabwe, Byron Caminero-Santangelo reminds us that African ecological practices were not always in harmony with ecology. He also offers a more careful analysis of Maathai's feminism and political positions than the tired critique of lost utopias. Maathai's views, he argues, may present a precolonial idyll, but her environmental politics cannot be reduced to that:

It is difficult to discount her contributions to the shift toward democratic change in Kenya and to increasing attention to the part played by capitalized monocrop agriculture and resource extraction in the environmental slow violence underpinning social problems in Africa. These successes were not necessarily achieved *despite* her deployment of pastoral tropes; among Kenyans, those tropes helped connect environmentalism with anticolonial struggle and refute its association with colonial conservation and with the foreign. Her stories of traditional ecological practices and values could encourage Kenyans to rally around them as signifiers of counter-hegemonic collective identity and to embrace the protection of the soil as a crucial part of that identity.[38]

South African soils are damaged. They are poisoned by pesticides; leached by fertilisers; torn by ploughs; at risk of drought; under attack from policies preferring apparently "climate-proof" seeds that are genetically modified to be resistant to the trademarked herbicide glyphosate, which means that the only plants that will grow are those under a corporate trademark. These trademarks fundamentally change the relationships between rural land and economy, for following a 1995 World Trade Organization agreement that seeds with the tiniest of genetic changes can be patented, seeds now come with a suite of laws that criminalise seed sharing, and even criminalise farmers for the genetically modified pollen that blows onto their lands. Planting loans, on which many farmers are dependent, often come from grain-trading companies that will specify to farmers what seeds they can plant, and lock them into sales based on future commodity price speculation, using both genetics and law to stop seed saving.

In what sense, then, is the "land" of "land rights" the same as the "land" that was the focus of Black Consciousness and African nationalism in the postcolonial era? At Robben Island in Table Bay, where Mandela and comrades were imprisoned by the apartheid state, Robert Sobukwe, the pan-Africanist leader, was imprisoned in isolation, not allowed to utter a word to his comrades as they passed him on their way to the quarry works each morning. Former prisoners (now tour guides) tell how he would greet them with a fistful of soil running slowly back to earth between his fingers: reminding them that they were sons of the soil. For the late professor of religious studies Gabriel Setiloane,

> land was not only the property of the living, but of the total community of the living and living dead. . . . When one understands . . . how . . . the place of [a] man's birth and up-bringing is "a holy place," because there he meets his ancestors, only then will one be able to comprehend the depth of insult . . . of the victims of wholesale removals of villagers and townspeople in Southern Africa. . . . For when the Bantu say of themselves or one to another that they are *"Mwana we mvu"*—son [or child] of the soil—it is so. They are tied to the soil, body, mind and soul. A child's umbilical cord is

buried into the soil, the same soil into which his ancestors are buried, thus linking him to them where they are. If he is removed permanently from that place the cord which ties him to them is broken.[39]

The relation of death and birth in soil something that Theophilus Hahn commented on in his account of Khoena religion: "Previous to burial, they place the body of the deceased in the same position which it once occupied as an embryo in the mother's womb. The meaning of this significant custom is, that the dead will mature in the darkness of the earth in preparation for a new birth."[40] The connectedness of soul to soil has long been part of Southern African land.

When German academic colleagues asked, of an early version of this chapter, "Do you not remember that this way lies the road to 'blood and soil,' the language of genocide?," they were speaking to the problem of nationalist and culturalist claims to soil as territory. The discussion left me asking: is it possible to reframe connection to soil without the nationalist language of being and territory, as so much of the early postcolonial Négritude movement did? Was that in itself a translation of connection to soil into the terms of the European colonists of Africa, in order to oppose them? Could the phrase and the sense of belonging that it contains be translated into the language of household presence-to the soil and its creatures?

Among the most pressing questions in political philosophy has been the work of rethinking "being" in such a way that it is not predisposed to violent ethnic nationalisms with their associations of territorial rights and mastery. Jean-Luc Nancy, for example, whose work was the focus of Jacques Derrida's final manuscript,[41] sought to rethink the way of relating to the world which had arisen as part of European modernity (and its hundreds of years of nationalist wars), and in which cultural or nationalist concepts constituted people's identity: their tribal being, as a social whole. Apartheid was justified on the basis of this concept; the assertion of someone's identity as "tribal" was what Aimé Césaire rejected (using the memorable phrase we have already encountered, "a screaming man is not a dancing bear").[42] For Nancy, if social identity is taken as a form of being that must be defended, it is inevitably drawn into a violent relation with those designated as its "Others." If, by contrast, one works with the notion of being "present-to" the relations and objects that are valued, relationship brings values and processes like identity making and dignity into social life. Nancy's approach links processes of relating and knowing.[43] The "being" approach claims that "culture" is unchangeable: it is something that people "are."

Curiously, this way of connecting people culturally to soil has much to do with the origins of the word "culture"—before it came to mean something nationalist or ethnic. The *Oxford English Dictionary* tracks the shifts in the meaning of the word from earth or soil to mind or spirit.[44] Whereas "culture" originally described habits and practices of tillage and animal husbandry, its meaning shifted in the early modernist thought of the 1700s to focus on ideas and customs, or literary and artistic development: splitting "human nature" from earthly practices. Over time, "culture" took on a meaning associated with practices that, in Europe of that time, indicated that specific nationalities and nationalisms had mastery over a territory.

What this detour into the history of an idea suggests is that reclaiming practices of care for soil, and connection with ancestors buried in soils, does not have to be translated into a territorialist narrative of "blood and soil."

Is "sons and daughters of soil" necessarily an argument for cultural nationalism? Or are there other ways of interpreting kinship with soil?

A great deal of post-apartheid intellectual leadership has been provided by the South African Constitutional Court, where a widespread Southern African philosophy of personhood, or "ubuntu," has been taken up as a resource for rethinking jurisprudence.[45] Constitutional Court justices, including Yvonne Makgoro and Albie Sachs, have sought to draw on ubuntu as an African ethical ideal for relations: that a person is a person through other people.

In post-apartheid South African scholarship, postmodern academics have focused their attention largely on the misplaced question of whether ubuntu does or does not exist. (Note that the question is framed in terms of being— that is, does it exist or does it not exist?—even though an ethical ideal doesn't operate as something that exists. The same litmus test of existence is not applied to more fashionable concepts such as corporate social responsibility or the market, for example.) The Constitutional Court justices, however, were focused on relationships rather than the somewhat absurd requirement to prove the existence of a concept, and got on with the task of articulating a South African jurisprudence based on the ethical ideals associated with the widespread sub-Saharan notion of ubuntu.

Might ubuntu as a theory of relation be useful for rethinking human relationships to land and to animals?

Linguist Nkonko M. Kamwangamalu makes the case that the idea of ubuntu is found in languages across sub-Saharan Africa, tracing variants of the term, which translates as "personhood" or "humanness," to Kenya, Tanzania, Mozambique, Zimbabwe, the Democratic Republic of Congo, Angola, and Lesotho.[46] "Morphologically, *ubuntu*, a Nguni term which translates as 'personhood,' or 'humanness,' consists of the augment prefix *u-*, the abstract noun prefix *bu-*, and the noun stem *-ntu*, meaning 'person' in Bantu languages," he says, adding:

> The concept of ubuntu is also found in many African languages, though not necessarily under the same name.... This concept has phonological variants in a number of African languages: *umundu* in Kikuyu and *umuntu* in Kimeru, both languages spoken in Kenya; *bumuntu* in kiSukuma and kiHaya, both spoken in Tanzania; *vumuntu* in shiTsonga and shiTswa of Mozambique; *bomoto* in Bobangi, spoken in the Democratic Republic of Congo; *gimuntu* in kiKongo and giKwese, spoken in the Democratic Republic of Congo and Angola, respectively.[47]

To his list can be added *hunhu* (Shona, spoken in Zimbabwe) and *botho* (SeSotho, spoken in Lesotho); some would also link this to the Shona word *ukama* (relatedness) and the Kiswahili word *ujamaa*, or "familyhood," which became the rallying cry of Julius Nyerere in Tanzania seeking independence from the British in the 1960s.

Connecting the philosophy of ubuntu or hunhu to humans only is, in my view, an effect of the history of the modernist division of nature from society. One of the few African philosophers writing about this is Munyaradzi Murove of the University of KwaZulu-Natal:

> Kama means "to milk a cow or goat." ... Those who are related by blood or by marriage are called Hama. When it is Ukama it becomes an adjective which means being related or belonging to the same family [but] not usually restricted to people who share the same blood.... Ukama is also based on the totemic species system.... It is not uncommon to hear [Southern Africans] referring to themselves [as]: "we are those who belong to the buffalo, we are those who belong to elephant, we are those who belong to Zambezi river etc."[48]

Writing about the ways in which Shona ways of world-making became a resource on which many Zimbabweans drew in surviving that country's political and economic crisis of 2008, Artwell Nhemachena describes the practices of reworlding the world in the aftermath of economic and political crisis in Zimbabwe.[49] His central conceptual enquiry follows the ways in which the notion of "chivanhu"—which is based on the root word hunhu (ubuntu) and translates as "the ways of human beings"—became a resource for surviving extreme state violence aimed at unmaking protesters' humanity and their worlds. Nhemachena's work reaches for the much broader ecological and cosmological set of ideas undergirding the concepts of hunhu, to use the Shona word, or ubuntu, to use the Zulu and Xhosa terms for humanness.

While it is not possible to recover a precolonial past, the sociolinguistic picture suggests an ethic of care for the relationships through which humans connect to one another—the living and the dead, humans and soil, seed, water, plants, fish, cattle, wild animals, and birds. This set of ideas *connects* world and person, nature and humanity, in ways that are very different to the concept of nature-apart-from-society that dominates contemporary conservation biology. If it is, for the moment, impossible to imagine all nature reserves integrated into human society, is it possible to explore the possibilities for African conceptualisations of human kinship with soil and its multispecies partners, in a time when the soils of the Anthropocene are so damaged?

Seeds, pollens, and soils are materially different and politically different to the soils conceptualised by Leopold Senghor's heirs, who reclaim their kinship with soil as its sons and daughters. Agricultural land is not acreage; it is relationships that are embedded into soil. Reclaiming land requires not only struggles for title deeds, but cosmopolitical struggles—struggles over what is rational, reasonable, and normal, and what constitutes common sense. These are struggles over forms of science that have been captured by economic, political, legal, and university structures; struggles over how to conceptualise the biogeochemistry of seeds and soils.

Audre Lorde cautioned that the master's house will never be taken down with the master's tools. The rationales that created the Anthropocene-Capitalocene are not going to be able to unmake it. What will today's graduates have needed in fifty years' time, if they were to have succeeded in decolonising the material relations of the Anthropocene?

The struggles over the farmland known as the Philippi Horticultural Area (PHA) in Cape Town exemplify the extent to which struggles over land are not only political, but cosmopolitical. The PHA Campaign's efforts to protect the last prime farmland in the city area from sand mining for glass and cement, and from developers, have come to encompass struggles against supermarket-led food systems, industrial agriculture, city financial policies, and social science consultancies, as well as struggles for soil health, composting, aquifer recharge, land restitution, small-scale farmer alliances, and the recognition of the necessity for urban planning to be based on urban ecological futures—not the financial futures of investment banks. The PHA campaign's work with damaged soils was the beginning of a different politics.

The leader of the PHA Campaign is Nazeer Sonday, who began his farming career as an "empowerment" farmer, growing tomatoes. In his first year, he says, he supplied up to a ton of tomatoes per week to the supermarkets. But he was not able to make a living from this. Having thought deeply about why his successful farm was a commercial failure, he links food-supply systems, land restitution, soil science, fertilisers, pesticides, composting, and the Cape Flats aquifer. The interlinkages together form a veritable curriculum in political ecology for a city preparing for a time of climate change, using an approach that links farmers in Khayelitsha and the Cape Flats with suburban residents, consumers, and garden farmers.

Sonday's farming project aims to show that small-scale, two-hectare, multi-crop farms can be commercially successful and ecologically enriching, even on soils that have been severely damaged. What he teaches farmers to do is tend the relations between plants to control insects; partner with roots and earthworms in no-till farming; and work with fungi, moles, bees, and composts. He describes working with these relations as part of a politics of escaping industrial-scale soil productivity with its high input costs for chemical products. "Why should I buy nitrogen fertiliser if I can get it free from the air?" Sonday asks. "I just need to work with the beans." (Beans fix nitrogen in the soil and are a vital resource for soil health.)

Sonday has taught me that damaged soils can be fixed without unaffordable industrial additives, through multispecies partnerships with soil microbes and molecules. Those microbial and molecular partnerships, for him, offer the political-economic liberation that will make successful farming and food systems viable, cheaper, and healthier.

The PHA Campaign offers leadership in a form of environmentalism for South Africa that can activate collective work, life, and thinking, based on reclaiming what it is to be a terran: a person living in partnership with the earth.

IV | A DECOLONIAL SOIL SCHOLARSHIP?

Climate crisis is centrally about cities: the ways they have grown; the material resources they need (cement, glass, fuel, water, food); and the "biogeochemistry" they generate. The PHA Campaign struggles with a neoliberal city and its plans for a top-down, market-oriented climate resilience strategy. The Western Cape government has been remarkable in being the first provincial government to develop a policy for "climate smart agricultural production"—although its water priority plans were caught woefully short in the drought extending from 2015 until 2018, and there is no mention of any conflict over seed policy in their mention of "climate resilient . . . crops."[50] Neither is there any mention of land restitution or the need to work out how to support emerging farmers who are taking up farming amid the looming climate crisis. But those who drafted its policy brief on the PHA, not dated but circa 2016, did not interview Sonday, despite his leading the legal and policy work on the area, and notwithstanding his written invitation to them to work with him and use his farm as a base for further research. Presumably the writers of the document (much of which appears to be based on the PHA Campaign website) felt he was "too political" to interview.

Throughout the Western Cape's climate-smart agriculture documents one finds the language of "resilience." Here, the insights of climate scientist Gina Ziervogel and colleagues in the African Centre for Cities and African Climate and Development Initiative are important:

> Many environmentalists . . . are offering solutions to build resilience that rely on green consumption, growth of capital markets, and other technocratic and financial fixes. [But] most of these fixes reflect an implicit optimism about market mechanisms and fail to embed ecology in social life. In the last two decades cities have increasingly become centres for reproduction and transformation of neoliberal ideologies, like the focus on individual agency and self-reliance, that collide with the concept of redistribution. The focus on resilience initiatives in cities is at high risk of mirroring the current approaches to mainstreaming attention to climate change in cities, namely a symptomatic treatment of the issue that often fails to trace the structural causes of vulnerability.[51]

What Ziervogel and her colleagues achieve in this paper is the beginnings of a transformation of the questions addressed by climate science in South Africa. A critique of the language of resilience is a first and crucial step to developing climate research and scholarship that opens up different kinds of questions in South Africa. But the need is to go yet further.

When a South African climate scientist objected to my questions about climate-smart agriculture—"you are using climate change as a lens for a humanities critique"—the phrasing of the objection testified to the urgent necessity for climate scientists to rethink how climate science as a field changes scholarship in much the same way as do epigenetics and neuroscience. Contemporary science is itself transforming the totality of scholarship. It is not possible to address the climate crisis or epigenetics or neuroscience if "nature" and "humans," natural and social sciences, science and capital and justice are conceptualised separately.

How else, other than with a human earth science, could climate advisors address the South African lead delegate to the Paris COP of 2015, Nozipho Joyce Mxakato-Diseko, when she said that climate negotiations were "the new global apartheid"?[52] Her comments reflected a global and critically important critique of northern-driven approaches to climate policy, with their assumptions of universal knowledge and managerialist, accountancy-based implementation.

I can understand that in the era of Trumpism, the need among climate scientists for "independent," "bias-free" research feels intense. But a retreat to value-free, mess-free, society-free consultancy is not a victory: it is a defeat. Reductionist science that is blind to its societal entanglements is more easily manipulated than when those entanglements are surfaced in open discussion. In the words of Isabelle Stengers in *Another Science Is Possible*: there is a necessity to "reclaim the art of dealing with, and learning from, what scientists too often consider messy, that is, what escapes general, so-called objective, categories."[53]

For example, the PHA is a landscape of damaged soils. Rather than assuming that the entire PHA consists of soils as they have been defined in their ideal state by pedologists, it is more useful to work with the different kinds of damaged soil in the PHA. Seth Denizen, for example, classifies different kinds of human-affected urban soils, including soils shaped by gardening and lawns; non-soils, like mined rock or asphalt; commodity soils, which are deposits of manufactured materials gone to waste; mortified soils that have been stripped; and soils that are considered undisturbed.[54] By attending to the specific biogeochemical histories of urban and peri-urban soils, Denizen offers the basis for a Capitalocene-responsive scholarship: the material beginning of a project with emerging farmers that is capable of—literally—the decolonisation of soil, by unmaking the residues of colonial and neoliberal policies that have damaged soils in the misguided pursuit of maximum extraction.

Climate change is unequivocally entangled with the histories of metabolic rift and settler-colonial environmentalism in South Africa. Contemporarily, climate scientists advising on adaptation strategies have to work with national government officials who see environment as the opposite of development, and a provincial government in the Western Cape that sees in nature a symbolic greenery that can bring economic benefit and electoral gain—or be sacrificed for political benefit where necessary. Yet neither approach—at the national or provincial level—is useful in mobilising publics to partner with earthly and local ecologies on the scale necessary to address the climate crisis that is already with us, and yet to come.

If climate scholars would recognise as a conceptual frame their own core finding—that the separation of the categories "natural" and "social" is no longer of any use—a climate-transformed agricultural scholarship could confront the legacy of metabolic rift, settler-colonial natures, and global extractive capital. Looked at in this way, climate scholarship and decolonial scholarship share common cause: the unmaking of modernity-coloniality.

PART III

FUTURES
IMPERFECT

Stupidity is characterized, first of all, by a certain fascination with false problems, with "infernal alternatives," a kind of laziness or tiredness of thought, that "naturally" presents itself in every situation: truth or belief, experience or representation, facts or values, subjective or objective, and so on. . . . [These are] terms that had been previously separated and constructed, but whose modes of construction are no longer put into question. . . . Behind these false problems, so innocently epistemological, there is an entire political organization of thought at stake. To take an interest in the fabrication of false problems, in their dispersion and in their manner of mobilizing evidence to present themselves, is to interrogate the construction of a certain image of thought, at once political and speculative, which has stupidity as its final expression.

—Didier Debaise, "The Minoritarian Powers of Thought:
Thinking beyond Stupidity with Isabelle Stengers"

One of Aesop's fables tells the story of a deer. Blind in one eye, she was unable to see from both sides. She considered her limitations and decided on a solution: to live at a place where the forest met the beach so that she could train her good eye on the forest to watch for predators. She lived there happily for quite some time, safe from leopards and lions. Then came the day that she was seen by fishers from the sea. They crept up from their boat and feasted on her for days.

The problem was not the deer's eye, but her analysis of her situation. She divided the world into land and sea, and confused her body's capacity for sight on two sides with that world. Had she understood that she did not live in a world of two different sides, she might have survived. But, frozen in the glare of her own light bulb of how to analyse things, she didn't.

When scientists separate out the world into land and sea, or nature and society, and leave the weaving together of their facts and probabilities to others, and when they fail to see that the questions they ask derive in large part from how they look at the world, and that their investigative technologies have not escaped the complicated, messy, entangled, and contingent world, then they are as vulnerable to capture as Aesop's deer was to hungry fishers.

––––––––––

The story of the deer reminds me of what Isabelle Stengers describes: "the rather horrifying experience you can get for instance when trying to speak with so-called neo-liberal economists, the stone-blind eye they turn against any argument implying that the market may well be incapable of repairing the destructions it causes."[1]

Homo sapiens has in the era of neoliberalism come to be understood as *Homo oeconomicus*. An unassailable, absolutist dogma appears rational and reasonable: all is in service of economic growth. The "anthropo-not-seen" is Marisol de la Cadena's description for political struggles against the such worlds defined and transformed by the marriage of state and private companies under neoliberalism. "The universal *Homo oeconomicus*" that frames the idea of the human ("Anthropos") in the "Anthropocene" excludes the experience and struggles of those who do not define their worth or well-being as humans in some sort of "econo-sphere" that defines the worth of all the other earth systems, yet is somehow miraculously separate from the planet.[2]

Dominant economic logics—and, within them, universities and sciences of all kinds—serve to render invisible the concerns of those who refuse the roles prescribed for them by the prevailing idea of *H. oeconomicus*. Stengers argues: "What presents itself as a logical consequence . . . has been fabricated by multiple processes of so-called rational reorganization that in the first place aimed at sapping or capturing the capacities for thinking and resisting of those who were apt to do so."[3]

Amid the worst of the university strife at Cape Town in 2016, when students opposed the rise of the neoliberal university and named it as a contemporary form of coloniality that "disappeared" their concerns about racism in campus life and curricula, Jennifer Ferguson, singer, poet, and a former African National Congress Member of Parliament in Nelson Mandela's presidency, performed in Cape Town. "Here in South Africa, we are always in the crucible," she said between songs. "There are never any shortcuts. The only thing we can do is to be present."[4]

Her words meant a lot to me in those painful months, echoing as they did the exhortations of my colleague June Bam to academics and graduates to listen and listen until we could hear what we had not heard before. The task of the moment was to find a way out of two powerful spells: the spell of white privilege in which black experience was rendered invisible, and the spell of neoliberalism in the knowledge economy, in which all value, and all that was felt important, was translated into monetary value.

What would it mean to draw these insights on being present to what was not being heard and not being seen, into the idea that the value of ecology and nature is best assessed via its supposed "economic benefits"? What might it mean to draw decolonial critiques in to wider debates of modernist practices of objectification, and be present to relations and experiences that were deemed not to matter?

The idea that "wildlife" can exist only behind a fence (which will pay for their protection) is limited, and limiting, for environmentalism. The value placed on some species of now-rare wildlife makes them a priceless target for poachers who operate outside of the formal economy. The situation initiates a terrible cycle of monetisation and militarisation: when animals are rare they become more valuable; therefore, they are more likely to be hunted and poached and sold in an underworld global network; therefore, ever stronger divides need to be placed between them and society; therefore, electric fences and militarised rangers are the solution; therefore, the prices go up on the underworld markets; therefore, the animals become an even more valuable target; therefore, militarised protection of animals at risk of extinction increases. And for animals without "market value" their destiny on the "human" side of the fence puts them at risk of hunters, culling, or factory-based extractions.

Can that script change? Is it the only possible future for wildlife in this Anthropocene, to be behind ever stronger fences, and patrolled by ever more sophisticated drones and technologies? Is a war between people and the wild the only possible suite of relationships that can avert extinctions?

Working with the management of peri-urban wildlife, the two chapters in this section question contemporary science-based policy that seeks to script the relationships of *H. oeconomicus* with wild species as monetised and/or militarised.

The chapter on baboons questions the idea that fences and fear-mongering ("Baboons Are Dangerous Wild Animals!") are the only way to manage the surviving primates that live alongside humans in the southern reaches of the city.

The chapter on marine management questions two organising ideas. The first, dominant in fisheries management and city management, is that the city has no impact on the sea. That idea comes directly from the belief that land and sea are separated (à la the deer). Second, the chapter contests the idea that financialising fishers' relationships with the sea through the quota system is useful. If marine life is reduced to rands and dollars per kilogram, the ocean is simply an ATM. Asset stripping will follow, and militarised interventions are both inevitable and unsustainable. Far more useful is an approach that builds on traditional fishers' care for the sea. What kind of fisheries scholarship might support that?

All of these questions reflect troubles with the current constitution of environmentalism in South Africa, in which the environment is protected to the extent that it harms humans or benefits them. What is harm? What is benefit? If they are defined only in financial terms, environmental managers are unable to support and protect the fragile remaining webs of care for species and spaces. The chapters of part III tease out the ecological implications of relationships proposed for *H. oeconomicus* and non-human living creatures. Both studies address versions of environmental science-for-governance whose ideas of nature take no account of the history and social context of definition of a human as *H. oeconomicus*, do not question their notions of property and territory and borderlines, and accept as truth the illusion of a division between society and nature.

The challenge of addressing such a deeply held social fiction is immense. In the words of ecofeminist and environmental philosopher Val Plumwood:

The misty, forbidding passes of the Mountains of Dualism have swallowed many an unwary traveller in their mazes and chasms. In these mountains, a well-trodden path leads through a steep defile to the Cavern of Reversal, where travellers fall into an upside-down world which strangely resembles the one they seek to escape. Trapped Romantics wander here, lamenting their exile, as do various tribes of Arcadians, Earth Mothers, Noble Savages and Working-Class Heroes whose identities are defined by reversing the valuations of the dominant culture. Postmodernist thinkers have found a way to avoid this cavern, and have erected a sign pointing out the danger, but have not yet discovered another path across the mountains to the promised land of liberatory politics on the other side. Mostly they linger by the Well of Discourse near the cavern, gazing in dismay into the fearful and bottomless Abyss of Relativism beyond it. The path to the promised land of reflective practice passes over the Swamp of Affirmation, which careful and critical travellers, picking their way through, can with some difficulty cross. Intrepid travellers who have found their way across the Swamp of Affirmation into the lands beyond often either fall into the Ocean of Continuity on the one side or stray into the waterless and alien Desert of Difference on the other, there to perish. The pilgrim's path to the promised land leads along a narrow way between these two hazards, and involves heeding both difference and continuity.[5]

Amid a fragile web of planetary systems, cities are like the bulwark of a castle, having expanded dramatically since the 1960s. Occupying more and more formerly "wild" spaces, they also affect oceans in dramatic and unexpected ways. Focusing on two contested sites in the greater Cape Town area, part III focuses on the way environmental governance problematises both a terrestrial and a marine situation. Inquiring deeply into each situation, these chapters seek to open up a line of flight from the irrational delusions of the knowledge economy in which so many environmental managers find themselves trapped like Aesop's one-eyed deer.

WHAT IS IT TO BE A BABOON WHEN "BABOON!" IS A NATIONAL INSULT?

5.1 | Baboon harvesting mussels on the rocks at Cape Point Nature Reserve.
Photo: Lesley Green.

In the environmental management meetings of the city of Cape Town, allegations that green budgets give effect to the interests of white privilege wave the red card of racism. "Transformation and equity gains have been reversed under DA [Democratic Alliance] rule," declared the African National Congress in 2011, "[and] the fact that DA councillors lobby vigorously for funds to combat the baboon problem in the South while turning a blind eye to the scourge of rats on the Cape Flats, is [an] example."[1] Among the last midsize mammals on the Cape Peninsula, chacma baboons—*Papio ursinus*—received a budget of R11 million for their management on city turf in the 2014–15 fiscal year.[2]

Baboons have not always been vermin in African histories. They were among the gods in ancient Egypt, and many a baboon stone carving can be found in museum collections. In !Xam rock paintings, they are creatures into which humans could transform: "therianthropes" is the word used for figures of human-becoming-animal. One of the !Xam words designating the chacma baboon was /nera-b, "the one who measures his strides," which was explained to an early researcher as "Have you not seen how the baboon imitates what a man does, how he measures his strides?"[3] And indeed the name *chacma* was originally the Khoe word for baboon, which is *choa kamma*, and the !Xam *!choa camaa*, both of which are based on the sharp "chaac chaac" bark which the baboons screech when there is a disturbance, or a fight breaks out in the troop's social hierarchy.[4]

In South Africa's Eastern Cape, Nguni tradition links the baboon with the healer. Many Xhosa sangomas (healers) are recorded as wearing a baboon pelt on their heads—a linkage made, at least in part, because baboons, like humans, reduce their vulnerability to illness by self-medicating with plant roots.[5] In the context of the frontier wars in the Eastern Cape and the deadly Mfecane struggles in Natal, the San, Khoe, and Xhosa were forced into alliances, and in rock paintings of the time, argues archaeologist William Challis, baboons became iconic of the capacity to protect oneself from both illness and the ills of war.[6] In Zimbabwe, some Shona have the clan name of the baboon, and are therefore at least in principle tasked with looking out for their well-being—including celebrated Chimurenga musician Thomas Mapfumo, who is affectionately called "Mukanya," the praise name of his Shona clan, referring to its totem, the baboon.[7]

Working partnerships between people and baboons are not hard to find in historical records. In Namibia and Namaqualand, there are reports of baboons being domesticated by the Namaqua and serving as goatherds and cattle guides.[8] There's even a record of a baboon having been employed by the South African railways: signalman "Jumper" Wide lost his legs managing a railway crossing, and at some point thereafter saw a baboon (Jack) leading a team of oxen. Signalman Wide bought Jack and trained him to change railway signals on his command.[9] After a passenger complained that train safety was being left in the opposable thumbs of a non-human, Jack and Jumper were dismissed, but both were re-employed after they passed the tests set for them by the railway company. Both received official employment numbers.[10] Jack's salary, apparently, was paid to him in tobacco. The story seems quite similar to the one depicted in an Egyptian tomb in which two baboons on leashes act as police assistants to catch a thief in the marketplace.[11]

5.2 | Signalman Jack, the baboon operating the railway signals in Uitenhage, South Africa, c. 1910.

The word "baboon" appears to come from the name of an early Egyptian deity, "Baba" or "Babi," who ensured that the dead would not suffer impotence in the afterlife.[12] Baboons were also depicted as one of the heads of Thoth (who could alternately appear as a baboon or an ibis). Thoth was the source of all knowledge and the counsellor to the sun god and creator Re (also known as Ra). Not surprisingly, some ancient Egyptian baboons achieved such social importance that they were mummified.

By contrast, the Greeks and Romans "adopted a kind of 'reverse Darwinism' in which the more an animal resembled a human, the more it was shunned, made into an object of ridicule, and declared to be fundamentally different."[13] In the Middle Ages in Europe, the rise of Christianity coincided with the fall of the apes as theologians used them

as living examples of what man would become if he turned away from God and gave way to his baser instincts. Monkeys and apes were no longer just devious sycophants—they now became creatures completely at the mercy of emotions, sadistic impulses, and lust. In the most extreme characterizations, they were depicted as the devil himself (or at least the devil's agents), sent from out of the land of darkness to perform Satan's work and lead people—particularly women—into sin. Baboons were the objects of particular revulsion and scorn, in large part because of their supposed deviant sexual appetites.[14]

It is unsurprising, then, that the revulsion for the sexualised baboon travelled to the Cape Colony, and under the British, the figure of the baboon became increasingly linked to offensiveness. Dirk Klopper notes in "Boer, Bushman, and Baboon: Human and Animal in Nineteenth-Century and Early Twentieth-Century South African Writings" that in South African writings in English, the rough-living Afrikaner Boer, the stock-thieving Bushman, and the nefarious Baboon came to form a triad of Others. To the English who sought to establish their own evolving status as "moderns," the three were dirty bush dwellers, one and all: more animal than human.[15]

In the 1940s, the *American Journal of Psychology* published a piece on the "Baboon Boy" of South Africa, who was said to have been raised by baboons.[16] A response in the subsequent issue declared the story a hoax, in which a man with a brain injury from an ostrich kick to the head had been presented as a feral foundling in the hope of selling a story to Hollywood.[17] To make the story plausible, farmers, police, and many in the wider community would have had to be enrolled in the fabrication.[18] That South African and American academics had been unable to read through the racist discourse that "simianised"—rendered as apes—black South Africans during the rise of English capital and Afrikaner nationalism is an early indication that South African primatology had yet to reckon with the need of those who deemed themselves modern humans, and enrolled races and primates in a sequential evolutionary history. To the extent that monkeys and apes connected humans and nature in the theory of evolution, in modernist consciousness they became the lightning rod for the efforts of racists to separate black and white, couched in the terms of "moderns" who separated themselves from nature. Simians—apes and monkeys—found themselves fashioned into instruments of insult, in respect of class and race, in the making of the modern "Anthropos."

While simianisation has resurged in the form of crude racist insults in the United States amid rising racism, simian slurs in South Africa never really ceased. When in 2016 one Penny Sparrow compared black people with monkeys on Facebook, it caused a national furore.[19] Sparrow's simian insult was far from isolated. Google the words "baboons" and "racism" with "South Africa," and be confronted by the spectre of a racism so powerful that the ruling African National Congress issued a statement in 2016 calling for its criminalisation: "We can no longer as a nation tolerate such dehumanising violations, where [the] Black majority are treated as subhumans and are referred to as monkey, baboons and other derogatory racist epithets in the land of their birth. . . . Racial bigotry and apartheid must be considered serious human rights violations that must be punishable by imprisonment."[20]

It's worth noting that American primatologists don't necessarily think of simian analogies as insults: a book on baboon medical research team leaders in the United States was called simply *The Alpha Males*: "They were all individualists strong in their own ways. They inspired loyalty, whether by tyranny or by kind... endeavors. They all were charming in social situations [and] impressed their superiors.... Most important was the single-minded belief in what they were doing."[21] Nonetheless, for most everyone else, to be simianised is a terrible insult. American racists regularly insulted President Barack Obama and First Lady Michelle Obama in simian terms.[22] More locally, Professor Malegapuru Makgoba caused an uproar in 2005 when he described university transformation: "The dethroned white male in South Africa is playing the same role as dethroned baboon troop leaders do."[23]

So in South Africa, Cape Town's chacmas—*Papio ursinus*—face a challenge: how are they to be conserved when their species has become an insult?

"In his search for another logic of metamorphosis, Achille Mbembe, in *On the Postcolony*... tracks the brutalization, bestialization, and colonization of African subjects in philosophy and history," writes Donna Haraway, going on to say:

> The readiness with which taking animals seriously is heard to be an animalization of people of color is a shocking reminder, if one is needed, of how potent colonial (and humanist) tools of analogy remain, including in discourses intended to be liberatory. Rights discourse struggles with this legacy. My hope for companion species is that we might struggle with different demons than those produced by analogy and hierarchy that link all of fictional man's others.[24]

This chapter searches for a way to think about baboon management, and baboon well-being, outside of the available frames, in the hope of finding a way to struggle with different demons—and restoring care for baboon lives and well-being.

I | CAPE TOWN'S URBAN WILDS

At the time of this book's research and writing, decision-making in Cape Town's council chambers (and the Western Cape province) reverses national and local norms: the nationally ruling African National Congress comprises the opposition, while the national opposition party, the Democratic Alliance, governs. Capetonian filibustering achieves a level of finesse bettered in few cities—and as a result, Cape Town baboon managers are often caught between a rock and a hard place in council budget meetings.

Most of Cape Town's chacma baboon troops live on the Cape Peninsula, where there are many versions of "the nature of the baboon" in play. To the environmentally minded, baboons are, for the most part, interesting companions who were there first and to whom humans ought to relate with firm respect—like the peninsula resident who described asking an elderly baboon whom she called "George" not to enter her home again after he had given her family a few scares on their small farm. She told me she went to see him at the compost heap to explain that he could eat from the compost heap if he would leave them alone inside the house. He did so, she told me, and did not enter the house again. But other baboon neighbours experience them as a threat to dogs, children, groceries, and property.

The rapid growth of the city has brought an increase in food waste; an increase in tourists looking for a monkey trick; a larger number of vegetable gardens and fruit trees; more fast-food chains with delicious disposable plates and cups; new agricultural activities, such as ostrich farms and vineyards; and ever more well-provisioned middle-class dustbins to raid.[25]

In the new rural suburbs, baboons learned new skills fast: how to lift a sliding door off its rail; how to open refrigerators and windows; how to open an unlocked car door; how to raid an ostrich farm for eggs; how to open a double-locked wheelie bin; how to raid an artisanal ice-cream counter. When bins were baboon-proofed with double locks, some baboons worked out that the lids would come off if the bins were knocked over. When some residents started to lay their baboon-proof bins down so they could not be knocked over, the largest baboons learned to jump on them until they split.

Baboons not only ate food from kitchens, but tipped their patrons with generous faecal deposits. An entire month's groceries could be spoiled in one raid, with refrigerators, cupboards, and boxes opened, sampled, and scattered. Window putty could be stripped until the panes fell out; gutters ripped off by rambunctious "juvies," or juveniles, who chased each other in large groups, screeching and thundering over rooftops. The costs to households were high, both in preventative measures—baboon-proof window bars, baboon-proof vegetable gardens—and in clean-up costs and grocery losses. Ensuing hostilities led to poisonings, maimings, and shootings.[26]

Within the city limits, baboon management involves a tussle over who has legal responsibility for baboons that cause damage: the city council (elected by ratepayers); Table Mountain National Park, where the baboons officially have their homes; Cape Nature, which manages provincial nature reserves; or the navy, which owns and controls large portions of land that are in legal terms excluded from the city's jurisdiction. Different laws apply in each of those spaces. But, like the proverbial chicken, the chacmas cross the road, or the fence, or the legal demarcation, without regard to the finer points of law. The ten local troops have made it clear they have no interest in staying within Table Mountain National Park, and with four opposable thumbs and a tail, they traverse fences with superhuman ease. Will their survival or extermination be decided by council ordinance or military ordnance? Or both, as council ordinances legalise culling of "problem individuals"? As city voters shift from one party to another, will their nature be determined by a politics of green, white, blue, black, or red?

The multimillion-rand city baboon management budget took a multi-year crisis to secure, along with struggles over which version of baboons was to be understood as "Science" (and often "scientism," as discussed in chapter 3).

Primatology, as a global discipline, offers several contending accounts of baboons' nature. Indeed, so contentious is the "nature" of baboons and other primates in primatology that feminist primatologists have rejected the accounts of male aggression and dominance that had characterised so many patriarchal studies of the nature of baboons and bonobos.[27] That conflict was one of the sources of what would later be called "the science wars" in the United States and Europe. In Cape Town, it is useful to see baboon management as an extension of that same science war—in which the leading voices in Cape Town zoology (who do the city's primatological research) have declined to engage with the feminist and decolonial primatology that has emerged in the United States and Europe since the 1980s.

Intrigued by the problems in human-baboon neighbourhoods as I'd cycled through them on weekends, I sought permission to observe monitors and tourists engaging with baboons in the area of Scarborough and Cape Point with five students in late 2014. The query went up the city's email ladder, and eventually the answer came back: permission was "reluctantly" given, we were told. Students could observe the monitors only on condition that there were no cameras, no recording devices, and no notebooks, and that the project was not to be continued in any way.

As an anthropologist schooled to attend to patterns in behaviour, my curiosity had already been piqued by the similarity of the actions of baboon monitors to the riot squads confronting student demonstrations in the 1980s. To receive these instructions for my student field trip was equally reminiscent of restrictions on reporting on apartheid state policing. The more I tried to understand the city's baboon management strategies, the more I felt like Alice in Wonderland: curiouser and curiouser. Though the students and I chose to comply, I was of the view that there was no legal basis on which city officials could issue these instructions, given that the monitors were city contractors operating in a public space.

The level of secrecy was troubling, and it was not confined to a single instance. In 2015, three weeks after a runaway mountain fire from which baboons had (surprisingly/not surprisingly) not fled in the direction of the monitors who usually shot them with paintballs, Jenni Trethowan of Baboon Matters went in search of the body of a baboon that she could smell on the mountain. She found the carcass—and was arrested and fined.

A key issue in the struggle over access to information is the question of baboon numbers, which is the basis of any assessment of a troop's social well-being. Knowing the ratios of males and females, of juveniles, adults, and babies, is key to understanding their health and social behaviour, and gaining an indication of whether troops are being shot or poisoned (by residents) or culled (by baboon managers). Trethowan's attempts to get official figures of the baboon population had been met with the argument that the information was part of a PhD in progress and was therefore under embargo, and baboon troop numbers therefore could not be made public. Yet if baboon management was part of the city's mandate, surely the city had the right and the duty to ensure the information was public sooner than a five-year PhD cycle?

In late 2016 Trethowan called me to ask for advice after presenting material at a South African Wildlife Management Association conference where she had compared official baboon statistics with figures that were available from researchers. Her figures were obtained, she said, from data published online by Human Wildlife Solutions (HWS), and from a dissertation available online through the university library, to which she had repeatedly been directed by the city in her requests for figures. After the conference, however, she received an email from Justin O'Riain, a senior zoologist and head of the Baboon Research Unit (BRU) at the University of Cape Town, advising her that she was guilty of violating copyright law for using those figures because she had not verbally attributed them to BRU (the attributions were on the presentation slides).[28] Her presentation had compared different researchers' troop counts, calling attention to the differences between the figures of researchers and those supplied by the city's contractor (HWS itself). Trethowan had questioned the categorisation of immature and mature males and females, and contested the claim that total troop numbers are an adequate means for evaluating baboon troop well-being. Because of the importance of the social lives of baboons, skewed age and sex ratios matter.

As a former research ethics committee member in my own faculty, it seemed to me to be counter to the spirit of scientific enquiry to be discouraging critical engagement by a member of the public. Interested in obtaining a baboon research permit to observe resident-baboon interactions more closely, I discussed the matter with a SANParks official who advised me not to bother applying for a research permit, as researcher access was apparently tightly controlled by academics and officials who kept out critical players.

Baboon management in the city is a battle zone: between those who claim that a pure science of primatology can exist untainted by social thought,[29] and those who live in the area and question the dissonance between what they are told is policy and what they see on the ground. They are the ones who dare to imagine that baboons and humans might be better neighbours than the metaphor of war will allow. But within the city, an authoritative version of science (that is, scientism) bolsters officials' claims to the right to have the final say over baboon-human relationships. Baboon neighbours are dismissed as irrelevant. For the advising scientists, management by numbers provides the illusion of clarity, as if a number provides the sum of all truth, and as if their methods enable them to provide incontrovertible numbers.

What the management numbers make *in*visible are the acts of observing and counting. Not everything that counts is counted. Not all evidence is evident. Trethowan's concerns are decidedly matters of science in primatological research elsewhere. Her contestation of what is measured in the official baboon counts has parallels with feminist and decolonial concerns in primatology, about which scholarship has been available for at least the past three decades.[30]

Eminent scholar Donna Haraway in 1989 published *Primate Visions: Gender, Race and Nature in the World of Modern Science*, in which she examined primatology through feminist thought and antiracism lenses, and explored the feminist primatology that was coming to the fore amid the decolonial concerns of the time. Primatologists Thelma Rowell and Barbara Smuts came up with baboon studies that completely contradicted the patriarchal, sex-and-aggression-and-dominance-based accounts of baboon social structures that had dominated the field at the time, describing friendship and care in everyday life.[31] Shirley C. Strum and Linda Marie Fedigan edited a six-hundred-page volume titled *Primate Encounters: Models of Science, Gender and Society* that emerged from a symposium trying to resolve the conflict. Dorothy L. Cheney and Robert M. Seyfarth's study *Baboon Metaphysics: The Evolution of a Social Mind* describes the social complexity and social intelligence of the species, describing baboons' empathy, emotion, sentience, and sensitivity in ways that contrast with the behavioural ecology and patriarchal sociobiology that HWS and the University of Cape Town's BRU advocate for baboon management in Cape Town. The remarkable exhibition *Ape Culture*, curated in Berlin by Anselm Franke and Hilda Peleg, sets out a clear rebuttal of patriarchal primatology that focuses on male aggression and domination at the expense of a grasp of the relations of care and friendship within troops.[32]

Yet "the nature of baboons" (as if their nature were singular and uncontested) as framed by the city's consultant researchers focuses on the kinds of questions that enable the city's urban management mandate to be fulfilled: on baboon injuries;[33] on troop territory (which is problematic given that juvenile males need to roam);[34] on possible human-baboon disease vectors;[35] on domestic refuse management.[36] All of these assume that the best possible form of baboon governance is to manage by numbers within GPS-defined space, and to keep humans and animals apart along x roadways for y amount of time. The possibility that research questions might address the multiple natures of baboons in primatology before proceeding is not entertained.

Nor is there consideration given to the ways in which research questions are framed by the city's governance mandate. Nor is there critique of the regional zoologists' belief that baboons must be kept "wild" and away from humans in the city: an assumption that shapes the research questions, and therefore also the research findings. The social well-being of the troops is not visible in the data being collected. In direct consequence, the impact of the loss of healthy baboon social structures on their behaviour is not understood—nor is the historical record of constructive human-baboon relationships. By the same token, the rhythms of different residential areas where employment or unemployment, among other demographic factors, determines when people are in the neighbourhood. The baboons' daily rhythms of presence and absence in relation to different neighbourhoods' street life and trash profiles is not part of this research. Instead, the presumptive nature of the baboon is as a problem animal.

"She cares deeply for animals, just like I do, but we have completely different approaches," O'Riain is quoted as saying of Trethowan. "She seems to want to hold and protect them. I want them wild and at a distance, doing their own thing."[37] This statement characterises the Cape Town baboon conflict as that of a man who wants wild and free baboons versus a woman who wants to hug the animals. That particular version of "wild and free" is an odd version of liberty, since it requires fences and paintballs and relations of animosity between people and animals. Its idea of wildness is that it is a "male" space, and it situates its opponents as women who operate (and belong) in a domestic sphere. But in all the time I spent with Trethowan, not once did I hear her express any wish to hug and hold baboons, or keep them in her home. What I did hear was a concern for the baboons' social well-being within their troops, and a wish to facilitate baboon-human neighbourliness.

For O'Riain, human-wildlife relations are necessarily conflictual. For Tre-thowan, they do not have to be so: neighbourliness is possible—but humans need behavioural guidelines, and baboons in urban areas sometimes need specific forms of care, such as when they contract mange, or get injured in a peri-urban wildfire.

The city's baboon managers rightly want to prevent citizens being "ter-rorised by baboons" (their language)—but don't seem to see that taught fear and experienced terror are linked. Here, as elsewhere, those who are concerned about the baboons are dismissed as sentimental and emotional women, even though their interest in the social well-being of baboons is ex-pressed in terms that are central to the international literature on baboons. But that version of baboon research is totally excluded from the "official" Cape Town research conducted for the city.

"Anthrozoopolis" is the word Haraway used to describe the Cape Penin-sula in an email responding to an early version of this chapter.[38] The word asks for attention to interspecies entanglements, and prompts exploration of different forms of animal relations. While taking seriously the concerns of residents and city officials who don't want kitchens trashed or people or dogs or chickens threatened by baboons, I also want to take seriously the concerns of the baboons, who, like other animals in this locale, are trying to adapt the terms of their survival in a time of increased human settlement, increased hazards like mountain fires, more domestic dogs, more traffic, more garbage, and different legal frameworks that change as you cross invisible lines on land that are referenced to "the map."

Table Mountain National Park runs the length of the western seaboard of the city: some fifty kilometres as the crow flies from Cape Point in the south through to the city bowl that surrounds the harbour at Table Bay in the north. Surrounded by urban development on the lower slopes, the city's mountains continue to support midsize mammals both outside and inside the settlements on the mountain's edge, including otters, porcupines, cara-cals, a range of small antelope like the duiker and the klipspringer, and ten troops of baboons, as well as many of the larger bird species: eagles, owls, har-riers, kites, goshawks, and flamingos—along with the more familiar urban bestiary of rats, cats, and canines of all sizes.[39]

The lives of larger mammals—leopards and lions and the large antelope—were extinguished within the city's limits in the 1800s. Lion roars were last heard on Table Mountain's slopes as recently as 1974 at the University of Cape Town's zoo, continuing Cecil John Rhodes's vision for roaring captive lions adjacent to the official residence of the country's premier. His dream ended when a drunken student rugby player thought to score a try in the lion enclosure, confronting the university with the need to limit its liability for acts of terminal human stupidity.[40]

On the southern end of Table Mountain National Park is the peninsula of Cape Town, where along the fence of the Cape Point National Park, creatures of at least 351 kinds—including humans—share land, water, air, and seashore.[41] The headline grabber, though, is the baboon.

Until the 1900s, baboons were shot on sight, even though they were a protected species, because under various nature conservation ordinances in the 1960s, animals that were suspected of causing damage were allowed to be killed. At that time, the renowned director of Nature Conservation in the then–Cape Province, Dr. Douglas Hey, recommended the use of thallium or lethal doses of sodium cyanide in small gelatine capsules to control problem baboons.[42] It was conservation on a genocidal scale, using a variant of the toxins that had been used at Auschwitz. And indeed, these same toxins were reused on baboons by the apartheid state's chemical warfare team, named Project Coast, led by Dr. Wouter Basson, whose medical specialist training as a cardiologist would have exposed him to the use of baboons in medical research. According to testimony given to the Truth and Reconciliation Commission by the leader of one of the teams on Project Coast, baboons were used extensively for tests of toxins, including in gas chambers.[43] Basson denied the testimony of his former colleague on this, but it is difficult to imagine a reason for a professional to concoct such stories.[44]

The historical relationship between science and baboons, in South Africa, has not been a kind one. In cardiac surgery, globally pioneered in Cape Town, xenotransplantation—transplanting an animal heart into a human—was attempted as early as 1977;[45] the chosen heart was that of a baboon. Without the chacma, South Africa would not have become the global pioneer of heart transplants—a technology which "intimately intertwines animals and humans, [challenging] boundaries such as animal/human, subject/object and us/them."[46] Those baboons came from Cape Point in the 1960s, and the Silvermine, Kalk Bay, and Chapman's Peak troops were removed for the same purposes in the 1970s and 1980s. Chacmas continue to be used for medical research, for example, in calculating body sizes in relation to pacemaker strength, and in studying genetics and epigenetics of human disease.[47] Many of the baboons used in medical research were taken from the Cape Peninsula.

The practice of removing troops was only outlawed in 1997,[48] but baboons remain regularly used in South African medical experiments, where their mortality is not infrequent.[49] The lack of critique from approved baboon researchers is evident in the discourse of "harvesting" Cape baboon troops for medical science found in BRU descriptions of their recent history.[50]

A reading of baboon research in South African science suggests that while human-baboon bodily entanglements are foundational to the development of heart transplants in Cape Town's cardiology, their cognitive and sentient connections to people are passionately denied in Cape Town's primatology.

In the late 1990s, as baboon troops grew, environmentally minded residents in the area formed the Kommetjie Environmental Action Group (KEAG) in the suburb of Kommetjie on the Atlantic edge of the Cape Peninsula. A zoologist who lived in the area, Ruth Kansky, initiated baboon counting in 1998 with funding from the World Wildlife Fund and SANParks and, from that experience, learned that baboons tended to stay on the far side of whatever spot she was observing from. Realising this, she (and others in the area who were affiliated with KEAG) initiated a baboon-monitoring programme in partnership with the city, with the goal of keeping the baboons as high up on the mountain as possible, and away from human settlements. But as the human population expanded rapidly in the South Peninsula in the 1990s, conflicts with baboons grew.

Initially, the baboon monitors were managed by local residents and animal activists on a shoestring budget, with a modest contribution from the city of Cape Town to cover the costs of the baboon monitors. Tensions grew, however, between the city and SANParks over legal responsibility for boundary-crossing nomad baboons. Under the leadership of Jenni Trethowan, and in partnership with the city, the KEAG baboon-monitoring project grew, along with a project of public education built around walking tours on the mountain to encounter the animals in everyday life. The predominant approach was informed by animal welfare thinking, with which the Baboon Management Team appointed by the city appeared to concur.[51] In the same period, the baboon troops themselves were undergoing changes, as some troops split and began to move independently. The city's biodiversity management team has particular difficulty managing the land along the Table Mountain National Park fence, which demarcates the point where urban law of property meets the Protected Areas Act 57 of 2003. But struggles over laws and financial responsibility for damage-causing animals did not make the headlines. Rather, it was the deteriorating relationship between baboons and urban citizens that caught journalists' attention.

That the welfarist approach of the earlier Baboon Matters team had its limitations is clear. But it is important to note that they were developing their protocols in a new and fast-changing situation. They had a fraction of the budget of those who came later, and did not have the resources to put monitors in place to respond to all the baboon troops. As residents in the area, members of Baboon Matters also found themselves in the difficult position of having to respond to their own neighbours' aggression toward the baboons. The moral outrage of Baboon Matters supporters on social media undoubtedly strained relations between neighbours yet further when supporters tended to idolise baboons as innocents and castigate humans as demons, with the result that at least in their early days, baboon welfarists came to be caricatured as "bunny huggers" and "emotional" and "troublesome women" who had to be stopped by the hard power of cold science.[52]

II | BABOON MANAGEMENT 2.0: PROPERTY, TERRITORY, AND SPATIAL PLANNING

Around the time of Cape Town's 2009 Baboon Expert Workshop, Baboon Matters lost the monitoring contract for the city. A new group of monitors was appointed. Baboon Matters was still concerned for baboon welfare and critical of decisions to cull "problem animals," and was portrayed in the media as naively fighting for innocent baboons against evil humans. Its public image suffered.

The issue of baboon habituation to urban areas came to a head in 2011. US-based primatologist Shirley C. Strum, who had worked extensively with baboons in Kenya over many years, visited Cape Town for two weeks at the invitation of the BRU. At that time, baboon management difficulty centred on a block of flats on South African Navy land that was not under council jurisdiction. Moreover, the navy apartment managers were notoriously uncooperative in ensuring the management of waste in baboon-proof bins, with the result that the baboons were difficult to control in that area.

In a *Cape Times* article published a year later in July 2012, a few days before the baboon management contract would once again be up for renewal, Strum described the monitoring as "anti-science" and motivated for the baboon management policy to be replaced by "science." And, crucially, she declared that the activists who were calling for the baboons to be treated humanely were the problem: "If there were to be a trial for the 'murder' of specific baboon males," she wrote, "I would testify as an expert witness that the very same people bringing the charges are the ones who should be on trial because it was their objections that prevented methods that could have saved these baboons. . . . If deterrence had been used successfully earlier, there would be no need to kill any baboons today."[53] Even though there were by then a number of players in the baboon management arena, Baboon Matters took the brunt of the public opprobrium that followed.[54]

Strum's article acknowledged none of the history of their work, nor the activists' research base about the different troops, nor their publications and scientific researchers and advisors, nor the complex legal, political, and funding landscape in which their work had proceeded while navigating navy, city, and national park legal regimes. It made the mistake that anthropologists caution one another on: "the error of the ethnographic present." This refers to taking what one sees at *one* moment as representative of the situation everywhere, all the time. Simplifying that generalisation into two categories, good and bad, the article gave no deeper insight into what was at stake; it simply switched the moral labels on existing categories that were already a part of the problem. Good and bad: welfarists and scientists simply traded places. It was powerful rhetoric.

Within a few days of Strum's article, HWS was awarded the city's baboon management contract with roughly ten times the budget of the previous incumbent: some R10 million per annum (more than a million US dollars per year, at 2009 exchange rates) to manage all ten of the baboon troops in Cape Town. Thus authorised, HWS, as they are known, explicitly took a paramilitary, patriarchal "we'll sort this" style *contra* the group of interested, active, diverse, and women-led residents interested in fostering interspecies neighbourliness. Under the new protocol, the welfarist approach was replaced by a new masculine regime based on shooting baboons with paintballs. As any paintballer knows, these shots are extremely painful, causing lasting bruising, especially at close range.

"Baboon Management 2.0" began not on a particular date, but as a slow paradigm shift that commenced in 2009 during the baboon crisis in the city. At that time, Baboon Matters and other animal welfarists and contractors had been arguing for increased resources to monitor the baboons. But at a crisis meeting, a zoology lobby led by O'Riain argued that "baboon management can be distilled down to two things: population and space." Regarding the protection of space, O'Riain suggested that the home ranges of the baboon troops should be put onto the Cape Town city maps, and that "new developments must factor in the cost of annexing existing troop home ranges" since "all space is not equal." "With monitors," he is minuted as saying, "the troop could be kept out of the urban areas and . . . effectively herded and kept within natural foraging areas." He concluded, according to the minutes, "When we are managing a closed system with decreasing home ranges, there are inevitable and difficult decisions to be made. BRG [the Baboon Research Group, later the BRU] is gathering data to inform management as we need to inform whatever action we decide to take in managing the populations on the Cape Peninsula." Piet van Zyl, of the city of Cape Town's spatial planning unit, responded enthusiastically, "The presentation by Justin O'Riain indicates that we now have a robust piece of research and we need to integrate this into the city's planning, specifically the District Spatial Plans currently underway and specifically for the Southern District, [which] should have a baboon overlay to inform land use decision-making."[55]

Reading the record of arguments from that time, it appears that in that moment of accepting the then-BRG's research methodology,[56] baboon management was reconceived as a spatial planning problem that could be monitored via a "baboon overlay" in a geographical information system. With the city's relationships with its baboons reconceived as a spatial data management project, the natural and inevitable consequence involved the technologies of territoriality: maps, lines, weapons, and policing.

This moment was, in the million-year history of the baboons of Cape Town,[57] of signal importance. Yet for Ruth Kansky, this "drawing the line" approach was based on research that, in her view, was incorrect:

> They [the advising zoologists] concluded that baboons were not to be kept up in the mountains and far from the villages because in their view most of the good quality forage was lower down. So they told the [Baboon Matters] monitors . . . to not chase them up the mountains as they had done, and rather let them come closer to the village, but "keep the line" around the villages. After applying this strategy, things got out of hand because the monitors could not keep them out at all, resulting in the baboons in the villages all the time. After a year or two of this strategy, [US-based primatologist] Shirley Strum came and found the dire situation of baboons so habituated and in the villages all the time.[58]

Kansky contests the view that baboons will be unlikely to find adequate forage higher up the slopes, arguing that there is indeed adequate forage for them higher up. This question is fundamental to the contest over baboon management. After all, how do you persuade a chacma *not* to cross a road?

The "baboon spatial overlay" in the city's Spatial Planning Division placed the baboon "troop territories" at the edge of human settlements that were demarcated by roads on which rangers' vehicles could patrol. Specific baboons were to be monitored with radio collars.

With the information from the collars databased in geographical information systems in line with the technical protocols of city spatial management, a new science was born. While baboon spatial areas had always been understood and agreed upon, in the moment when Cape Peninsula baboons were mapped in specific areas in order to police them, Cape chacma primatology evolved into something quite different, becoming a kind of zoo-demography. The monitors also changed their purpose and function, no longer shepherding the baboons to the upper slopes of the mountain, but policing a line between humans and baboons.[59] Crucially, Kansky notes, the crossing of that line became a key performance indicator for the contractor, *and therefore the key event to be entered as data*. With the tabulation of individual baboon transgressions arose a new science of zoo-criminology.

Perhaps the most important question is whether the idea of the line is appropriate: the possibility that the baboons can forage and live well higher up the mountain slopes should be the focus of further investigation. The idea of having "the line" along a road may be useful from the point of vehicle-based management—but it creates its own problems, and inevitably necessitates violence. The idea behind the data "object" that can be counted—in this case a baboon crime such as crossing a road—is not intrinsic in the world, there to be counted. Rather, a baboon crossing a road becomes an event only when a specific question arises in relation to available technologies of policing, such as when the question, "Did HWS execute its mandate to keep the baboons out of the urban area?" is ascertained by way of reference to a road, monitored by vehicle patrols, rather than a region higher on the mountain beyond the reach of vehicles. Without that particular technology—road, vehicle, urban spatial overlay—the event of the baboon crossing the road would not be rendered as an object to be counted and turned into a measure of an individual baboon's culpability, cullability, and killability. The case of baboon management science in the Cape Peninsula speaks to the question at the heart of science studies: How are objects conceptualised? Once "made," what makes them countable? Are the most useful things being counted?

Geographical information systems technology provides a powerful meeting point for the city's bureaucracy and local scientists; it provides both with the illusion of making "the truth" visible. And the technologies of spatial governance in Cape Town—probably the most racially divided city in South Africa, where particular roads marked racial boundaries under apartheid, and the painful geography of racial separation has persisted—lend themselves seamlessly to techniques of keeping baboons and humans apart. But none of the studies on which the new policy was based took account of the huge variability in different areas, specifically with regard to presence on streets, garbage handling, and types of food waste. Social science is wholly absent in the major studies that advised the city on how to manage its "human-wildlife conflict." Without any attention to variable human activity and its effects on baboon behaviour, the studies introduced a blind spot into urban baboon policy: people.

Baboon Management 2.0, with its baboon criminology, rests on practices of urban spatial management that are uncomfortably familiar, involving the conceptualisation of a line of territory holding populations apart, policing, and the criminalisation and removal of "problem individuals." Its logics of deviance subsist in metaphors that derive from racialised fears of suburban invasion. That those ideas have framed the scientific assertions of how baboons are to be managed is indicative of the impoverishment of science that occurs when critical social sciences and humanities are not part of the discussion. The result is a form of urban wildlife governance that believes violent strategies of repulsion and expulsion are the means of choice to prevent extinctions in a growing city built around a nature reserve.

III | ZOO-CRIMINOLOGY

Two years after taking on the baboon management contract with the city of Cape Town, the HWS report on that period reads as follows:

> During the recent cold and wet months, the troop moved to the cliff site just above the Red Hill Settlement (RHS), which is not unfamiliar to the baboons as they had slept there several months ago.... However, RH2 [Red Hill troop baboon number 2] and WF2 [Waterfall troop baboon number 2] started raiding a farmhouse in the area, and also made several failed attempts to raid the settlement as rangers were able to block their efforts....
>
> On 13 August, the entire troop attempted to raid the RHS however rangers were very quick to respond and blocked entry. Although the sleep site is more sheltered for them during the cold and wet months, the presence of the baboons has led to more conflict with farmers and RHS residents. This situation will continue to be monitored closely.[60]

On the Da Gama troop, adjoining the naval base, where fast-food packaging litters the streets and there are problems with trash management, the report notes:

> DG10 [Da Gama Troop baboon number 10] continued his raiding activity and was euthanised on 1 August.
>
> DG11 and DG12 continued to be the main culprits in further raiding activity throughout the month. Both of these males are young sub-adults that had previously followed DG10 and raided with him. DG11, being of similar age to DG10, took over leadership of the small raiding party and, during the period of merely one month, was involved in seventeen incidents of raiding and/or being in town.

However, following the euthanasia of DG10, the troop has been sleeping together as one unit again which greatly relieves pressure on the rangers in the mornings.

On 15 August, DG11 and SK3 [Scarborough baboon number 3] led the troop into Scarborough where they spent five hours raiding bins in town and eating fruit from trees. They were eventually pushed out of the area and slept close to Scarborough. Subsequently, extra rangers have been put in position to prevent the troop from sleeping at the Scarborough-facing side of Grootkop.

The Da Gama Troop is a habitual raiding troop and will take every opportunity to raid. It has been shown during the past year that, regardless of which sleep site they use, if town or the residential area is visible to them, they will attempt to go there unless rangers set up effective lines to prevent them from doing so. . . .

On 16 August, GOB9 [Groot Olifantsbos 9] and MCF2 [Misty Cliffs 2] did not return to the troop and instead made their way to the Misty Cliffs roost site and spent the night there. They both returned to the troop the following day. MCF2 was not in oestrus at the time and it is therefore likely that these two pair together for the sole purpose of raiding in town. MCF2 has a history of spending hours alone hiding in properties and GOB9 has learnt to do the same.[61]

As noted earlier, to an anthropologist the metaphors and analogies used in policy and law are important because they serve to script relationships, and as such they render "natural" and "inevitable" actions that are a matter of social and political choice. In the language of these early HWS reports, the paradigm shift to Baboon Management 2.0 is clear. Individual baboons are "driven back" by "rangers" (no longer "monitors" and a term much closer to "police") who are armed with paintball guns that they fire at will. Individual baboons are no longer "George" or "Eric" or "Fred," as they were known by their human neighbours at Baboon Matters and in early HWS reports, but "RH2" and "WF2." While such codes are commonly used by primatologists—women primatologists like Shirley Strum, Barbara Smuts, and Jane Goodall being among the exceptions—they are, in the context of the South Peninsula, also rhetorical: they stake a claim to "scientific objectivity" by overriding and negating relationships that have been built up between human and baboon neighbours. The creation of numbers and codes is a statement that the baboons are now objects in a databank that, not coincidentally, doubles as the means of producing "facts" about baboon "crimes" that will lead to individual culling.

The defence of the use of codes offered is that a name is anthropomorphic, but a databank code is not. A name, however, speaks of a relationship between observer and animal, and acknowledges the relationships in a troop. A code that replaces a name is a negation of neighbourliness. Both have their origins in specific philosophies of animals: in one, they are subjects; in the other, they are objects. For the city's primatology advisors, naming a baboon "George" or "William" is considered anthropomorphic, but numbering George as DG11 is not. A name or a number: the one allows a relationship; the other denies it. Their difference is the rhetorical effect: the number renders the animals as objects, making a subject-object relation the only possible one between humans and baboons. It naturalises policing and facilitates killability. There is no difference in the levels of anthropomorphism.

What is critically important to note is that *making an animal into an object is no more free of social influence than naming them is.* Rendering an animal as a "code" (i.e., a thing) replicates a form of relationship that arose within modernist thought: that nature is studied by science as a series of objects without any influence from culture or society. But that notion of the subject-object divide does not inhere in nature any more than the nature/society divide inheres in it. To regard any animal or living being as an object is not to choose to avoid a social relationship: on the contrary, it enacts the dominant colonial relation of "thingification" identified by Césaire (see the introduction to part II for a discussion). Moreover, the approach denies the relationship of the scientist with the field of study. In this case, the conceptual separation of nature from society (which, again, is not an empirically observable fact, but a modernist fiction) allows the scientist to imagine that it is tenable to study baboon-human conflict by studying the baboons alone.

Code or no code, it is the databasing of individual baboon behaviour that leads to a decision on which individuals are troublesome enough to cull. Under Cape Town's Baboon Management 2.0 regime, citizen-baboon conflict requires force that is to be applied judiciously by first collating data on individual baboon behaviours. In 2014, day by day, the behaviour of individual baboons was being logged by class of behaviour, as follows:

1 | In urban area
2 | Raid bins
3 | Non-malicious damage
4 | Attacking pets
5 | Raid unoccupied house or vehicle
6 | Raid occupied house or vehicle
7 | Threatening behaviour/attack/stealing food
8 | Breaking and entering/damaging property
9 | Nocturnal raiding.[62]

Criminology meets animal demography here. Each baboon designated a "problem" has her or his own line and code. The number of "crimes" is tallied up by monitors on a calendar-month sheet that corresponds with the rangers' pay-runs, not the month itself. In the August 2014 report cited above, GOB9 is found guilty of seventeen counts of adverse behaviour. The troop as a whole has their behaviour coded too, on a day-to-day basis—although whether it was three members of a troop or twelve is not recorded. The troops' nightly sleep ("roost") site is noted—on Misty Cliffs or in the Cape of Good Hope reserve, together with the amount of time that the troop spent in town, based on a twelve-hour day, seven days a week.

With this log, baboon movements in property and territory—not baboons as social creatures—become visible to the city of Cape Town's policymakers. The database is designed to address two questions: at what point do individual baboons become legitimate cull targets,[63] and is the service provider meeting their Service Level Agreement with the city of Cape Town to keep the baboons out of "urban areas" 80 percent of the time? On the latter answer depends the continuation of the contract.

According to the record from which the above extracts are taken, the Misty Cliffs troop as a whole is reported to have spent between 40 and 240 minutes in the adjacent village of Scarborough for six days of that month, and the "problem animals"—the males and females listed by individual ID— are noted to have spent between forty minutes and five hours in town that month. The ID features of GOB9 are listed here, for closer surveillance, in the event that this baboon will need to be captured and culled. Effectively, the crime sheets provide a "killability index," determining "what counts as a liveable life and a grievable death."[64] Some sixty-five baboons were culled under this protocol between 2009 and 2017:[65] an astonishing number in a population of some five hundred.

This variant of zoological criminology is a security operation: policing a geographical line between human and animal, society and nature. Yet the only possible offender in this account is the baboon. Humans' culpable behaviours, such as leaving trash cans open, or a car window open with food inside, are not recorded. In HWS's daily index of individual baboon criminality, there is no accountability for human actions, nor is there an accounting of observable differences between baboon troop behaviour and human activity in the specific terrains they inhabit. "Why did the chacma cross the road?" is simply not asked. A troop living at a popular picnic site, for example, is databased on the same terms as a troop living on a relatively undisturbed part of the peninsula; a troop living near navy quarters with polystyrene-and-tomato-sauce detritus blowing in the wind is dealt with on the same terms as a troop living alongside an electric fence. The line of human-baboon engagement *only applies to baboons*. Humans, here, are innocent of baboon crime, and protected by law. Yet the metropolitan space on the edge of a vast reserve, with a fence-hopping wildlife population, clearly cannot be conceptualised as line-bound; clearly it will be a space of human-animal engagement. It is an anthrozoopolis.

The assumption that baboon governance requires that they be kept separate from humans mobilises all the techniques of forced separation in which apartheid specialised, beginning with surveillance in relation to territory/property. That a road serves, for the most part, as the critical dividing line is a function of the logistical approach which, in order to manage a working day effectively, employed rangers, need to have vehicular access to whatever edge has been decided upon. As discussed in chapters 1 and 2, once property ownership is the defining relationship with land, a relation of violence is an inevitable consequence once "boundaries" are breached. To cover up that violent territorial relationship—which is in its exercise unpalatable to most of the public—there is a bureaucratic anxiety that leads to clampdowns on photography, recording, and investigations by activists, photojournalists, researchers, and students; a secrecy around official statistics; and the exclusion of different approaches.

———

By their own account and that of others, HWS scored significant success in the first year of their operations, with a notable reduction in the number of baboon crisis calls from residents in the area. In evaluating the scale of the improvement, however, it needs to be borne in mind that the data on the baboon presence in settlements are being provided not by an independent evaluator, but by a ranger reporting to the contractor within the framework of concerns generated within the employer-employee relationship. At neither level—that of the ranger nor that of the organisation—can the figures be considered those of a disinterested science. In addition, HWS has far more resources than the earlier monitoring groups did—more vehicles, more rangers, more walkie-talkies, and more means of monitoring the baboons through the use of radio collars. Perhaps most important, they have the benefit of less infighting among the city, the navy, and Table Mountain National Park over their respective jurisdictions.

During fieldwork in Scarborough in December 2014 and January 2015, almost everyone I met was of the view that the baboon issue was less of a problem than it had been. A year later, however, the situation had changed: the Misty Cliffs troop of baboons was greatly reduced in size after repeated culling of "problem males" by the authorities, and at least one by a resident.

A surviving pair of females were taking care of three infants, one of whom had been orphaned and adopted by one of them. The infants were at risk of infanticide from the alpha male in the neighbouring Groot Olifantsbos troop, and "the girls" as they became known, had discovered the balcony of an unoccupied mansion where they were out of the line of paintball fire from the road. It seemed that they would go there after the rangers had left for the day, moving in the dark with their infants from the sleeping place to which they had been herded. As a result, residents were awoken at first light by barking dogs and rapid-fire paintballing. "'I Am Living in a War Zone,' Says Resident as Baboons Take Over" was the title of a news article that appeared in mid-January:

> Alison Brown is shaking with fury. She is wearing a bikini covered with a beach towel which she clutches tightly to her chest. "This is one of the most spectacularly beautiful places in the Cape but I am living in hell. It's nothing but a war zone," she says.
>
> Brown lives in Scarborough, between Misty Cliffs and Cape Point.
>
> "It starts at the crack of dawn when the baboons start foraging.
>
> "Then, when the baboon monitors pitch up for work, the paintball guns start going off—not one shot at a time, more like 15 to 20."
>
> Her neighbour and Kommetjie Environmental Awareness Group member Russ Weston says: "The baboon monitoring programme is not working."
>
> When *The Times* visited Scarborough the chacma baboons, one of the peninsula's remaining 15 troops, were raiding rubbish bins while a monitor was napping metres from the scene.
>
> A few weeks ago a female baboon was shot and killed by a resident with a high-velocity pellet gun. Last month the Misty Cliffs troop's last remaining male was killed.
>
> "These horrific killings are the direct result of the monitors not doing their job," Weston says.
>
> He conceded that residents were also at fault. Many did not secure their bins.[66]

Paintball-gun shots are painful, and their use by the monitors has led many members of the public to buy paintball guns or even pellet guns to control baboons. The monitors (or rangers) are trained in how to target baboons reasonably safely and have to pass accuracy tests, according to the paintballing protocol; members of the public do not. In this particular case, in trying to manage these remnant females, the rangers would have been shooting at or near the mothers with infants, which is expressly forbidden under the protocol.

The continued use of euthanasia is disturbing, as the removal of an alpha male may lead to a spate of infanticides as the incoming alpha kills his predecessor's youngest offspring. The result is that while specific baboon troops may be numerically the same as before, the age and sex ranges are skewed: adults and juveniles without subadults or infants, and one adult male to many more than the norm of four adult females.

Such painful and socially disruptive methods are questionable: if an aggressive relationship based on pain and death would aggravate the fear-aggression response of a dog, child, cat, or hamster in a domestic space, will it not, in the long term, aggravate the possibility of aggression of the baboons in a suburb? Certainly, at this point, Cape baboons seem to me far more skittish than those that I've observed in nature reserves elsewhere.

In short, the Cape Town anthrozoopolis feels as if it is run by a kind of zoo-police. This would not be the case if a feminist and relational primatology had driven the intervention. Such an approach would have problematised patriarchal assumptions about baboons; situated the facts produced by science in the wider context of their production; attended to the ways in which specific concepts and concerns generate what can be called a fact; and paid careful attention to the relationships both within troops and with their human neighbours. Primatology like that of Donna Haraway, Barbara Smuts, Thelma Rowell, or Dorothy Cheney and Robert Seyfarth provides plentiful scholarly grounding for this approach. It is even difficult to imagine Strum's work on "the Pumphouse Gang"—the name of the baboon troop she observed in Kenya—informing a primatology that seems so disinterested in baboon social lives in a context where baboon troop discipline is paramount. The question, then, is: What is the version of the nature of baboons that is at work in the science that currently dominates the city of Cape Town's baboon policy?

IV | CONTESTING BABOON "NATURE"

For Cape Town's current baboon managers, "fortress conservation" is thought of as the most appropriate strategy in the middle of a city. But a militarised conservation that fosters the unmaking of all multispecies neighbourliness in an area in order to manage property-line transgressions establishes fear and violence as the sole possible form of relationship between humans and baboons. In this regime, all relations of attunement with a neighbouring baboon are defined as part of the problem and attributed to "bunny huggers." The baboons cannot exist in the city, the fortress conservationist propounds, unless they have a relation of fear with humans. With sufficient fear, they are allowed to live, but without sufficient fear, they are required to die. The "cheeky bugger" who doesn't know his or her place, to use South African slang, has to be culled.

Modernity and coloniality have depended on the work of the territorial imaginary in order to map spaces of nature and culture. Apartheid South Africa, which took modernist divisions to the extreme, relied on the twin project of creating the nature reserve and the native reserve, with the former justified as the protection of nature, and the latter as the protection of culture, while paradoxically animalising people of colour. Apartheid city planning has been mobilised around the baboons, a mode of thinking where territory is conceived of as single-use: either baboons or humans, nature or society, but not both, have access to the land. Alternative forms of attuned relations with baboons are exorcised from public discussion by the claim that they are not objective science—even though they have plenty of scholarly backing.

Trethowan's activism is based on her deep insights into the baboons' social lives within the troop. Their social suffering is painful to see under the regime which has culled alpha after alpha, with frequent infanticide when a new alpha comes in, and skewed troop demographics by the overwhelming elimination of males.

In January 2016 I accompanied her along a suburban avenue to find the female baboons who were the survivors of the Misty Cliffs troop after a series of cullings (it was this situation that had been described as a war zone, in the news report cited earlier). I had seen them in the upper streets of Scarborough: one carried her own baby on her belly and an orphaned baby on her back. She was, Trethowan said, trying to avoid the alpha male from the Groot Olifantsbos troop in the neighbouring Cape Point reserve into which the rangers were trying to herd her. The new alpha would almost certainly kill the tiny offspring of the dead alpha male. It was January; there was a heat wave, and a recent mountain fire meant that in the sunlight, the bare soil and stones on the hill were even hotter than usual. A few days before, Trethowan had seen the female and her peers trying to get down to a stream near Misty Cliffs to drink, only to be herded back by six HWS rangers all firing paintball guns to force them across the stream and back up into the mountains where the vegetation had been burnt. But to do that, she said, the baboons had to run the gauntlet of several large dogs. As she and I drove around a few days later looking for the "girls," as she called them, a lone ranger told us that the mother with the baby and the adopted orphan had been pushed into the reserve, but that the other two females with two babies were in the bushes amid the hilltop Scarborough houses. Trethowan was immediately worried. "How can she defend herself and her babies from that alpha when she is alone in the reserve?" she asked.

The difference between Trethowan's version of baboon nature and that of HWS was clear: her accounts of what was happening to the baboons attended to relationships—to troop dynamics, to their needs, and to the hazards they faced. By contrast, HWS was attending to loggable incidents of individual behaviour by GPS points, as contractually the terms of their success involved their keeping baboons out of suburban areas 80 percent of the time. Their November records showed that in the case of Misty Cliffs, that rate had dropped to 82 percent—perilously close to their minimum success rate—and this was in terms of their own records. The pressure was on to get that figure up in January. Their conceptual frame derives directly from the kind of thinking that science attends only to measurable space, within which relationships are irrelevancies. But would there be a world, without relationships? In what respect does stripping relationship from reality give access to reality? It cannot: it is an impoverished approach to science. It is neither empirical, nor is it consistent with leading primatology elsewhere, including that of Shirley Strum.

Resident primatologist Ruth Kansky asks, "Is it a reasonable management objective to keep baboons away from humans in an urban park?"[67] Baboons are not generalisable, she says; they evaluate humans in every situation, and respond in particular ways.

Kansky knows this well: she and the late Wally Petersen initiated the baboon monitoring in 1998, and she has records of baboon troops in the region going back till then. She and other zoologists speak of specific troops having specific "cultures"—specific adaptations to habits and daily rhythms in different areas of the peninsula, where settlements span the full range of South African demographics, from a shack settlement on Red Hill, where people spend most of the day outside their homes; to a naval base with a migrant population above Simonstown harbour; to the Brooklands ruins from forced removals in the 1970s; to low-income families in affordable housing and apartheid-era blocks of flats; to organic-eating families and wealthy landowners with beachfront mansions that empty out in the mornings and fill up again in the evenings.

"Different people want different relationships with the baboons," she notes. "So what is a democratic response in that context?" Some want a mutualistic relationship with the baboons, she explains; others—such as many on the naval base—think of undomesticated animals as best viewed in a gunsight. The current science base for baboon management, Kansky points out, assumes that everyone wants a "zero-tolerance" approach to baboon presence, when there is a wide range of approaches that should be part of democratic debate about what to do with the baboons—including that of people who want a mutual relation with baboons. "The current approach to science is making value judgments based on deeply held philosophy," noted Kansky. "But they claim that the current baboon management approach is 'pure science,' while others are making value judgments."[68]

Haraway's *Primate Visions* argues:

> Scientists themselves . . . keep pointing out that they are, among other things, watching monkeys and apes. In some sense, more or less nuanced, they insist that scientific practice "gets at" the world. They claim that scientific knowledge is not simply about power and control. They claim that their knowledge somehow translates the active voice of their subjects, the objects of their knowledge. Without necessarily being compelled by their aesthetic of realism or their theories of representation, I believe them in . . . that my imaginative and intellectual life and my professional and political commitments in the world respond to these scientific accounts. Scientists are adept at providing good grounds for belief in their accounts and for action on their basis. *Just how science "gets at" the world remains far from resolved.* . . . Evidence is always a question of interpretation; theories are accounts of and for specific kinds of lives.[69]

For Trethowan, just how HWS "gets at" the world of baboon behaviour is key to formulating a critique that might improve baboon governance. Besides the room for error in the monitors' accounts of baboon behaviour (given that their performance evaluations depend on low numbers) and at the level of data handling, and besides the amplifications inherent in the way the count is done (if an entire troop of twenty baboons is in a village, she says, every baboon gets a mark against it, and it is recorded as twenty individual transgressions), the key issue is that there is no accounting for the interpretation of the context of recorded transgressions. In several conversations, she mentioned many kinds of factors. Human transgressions—such as having poor garbage can design, or leaving household garbage accessible, or not barring windows—do not make an appearance in the reported numbers. They may appear in the narrative of the monthly HWS report, but not in a way that mitigates the charges the numbers tally. But for the baboons, those numbers are a matter of life or death. Even more important, the troops' social lives do not make an entry into the counts, such as the consequence of inadequate juvenile socialising when alpha males are successively culled; when baboon mothers seek to protect their infants from the incoming alphas; or when young adult males leave their troops in search of another troop. Trethowan's objection is that HWS is de facto managing individual baboons via a single-species assessment in a multispecies zone, without taking account of the complex social lives of baboons.

For Haraway, writing in California in the 1980s, the feminist critique of primatology was that it was a field deeply and demonstrably informed by prevailing norms of sex, gender, and race. "The women and men who have contributed to primate studies have carried with them the marks of their own histories and cultures [which] are written into the texts of the lives of monkeys and apes, but often in subtle and unexpected ways," she wrote.[70]

Similar questions are needed of South African baboon management: how are South African social concerns carried into its science? Where territoriality is tied to techniques of urban spatial management, its policing derives from the notion of the line; notions of deviance and crime that need to be controlled; the forms of compliance that are expected; the right to assert separation without negotiation; the production of objects and subjects; the rule of law by fear. The securocratisation of urban South Africa through private security companies like ADT has been paralleled in this HWS operation, culminating in their right to determine killability—the right to cull. Working with this language, one can begin to grasp the scale of the shift that has occurred: from being a conservation-oriented entity that addresses human-wildlife difficulties, HWS has become a multispecies security company. My impression, on my bicycle in 2012, that I was cycling past a version of the riot police was not far off.

One of Haraway's earliest challenges to primatology was written in 1978, as feminists questioned the focus of primate scientists on aggression:

On one hand, we may reinforce our vision of the natural and cultural necessity of domination; on the other, we may learn to practice our sciences so as to show more clearly the now fragmentary possibilities of producing and reproducing our lives without overwhelming reliance on the theoretical categories and concrete practices of control and enmity. . . . It is not an accident of nature that our social and evolutionary knowledge of animals, hominids, and ourselves has been developed in functionalist and capitalist economic terms. . . . We are . . . engaged in a political-scientific struggle to formulate the rules. . . . The terrain of primatology is the contested zone. The future is the issue.[71]

In the book accompanying the exhibition *Ape Culture*, Anselm Franke offers a tale of the origins of a primatology that told a story of humans:

This tale is set in London in the mid-1920s, where the newly established "Monkey Hill" in London Zoo was enormously popular with both visitors and the sensationalist press. It was perceived as a "window" into nature. . . . What visitors to London Zoo observed confirmed their expectations of the raw, wild, and untamed primate nature beneath the thin "veneer of what was understood as culture." It was exactly what an enlightened late Victorian expected from the state of "nature" at the height of the British Empire: a fierce and violent war over resources, power and dominance ... offering, amongst other things, seemingly plausible explanations of male dominance and female dependence.... Thirty to forty years passed before [that approach was] finally refuted.... Field researchers demonstrated that baboons in the wild ... showed few of the behavioral patterns they had displayed [in captivity] in London Zoo. The "nature" that had been constructed there was entirely unnatural.... It was a symptomatic product of the social situation and milieu, which created an object of Knowledge devoid of social relations and reflexivity.[72]

The development of baboon zoo-criminology in Cape Town is fascinating because while it is defended in the name of "Science" (Enrique Dussel would use the term "scientism"),[73] it testifies to the entanglement of Cape Town baboon science with Capetonian society. Recent South African baboon journalism offers an almost infinite supply of evidence that media descriptions of baboon behaviour are dominated by the language of baboon criminal deviance. Media accounts—South African, British, and American—of the Cape Town baboon story are saturated with words and phrases such as "gangs," "living under siege," "ransacking," "looting," "hijacking," "rogue males," "robber baboon," "burglar baboon," "kleptomaniac," "running wild," a "gang of highway baboons," and "baboon death sentence," which pop up even in generally reputable media fora. These are very interesting words, for they reflect primary South African suburban anxieties: crime and invasion. These metaphors do what Haraway says primatology does: they use primates as a screen upon which to project unrelated human dramas, and try to resolve them.

Gangster metaphors for baboons, like any metaphor in the sciences, may be useful, but are also limiting. Where troops living on the urban edge become "gangs in suburbs," the response to them is already scripted: they are to be counter-attacked as marauding and invading nomads—when perhaps they are just doing what baboon troops do, and have always done, on the edges of urban settlements.

The production of hard boundaries between nature and society generates their breach since they are an unnatural, unreasonable, and illogical protocol for governing ecologies. Fences keep in or out neither baboons nor molecules nor pollens. For Nelson Maldonado-Torres, the very idea of a boundary between society and nature, human subjects and natural objects, institutes the relation of war that is at the heart of modernity.[74]

As long as the modernist regime of things and humans serves to divide environmentalism from society in a territorial imagination, environment will be something that needs to be policed; with which people are not expected to share.

Cape Town's baboon politics captures baboons in a web of apartheid metaphors and urban policing practices and racialised projections. Yet human history in Africa has long included a wily interspecies partnership and a great deal of care and respect for baboons, along with annoyance and irritation at times. The binary conceptual framework of modernity-coloniality has ruptured the history of attunements between animals and humans, replacing it with an approach in which animals are objects.

That web of apartheid-style language practises has survived because in a society characterised by a deficit of humanity that is epitomised in the figure of the baboon, the dominant means of characterising the "other" remains an intractable habit. The necessity for "moderns" to distinguish the self from the animal undergirds the history of primatology; it undergirds the fascination of international newspapers with Cape baboon stories; and it undergirds the "going viral" of baboon invasion stories. In the words of Franke, "Apes and monkeys have been held prisoners of the mirror that they represent for us."[75] What is the prospect that they can emerge from behind this screen?

V | LIFE SKILLS FOR ANTHROZOOPOLITAN ZONES

Jenni Trethowan tells a story of talking with "Eric," the one-time alpha in the Kommetjie troop, asking him what he wanted her to tell those who were gathered for a talk which Jane Goodall would be attending. She sat with him for a long time, she said, thinking about what he would say about humans, and went back to the meeting with the message, "You humans are so wasteful." Her time of "thinking together" with him enabled her to think about different patterns of waste generation in the middle-class and elite suburbs, where a great deal of food is thrown away, and in areas where food scraps are few in poverty-level households. In doing so, she began to think about the ways in which baboons relate to people, and the ways in which the different baboon troops respond differently in different settlements. Her dialogue with Eric finds a parallel in the city of Sydney, where Deborah Bird Rose and Thom van Dooren call for "an ethics of conviviality that is urban-based, emplaced, embodied, and enlivened through multiple stories enacted and expressed by multiple species."[76]

In February 2016 Scarborough resident Ushka Devi posted a photograph on Facebook of one of the Misty Cliff "mothers," with this comment: "Here she is in my kitchen! One of 2 females who with their babies are the remnants of the Misty Cliff troop. She grabbed the rye bread took a bite and threw it down. With no apology to our illustrious bakery. She went for the bananas as I shooed her out. Why all the fuss with rangers and paintballs and bear bangers? We handled this elegantly, woman to woman."[77] Shared hundreds of times in a few days, the post yielded several similar stories of multispecies neighbourliness: stories that have parallels in South African history across all sectors.

––––––––––

Is it possible to draw on different Southern African figurings of human-baboon relations to imagine different kinds of futures—futures that are defined neither by the militarised, territorialised, and violent relation that has come to life in policy and practice, nor the romantic figure of the baboon as the innocent child? Baboons clearly have the capacity to cause immense damage to homes and food sources—but this in itself is hardly a new phenomenon. Humans have always had to learn to conduct themselves appropriately in the presence of baboons. Why not now?

How might Capetonians reimagine baboon-human relations in the anthrozoopolitan zone that is this city built around a mountain? What approach to baboon conservation is possible other than one imagined only in terms of fear, violence, and territory?

A number of stories in the Bleek and Lloyd Archive sketch the kind of wily interspecies attunement that was needed for themselves, reluctantly or otherwise, the neighbours of baboons. They provide a great deal of advice, including on the proper wariness: "Baboons [are] not to be answered, when they address a Bushman in the early morning on his way to the hunting ground. They must, also, be alluded to in a very guarded manner, lest they should know that they are being spoken of." Elsewhere, the warning was this: "Baboons are not 'good people' and do not 'merely' want to sit and talk as people pass. The baboon gets angry and teases. It mocks and deceives people as they pass by."[78] We learn that baboons imitate humans: "The Baboon . . . understands like a man, and speaks, sounding like one."[79] And we are told that baboons have ways of knowing many things, even secrets: "The name of a Bushman seems to be known to a Baboon, even when the latter beholds him for the first time. . . . It tells the name to the others as the person passes by."[80]

5.3 | Artist Patricia Vinnicombe's rendering of a rock painting in the Underberg district of KwaZulu-Natal, South Africa. Source: The African Rock Art Digital Archive. Used with permission.

There is no romanticism here: on the contrary, the texts offer a clear-sighted view of a quick-witted and intelligent competitor for food around whom one needs to live with equally quick wits and an equally intelligent assertiveness. Baboons ask for an intelligent relation—not war.

The Cape Town baboon situation is paralleled wherever authoritarian and militarised forms of conservation science go unchallenged. These survive in the name of "objectivity," rendering as "bunny hugging" residents' attempts to attune to urban animals. The disregard with which city officials treat Baboon Matters is evident in correspondence that refuses to address any concerns.[81]

An understanding of the range of historical relationships between people and baboons offers resources for rethinking human-animal relations in the megacities of the Anthropocene. To do so could enable the recomposing of baboon management protocols in terms of an interspecies attunement that includes the social dynamics of baboon troops in relation to those in specific human settlements, with their different daily rhythms of movement, weekly routines, and different trash-producing profiles.

Learning to relate well to dogs is a taught skill—why could the same not apply to baboons? I can imagine many an "infernal retort" that would aim to make this suggestion look ludicrous. Can critics imagine an improvement on the huge road signs that, with the wording, "Baboons are dangerous wild animals," set all urban human-baboon interactions on a foundation of fear and control?

Rethinking the war of subjects and objects is central to the legacy of coloniality-modernity in sciences, too. A first step is to address the contradictions in contemporary primatology in the Cape in order to offer "a new proposition, articulating what was a contradiction leading to war."[82] That contradiction is the separation of nature and humans—and the forms of technoscientific governance that strip out relationships (as "society") and assume that only objects in space constitute science that is worthy of being used for governance.

Ensuing forms of technoscientific governance have been so frustrating to citizens worldwide that a backlash has ensued toward nationalist forms of politics that promise to restore relationships, however exclusionary and unjust, to political life. A whole new progressive politics is needed that refuses objectification of animals or people, and reclaims relationships wider than nation, territory, and humanity, to encompass neighbourly forms of political life and governance.

Addressing the era in which we live compels rethinking the co-presence of humans with other species. In urban conservation, there are no wild spaces that are free of humans. Accepting that, and working out the terms of a peace agreement within it, generates a different kind of research problem—one that is capable of "daring to add new dimensions to it, so that its answer is not that which, sadly, seemed decreed in advance."[83]

OCEAN REGIME SHIFT

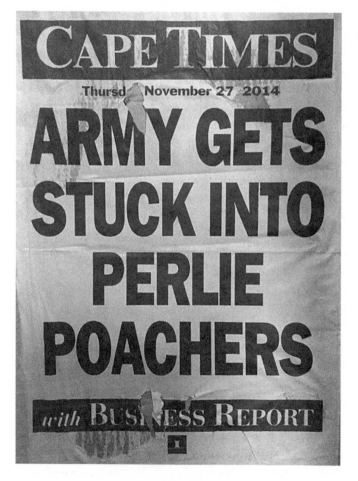

6.1 | Cape Times billboard, 2014. Photo: Lesley Green.

The first alarm bell was red.

It was 1989, and for decades the apartheid government had fomented fear of communism—"die Rooi Gevaar" (the Red Peril). And then came a "Red Tide," washing up on the shores of a Cape Town marine nature reserve named after the chief architect of apartheid, H. F. Verwoerd. An estimated forty tons of abalone died along the southern Cape coast east of False Bay,[1] along with thousands upon thousands of creatures in the kelp forest: crayfish, mussels, sea urchins. It was not the first red tide, but it was the first time it would be so extreme.

The second alarm bell was also red. It began in the early 1990s: beaches would turn red as a result of mass strandings of the rock lobster, *Jasus lalandii*, also known in Afrikaans as kreef, and in the fisheries management literature as WCRL, or West Coast Rock Lobster. They simply walked out of the ocean, out of the official "take zones" that had been set up by fisheries quota managers to manage their extraction and prevent their extinction. They would beach themselves on the sand by the hundred thousand—sometimes it seemed millions. When a lobster stranding was particularly bad, the army would be called in, and military helicopters would ferry the now-dead lobsters by air to a landfill.

The third alarm bell was heard in the words "ecological regime shift" in the mid-1990s. The phrase derives from invasion biology, and perhaps also from the Gulf War of the time, when "regime change" was the objective. Marine ecologists deployed the words in South Africa to describe a mass migration of lobsters from Cape Point, under the sea and across False Bay, to a rocky shore about a hundred kilometres away, called Cape Hangklip. This was not just a case of more eggs and better larval recruitment on-site—it was an actual mass migration of adults. Studies of their stomach contents showed they had changed their diet: eating different species, and setting off different trophic cascades (changes in the food chain), one of which added to the extinction pressures on *Haliotis midae*, the local species of abalone. *J. lalandii* was changing the ecology of the kelp forest.

The fourth alarm bell rang for sixteen years, starting in 1997. Local fishing communities that lived off the sea were classified as subsistence fishers under the marine environmental laws developed during the presidency of Nelson Mandela. The definition denied them the right to sell their catches. Over the years, they chained themselves outside Parliament, won battles in the Equality Court, and mobilised nationwide to develop a policy for the small-scale fisheries sector, in which they sought to be legally recognised as a specific sector of fishers with rights to national fishing quota, to reject individual quotas for specific species, and to cease being criminalised as poachers. Social science research showed that fishers were so distrustful of the controlling hierarchy of marine science that many made it a point of pride not to comply with fisheries regulations.[2]

The national Policy for the Small Scale Fisheries Sector (PSSFS) was adopted in 2013, allowing the catch of a "basket" of species instead of a single-species quota governed by the algorithms of sustainable commercial fishing quotas under the acronym TAC (total allowable catch). While the PSSFS said very little about ecology, it was expressed with an emerging discourse of care for the sea and for community members, so that they could live well, without hunger. But its implementation was handed over to the Department of Trade and Industry, which required every fishing community to form a proprietary company. The extraction of financialisable marine biomass, calculated with reference to the applied mathematicians' TAC, became the defining concern.

The fifth alarm sounded in 2014 when the army was called in to fight against the extinction of *Haliotis midae*, known as perlemoen or abalone, as part of a newly thriving global illicit trade in wildlife products believed to be aphrodisiacs.

The situation was surreal. Fishers were in their rights to ask: why and how in our democracy can a perlemoen have better representation in Parliament than us? Besides that: was it not surreal to mobilise a war machine to protect a snail? Could the army protect the perlemoen from the lobsters invading the kelp forests? Would the army restore practices of care for the ocean?

The sixth alarm came in 2014 and 2015 from kayakers and divers who reported huge floating ponds of human excrement close offshore from central Cape Town. Many of them were kayakers with whom I paddled regularly. When several contracted severe gastroenteritis after practicing their kayak rolls, they began questioning the city's claim that the marine sewer outfalls were not harmful. A microlight pilot, Jean Tresfon, provided photographs of the plume emanating from the outfalls, and local filmmakers made a short film on the city's sewage outfalls, which pump up to fifty-five million litres of effluent into the sea every day. The city's response was to issue a "cease-and-desist" letter to the filmmaker and the aerial photographer, and to argue that the two scientists who were interviewed for the film—epidemiologist Dr. Jo Barnes of Stellenbosch University and environmental nano-chemist Professor Leslie Petrik of the University of the Western Cape—were unscientific. The city's claim was that the occasionally high *Escherichia coli* and *Enterococcus* counts, both of which could cause serious illness, emanated not from the sewage pipe, but from stormwater runoff.

The seventh alarm rang in 2017: changing winds and rising air and sea temperatures led to unprecedented drought in Cape Town and its surrounds, after three years of low rainfall. The city's dams were about 35 percent full at the end of the 2017 rainy season, with rainfall levels among the lowest in recorded history, and far lower than even the most highly regarded climate scientists had predicted. By August 2017, dams were estimated to run dry sometime in the first quarter of 2018, before the first winter rains would arrive. "Day Zero," as it was called, would be a day when the city's taps would no longer deliver water.

The City of Cape Town's solution was to propose emergency desalination plants, situated next to the same marine sewer outfall that the kayakers had (literally) run foul of. The announcement came in September 2017, around the time that Leslie Petrik, Jo Barnes, several graduate students, and I were sending seawater samples to the labs to test for molecular compounds that could only come from sewerage and not from stormwater, in order to contest the city's argument that the pollution we were finding was caused by stormwater runoff, not the sewer outfalls. Our lab tests found every single one of the high-schedule drugs for which we had tested, and every one of the multiple household chemicals, in the seawater. We had hoped not to find them, for among these chemicals and pharmaceuticals are endocrine disruptors and carcinogens, with who-knows-what effects when mixed together in a seawater cocktail.

All fifteen compounds we tested for were bioaccumulating in the seaweeds and creatures at the shoreline: starfish, sea urchins, seaweeds, limpets, mussels. Our team of six researchers rushed a paper through to the *South African Journal of Science*, making the case that the molecular compounds being piped from the sewage outfalls into the seawater represented a public health hazard especially if desalinated sea water were to be piped into the city water supply. The paper made headline news nationally, and we were invited to make representations to the Western Cape head of local government, environmental affairs, and development planning, Anton Bredell.

The eighth alarm rang two days prior to the release of our article, when the City of Cape Town released its own scientific study on sewers and seawater, which it had commissioned in 2015, and which had been completed in June 2017, but which they had until then refused to release, even under pressure from ratepayers.[3] The press release stated, "A study by the Centre for Scientific and Industrial Research (CSIR) into the City's sea marine outfalls has confirmed that they pose no significant risk to human health and do not measurably affect inshore water quality or the wider environment."[4]

But the actual study had concluded, "It is . . . illogical and indeed irresponsible to imply that effluent discharged through the outfalls is not impacting on the marine receiving environments or posing a potential human health risk. Indeed, the notion of no impact to a marine receiving environment in the context of effluent discharge is unfounded."[5]

None of the authoring scientists at the Centre for Scientific and Industrial Research (CSIR), however, challenged the city's press release. The CSIR was under financial pressure, and about a year later, it was announced that job cuts loomed, specifically affecting their environmental research division.[6] According to a source, the environmental division was to be completely restructured. Staff scientists would need to reapply for their posts and recruitment restricted to "positions directly linked to secured contract funding."[7] Where government-funded scientific services turn into semi-autonomous scientific consultancies that must pursue contracts to generate their salaries and running costs, their claim to be independent and impartial must be questioned—as argued in chapter 2.

In November 2018, none of the city's three desalination plants was able to operate because the presence of unusually high harmful algal blooms was fouling their reverse osmosis membranes. In April 2019, a company called QFS (Quality Filtration Systems) that had won the tender to install a desalination plant at Cape Town's waterfront in 2018, near our 2017 sampling site, instituted legal action against the City of Cape Town for non-payment.[8] The battle concerned the seawater quality data, which, QFS claimed, was up to 400 percent higher than the COCT figures had shown.[9] Astonishingly, the city's tender documents had included official seawater quality data but also made it the legal responsibility of the tendering companies to supply their own. Even more astonishing, City of Cape Town Mayoral Committee Member for Water and Sanitation Xanthea Limberg later argued that that data ought to have been supplied across the seasons—yet the terms of the tender, amid the urgency of a city about to run out of water in 2018, were that the desalination plant had to be up and running within a matter of weeks.[10]

At the time of writing, it is expected that the case will reach the Cape High Court.

———

These situations paint a picture of both an ocean ecology in crisis and a science in crisis. In respect of a crisis in South African marine sciences, this is not because its practitioners are not good at what they do: on the contrary, the passion of Southern African marine scientists for the species that they represent—in scientific journals and in Parliament—is infectious. It has to be said that their journals struggle to include the social sciences. The focus is natural sciences, and in that, they offer excellence, study after study.

But therein lies the trouble. For absent from marine sciences in the Cape Town region is the urban-marine interface, with its sewer outfalls and its sanitation crisis; its chemical outputs and its agricultural fertilisers, pesticides, abattoirs, and manufacturing wastes; and its filthy, unmanaged urban rivers. As with much of the Cape baboon research (chapter 5), the research emphasis seems to be overwhelmingly on imagined pristine states: pure nature, existing outside of society.

So, while alarm bells have been going off about harmful algal blooms, lobsters, abalone, kelp forests, fish stocks, ocean warming, ocean acidification, and ocean plastics, a science has not yet emerged that is able to work integratively—that is able to address the anthropogenics of the situation as more than something that has "social dimensions" that require "regulation," "compliance," and "enforcement." Where the social does appear in marine research literature, it appears as demographic quanta of predation—recreational fishers, commercial fishers, small-scale fishers—and even though the annual human catch is absolutely central to the science behind the TAC, the annual international workshops to peer-review the algorithms do not include any social scientists. The result: a catastrophic and basic methodological error in calculations of lobster catches by recreational fishers, made by a social science consultancy whose methods were not reviewed, and went undetected for five years, resulting in a major revision of figures several years too late.[11]

Sciences that are split along the nature/society divide cannot conceive of their work as a field that integrates capital, global warming, biology, and chemistry—a suite of interactions that is only partially described by the emerging field "biogeochemistry" and would arguably be more useful under a name like "capitalothermobiogeochemistry." With attention to the relations of capital and government and ocean warming, current marine sciences may be better able to confront the terraforming neoliberalism that shapes imagination, action, and policy.

In what follows, I want to use this complex situation to argue that responding to the ecological regime shifts of our time requires a regime shift in ocean scholarship—an ocean regime shift. To sustain the seas, we need to reclaim science as something that is not necessarily reductionist; does not need to claim to be "apolitical"; is more empirically useful if it does not distinguish between nature and society. What range of scholarship will enable Capetonians, both publics and policy-makers, to think, act, and desire a "living together well with the ocean"?

What would this entail? For me, "living together well with the ocean" entails considering the full range of flows of matter regardless of imagined boundaries; property and the commons and the undercommons; extinctions as well as exclusions; living beings along with rocks and water.

Why is this important? Neoliberal governance and market logics actively seek to capture the authority of science. Reductionist science (and reductionist social science) offers points of data that can be narrated by others: it has surrendered the power of story. And when natural science is pursued without equitable and respectful partnerships with colleagues in the social sciences and humanities, it is at risk of making not only major methodological errors, but also major conceptual errors. And it lacks the tools needed to question the authority claims of politicians, or the truth claims of consultants in the name of "science."

Thinking with and alongside various actors and modes of existence in the Cape ocean ecology requires thinking from a multiplicity of perspectives, across disciplines, situations, and species. For anthropologist Trinh Minh Ha, the challenge is not to think *about* but *alongside* those whom one studies[12]—an approach very similar to the work of Elspeth Probyn in her book *Eating the Ocean*.[13] While Trinh's words were spoken with reference to the formulation of a postcolonial anthropology (in other words, about people), I find her words resonant with the attempt to move beyond the subject-object divide in scholarship on other living creatures. Philosopher Vinciane Despret, for example, titled a recent book in animal studies *What Would Animals Say If We Asked the Right Questions?*[14] In finding his line of flight from colonial thingification, postcolonial thinker Aimé Césaire found a line of flight from the prevailing gods of reason in the approaches of the surrealists.[15] From him, I have learned the power of posing questions about "realism" in ways that make the surreal evident. Doing so makes it possible to encounter the cosmological precepts in scholarly research and policy—and reveal their irrationalities for what they are.

Questioning authority is key. A deep tradition of storytelling, across sub-Saharan Africa, actively declines the rhetoric of *authorial* authority, offering instead the dilemma tale.[16] In such a tale, the art of authorship is not, as in the essay form, to persuade your listeners that you are right, but to stage a discussion of what is ethical, or what each actor might do next. Dilemma tales offer a mode of engagement very similar to Amazonian perspectivism:[17] understanding that knowing is not simply a question of "understanding the information of the world" but of "understanding the world in-formation." The different form of authorship here is not the authorial "authority over," but the authorial capacity to offer listeners the experience of a "presence-to" the complexities of a situation.[18] The knowledge they honour is less about the knowledge of the "being" that is each creature than about the "becomings" of a situation: who will do what next? Who acted ethically?

Being able to understand what will unfold next is a key part of the art of knowledge in Chinese Han dynasty thought that attends to the propensities of things.[19] In Amerindian perspectives, in African dilemma tales, and in Chinese thought, we can begin to see that the attention of coloniality-modernity to things and direct causal relations is something of an anomaly among many intellectual heritages. With regard to the perlemoen, the lobster, the fishers who are blamed for stock declines, and the problem of harmful algal blooms, the form of the dilemma tale offers a way of staging an encounter between perlemoen, lobsters, fishers, poachers, environmental managers, the army, and marine biologists. In these ways, the challenge of decolonising knowledge is not about offering a new kingdom of thought to replace the disciplines, nor generating a new field of study: it begins with a transformation of how we think about what it is to know.

I | BLUE FLAG SCIENCE

The waters of Table Bay sparkle as you drive into the city bowl, looking toward Robben Island where Nelson Mandela, Robert Sobukwe, and others were imprisoned. Follow the highway into Strand ("Beach") Street, which runs alongside the Castle. Nearby, in 1654, the Dutch colonist Jan van Riebeeck had, under the flag of Orange, taken fourteen wheelbarrow loads from a shoal caught by the sloop that had anchored offshore, which had filled its nets with one cast.

The beaches of Cape Town are now under a different flag: the Blue Flag, which describes itself as "a world-renowned eco-label trusted by millions around the globe."[20] The brand is held up by the city council as international proof that its beaches are safe enough for tourists, and ecologically healthy.

Are they?

When kayakers with whom I had been paddling in 2014 began to get ill, they began to join the dots: when the rain-bringing north-west wind blew, as it does in winter from April to September each year, that was when sewage was most likely to appear on the surface. My colleague, environmental chemist Leslie Petrik, explained what was likely happening: in winter the seas are colder, and the contents of the warmer sewage outfall less likely to mix with dense cold water; they would rise to the surface, and be blown back to shore by the prevailing winter winds. Around that time, social media went viral with the message: the city of Cape Town was sending fifty million litres of filtered but untreated sewage out to sea every single day.

The rich irony was that at the same moment, the city was fighting activists protesting the lack of waterborne sewerage in the shacklands. With the leadership of sanitation activist Andile Lili, supporters (known as "poo flingers") had begun dumping the contents of domestic toilets in public places, including on the steps of the Western Cape provincial government buildings. Their actions were copied by the University of Cape Town activist who threw faecal waste onto the campus statue of Cecil John Rhodes, starting the decolonisation movement on campus.[21] In these cases, a "hazmat" company was called in to clean up. But the city itself was sending filtered but untreated human waste into the ocean every day, including into a marine protected area on the Atlantic coast.

At issue here is the form that scientific questioning takes. Blue Flag science asks whether the water can be demonstrably safe for a tourist to swim in. To retain Blue Flag status, the city must answer a specific research question: whether the seawater contains faecal and streptococcal bacteria above the specified limits twice a month. The answer to that question requires no attention to the differences that will come depending on the time of day you sample the water, how deep and how often, what winds are blowing, and what methods you use to get the sample to a lab.

It is a textbook case of a science that asks a question that addresses the concerns of capital—in this case to keep the tourist economy booming—rather than seeking to enquire about the effects of the sewer outfalls. It is also a textbook case of the problems that accrue when a corporate deal—the requirements to fly the Blue Flag—replaces best practice in environmental management sciences, in this case, beach management. When science in the service of economic growth provides elected officials with a Truth that starts with a capital T and ends with a tiny TM, it is the kind of science that is designed to shut down questions rather than enable enquiry. It is as important for science to separate itself from the knowledge economy as it was for science to separate itself from the church in the 1600s.

As noted already, Leslie Petrik and Jo Barnes raise additional concerns concerning the biochemistry and microbiota of the outfalls. It is inconceivable that there are no geological traces of the millions of litres of sludge and effluent settling underneath Table Bay daily, particularly as the surface flow of the bay is five times faster than that of deep water. The disposal of toilet paper into the ocean, albeit disintegrated, constitutes a massive geological shift in the regional ocean as wood-based cellulose—which would previously have fed soil as leaves or rotting wood in a forest—now enters the marine ecosystem. Then there are the microplastics from city's millions of washing machines; the soaps and detergents and drain-cleaning agents sold by every supermarket and corner shop in the city; every bottle of disinfectant and shampoo and sunscreen; all the detergents and paint thinners and oil-cleaning products from every home, hospital, abattoir, and factory floor.

Sewer outfalls change the ocean chemistry and ocean ecology in a way that is both so micro that the change cannot be seen without the help of a microscope and so macro that it can at times be seen from space. I'm referring here to harmful algal blooms.

6.2 | The Green Point marine sewer outfall at Cape Town. Photo: Jean Tresfon. Used with permission.

Management sciences are generally sciences that attend to profitability. They do not prioritise ecological concerns or relations. The data that they produce address economic and organisational concerns. To do this, the most useful approaches are minimalist, reductionist, and territorial.

Ergo: Regime-changing scholarship is integrative, extending the "now" into history; law into justice; place to flows; the micro—microbes and molecules—to the hyper-object. Regime-changing scholarship must challenge the claims, methods, and authority of science that takes place under the flag of economic growth.

II | HARMFUL ALGAL BLOOMS

Generally comprising dinoflagellates and sometimes diatoms, some harmful algal blooms produce neurotoxins. Others are individually harmless, but can bloom so massively that they absorb all the oxygen in the water. Scientific literature on harmful algal blooms notes their association with disruptions of the nitrogen cycle as a result of commercial agriculture and urban sewage discharge (in which urine provides vast quantities of NH_3 or ammonia, and solid excrement contains vast supplies of bacteria). Add to that ocean warming and wind changes associated with climate change, and the future is looking bright for harmful algal blooms.

As fascinating as they are destructive, dinoflagellates come in many more shades than red. Some are green, some are black, and, indeed, some are known as the breathtaking phosphorescence in the sea—the "ocean fire" by which local Western Cape fishers used to fish at night "before the lights went out" around the 1960s.[22] When they are viewed under electron microscopes, the variation in dinoflagellate bodies is astonishing.[23]

Prompted by unexpected and catastrophic red tides that bloomed in False Bay in 1988 and 1989 and again in 1995–96, doctoral researcher Lizeth Botes set out to understand what species were blooming, and why.

Botes's species sampling is surely an example of the best kind of scientific research, as shown by the care taken over every name and every claim. The study looked at *natural phenomena* to try to explain the algal blooms: specifically, sea-surface temperatures and winds in relation to the deepwater upwelling that occurs when winds drive warmer surface waters away, and the deeper, colder, and nutrient-rich bottom waters come to the surface. Indeed, in a world of nature-without-people, this process of wind-driven upwelling would be the sole basis of the regional marine ecology. But in the mid-1980s, a great many other factors were at work on the shores of False Bay. These included the rise of commercial agriculture with the extensive use of fertilisers; the rise of the chemical industry to provide that fertiliser; and the growth of new shack settlements on the Cape Flats, specifically Khayelitsha, that were without sanitation services, so raw sewage now flowed through stormwater channels. Did these three factors contribute to a perfect storm that spawned the unprecedented algal blooms of False Bay? It is difficult to argue the contrary. A 1992 dissertation by geographer Carl David Rundgren shows the major sewage, industrial, and stormwater discharges into False Bay in 1986.[24] The SomChem and AECI factories were both producing chemicals. The Mitchells Plain and Monwabisi outfalls, and the Sand Rivers, Seekoei, and Eerste Rivier systems, were serving a rapidly growing population with associated increases in fertiliser and pesticide use and effluent discharge.

Thus, into False Bay in the 1980s flowed a vastly changed soup of contaminants, including nitrogens associated with urine, fertilisers, and the meat industry.

Now, some algae can produce their own food like a plant (by photosynthesis) and also consume other species for their food. Put them in a situation where they have nitrogens that stimulate photosynthesis, along with a delicious soup of bacteria, and the conditions are right for a perfect storm. Dinoflagellates and diatoms—marine algae—are highly responsive to changes in ocean chemistry, climate, and bacteria. Even changes in atmospheric dust from stronger winds far away have an effect. When unseasonally strong winds blow in the Sahara Desert—where sands contain iron particles—they can foster algal blooms in the Caribbean and subtropical Atlantic.[25]

And yet, with the nature-culture divide in mind, and determined not to be political, the science tracking and studying algal blooms in South Africa could not see the geological effect of law in the apartheid state's classification, separation, impoverishment, military control, and wilful disregard of the sanitation needs of shack dwellers in Cape Town. The rise of toxic algal blooms was not only a matter of an altered biogeochemistry, but a direct effect of political decisions.

Extinction risks to abalone and crayfish are certainly aggravated by poachers, but why make poachers the scapegoats for a system that creates poverty and toxicity—and poachers?

Marine science that researches "nature," but not human presence in it, cannot see the molecular flows of the human—and is unable to explain political biogeochemistry.

If we hold this political biogeochemistry in mind, does it offer any insights into the lobster walkouts, and lobster-led ecological regime shift?

III | LOBSTER WALKOUTS

Jasus lalandii offers much to think with.[26] The best ecological science can model the new lobster-led ecological regime, but has yet to fully understand either why or how the lobsters have moved.

Lobster motivation is difficult to assess: there is no Sigmund Freud for lobsters, although there used to be. It was none other than a young Sigmund Freud who demonstrated that lobster neural structures were of the same order of complexity as those of mammals. That thread of biological thought was cut short by the rise of radical behaviourism.[27]

The nature of *J. lalandii* comes to us at present through the legacy of a behaviourist biology. In the South African scientific literature, there are at least two very different scripts for understanding *J. lalandii*.

One is the *J. lalandii* already described above: *J. lalandii* the regime shifter of kelp forest ecology, enrolled in the ecological imaginaries of invasion biology and conservation policy. Brendon Larson's critique of combat metaphors in invasion biology is important: he argues that they are problematic in that they lead to an inaccurate perception of invasive species, contribute to public misunderstanding of what is at issue, and naturalise militaristic thinking that is often counterproductive for conservation.[28] I like to think of this script as *Che lalandii*.

The other script of *J. lalandii* appears in the extractive science of fisheries management, where the single-species community plays the part of a capital asset in the national bank, and its growth or decline is projected via the agency of a regularly revised algorithm that in turn dictates how much of the capital asset may be withdrawn. This *J. lalandii—J. lalandii* the algorithm—enrols the lobster in the machine economy: the cold chain of its supply and demand.

While the two accounts of *J. lalandii* are very different, they share an insistence that they are not "political" accounts of nature. Both fisheries activists and government officials routinely argue the contrary. To maintain the fiction of the separation of nature from politics, both are constrained to maintain another fiction: that the lobster is a creature capable only of a mechanistic response to stimuli, and not a sentient, responding being.

Since the case of *J. lalandii* offers a situation in which the lobsters have chosen to do what we didn't expect—it was not in our script for them—it is useful to consider whether *J. lalandii*'s representation is part of the problem.

In Amerindian contexts of thought, to say that lobsters have refused the scripts of extraction and predation that define their fishing is not likely to be met with sarcastic riposte. Similarly, to think like a lobster, among South African fishers, is not an outrageous anthropomorphism, but a reasonable recognition that all creatures search out places and spaces that offer well-being. Lobster fishers along the West Coast insist that crayfish are highly responsive creatures that dislike lobster discards being thrown in the water: "no creature likes its own graveyard." The observation would not surprise an evolutionary ecologist whose question is: what new choices did creatures have to make in order to survive? Only in the framework in which non-humans are commodified objects for extraction is it possible to assert that lobsters are incapable of response to the conditions of the Anthropocene.

Reading more widely about lobsters (beyond the *J. lalandii* literature in South African debates), I was astonished to find research in ethology that suggests both interspecies and cross-species communications between fish and crustacea about the availability of food and likelihood of predation.[29] The complexity of lobsters' neurological systems—which so intrigued the young Freud—is evidenced in studies of lobster behaviour ("neuro-ethology").[30] Their complex social lives include rituals of aggression and dominance.[31] They have a capacity for magnetic navigation and an ability to migrate long distances and adapt to new environs.[32] They are highly sensitive to smell, using chemical signalling to communicate. They have a strong fear response, as evidenced in their rapid, long-distance movement away from tagging sites in an acoustic tracking study.[33] They can, on rare

WATER | OCEAN REGIME SHIFT |

6.3 | Military helicopter disposes of lobster walkout remains. Photo: Rodger Bosch. Used with permission.

occasions, choose to move en masse. Lobsters, this literature suggests, are far more sentient, and far more collectively responsive, than behaviourist fisheries sciences have imagined.

To return to the question posed by Vinciane Despret: what would animals say if we asked them the right questions? What are the quanta of well-being that lobsters are looking for, in a slowly warming ocean so close to a city's sewerage flows?

Can we form an ocean science that partners with the desires of creatures to survive and live well? How can the humanities partner with the sciences to shift the scientific and political representation of ocean species? How does ocean scholarship change if we conceptualise ocean species as capable of objecting to the terms of our interventions to prevent their extinction?

IV | QUOTAS AND THE ARMY

The current means of calculating the TAC and enforcing quotas produces an unworkable moral economy of extraction of life as objects: it establishes an ontology of unbecomings in the name of conservation. Where governance turned small-scale fishers' emerging language of concern into proprietary companies, the rekindling of care for the ocean was erased. What was mainstreamed was the closure of the commons, and the rise of a new class of extractive owners—features that inevitably require policing and militarisation. Is this a sustainable approach to conservation? Reclaiming care for the ocean requires a different kind of relation than ownership and extraction. Extinction risks are not manageable via business administration, but via care for the commons.

Those who have invented and taught in MBA degree programs in the past thirty years have done enormous damage. I would like to see the rise of a new degree series: the MCA, a master's of commons administration, or perhaps the DCC, the doctorate of commons care. Why?

The situations that I've described have something important in common. Fishers have struggled against single-species quotas conceptualised by and for a capitalist accounting of the sea. Lobsters appear to be attempting to escape the high-take zones, the sludge and chemistry of sewer outfalls, and the low oxygen of harmful algal blooms that arise more frequently in a context where officials argue that dumping sewage in the sea is better political economics than establishing waste-water treatment facilities. Perlemoen are struggling to survive poachers, who are trying to survive the predatory economy and the new lobster regime. Ocean-users are fighting the brand-management approach to water quality, which is hopelessly inadequate on a day-to-day basis. What is in common here?

6.4 | *Kreef* (Crayfish). Linocut by Zayaan Khan, 2017. Used with permission.

If we listen together to the dilemmas that each is facing, it seems that they have in common a battle against a particular kind of reasoning, a logic that has become dominant in the era of neoliberalism. Once again, we find the three divinities of reason here: technical efficiency, economic profitability, scientific objectivity—an inseparable trinity.[34] They are rock-paper-scissors types of closed systems, in which the only decision is which one of them will be the figurehead argument that will clinch a deal, frame a winning argument, and invoke the others. The risk of environmental science, particularly that produced for purposes of governance, is that it becomes so entangled in neoliberalism that it confuses that set of values for nature itself.

In my view, it is an empirical error on the part of scientists to assume that neoliberal values and relations are capable of unmaking the geological effects of the Anthropocene. A regime shift in scholarship is needed. For that to happen, researchers need to become active in building alliances and partnerships with communities that can contest the rationalities offered by the three neoliberal gods of reason, and are able to respond with ecologies of knowledge that are not captive to economics. Thus, rather than partnering with the army and the financialised economy, an ocean regime shift requires building alliances across ecology, economy, and ecumene (society).

A first step in partnering ecology, economy, and ecumene in the Cape would mean partnering with small-scale fishers whose national struggle for recognition as part of the economy and ecology goes back to the days when the Khoena were driven from the Cape, and fishers' ancestors were given the surnames of Greek gods (Apollis), philosophers (Plato), or the months of the year (September) in the centuries of slavery. Those addressing expulsions are allies whose on-site care for coastal ecologies can be mobilised to address extinctions. That process has begun, but there is still a long way to go.

V | JUSTICE AND EQUALITY STRUGGLES

The 2014 billboard "Army gets stuck into perlie poachers" plays on the means used to pry abalone (perlemoen) from the rocks in the kelp forests of the Cape: a screwdriver. With the marine compliance inspectorate being unable to stem the collapse of local perlemoen stocks, poachers were now in the sights of the military.

Perlemoen poaching is indeed a problem: the illegal abalone network is vast and global. Similarly, poaching and overfishing are among the reasons for the collapse of crayfish stocks. The militarisation of conservation has indeed been a last resort. But are small-scale fishers and coastal communities the right targets for activity by soldiers? We cannot forget that the apartheid state turned a blind eye to years of crayfish poaching by a Hout Bay fishing company, in a case that ended up in the New York Federal Court.[35] Would environmental compliance managers dare to send the army after corporates? We should also not forget that a fishing corporate made a play, in the era of "State Capture" under the corruption permitted by President Zuma, to take over the annual fisheries scientific surveys—which would be the fox policing the henhouse.[36] Where were the objections of the marine science community? Stock assessment science is already framed by the questions needed for commercial extractivism—which is not a science helpful for managing the ocean. There is very little state investment in sciences of ocean ecology. Indeed, marine researchers like those at the University of Cape Town struggle for funds: the Marine Research Institute is a research asset currently stranded by both the university and the government.

In her assessment of different forms of environmentalism in neoliberal societies, Elizabeth Povinelli argues that contemporary environmentalism takes shape around three figures that, like the rock-paper-scissors image used above, cancel one another out.[37] Her three figures are those of the desert, the animist, and the virus. The desert is the figure of the extractivist: the masculine hunter, miner, extractor who argues that sacrifices must be made for the greater good, and who creates zones of sacrifice. The animist is the figure that stands for romantic environmentalists who make arguments for indigenous knowledges or the magic and wonder of nature, and who try but regularly fail to win political arguments that might halt extractive activity because they do not offer "pure science." And the virus is the figure of those on whom policing efforts and attention are concentrated: the disruptors (often labelled "terrorists") who upset stability, and refuse the logics of both the desert makers and the enviro-romanticists. These three approaches, argues Povinelli, frame the dominant approach to "nature" in contemporary political life.

In South Africa, I see the poacher as a figure that plays the role of "the virus" in Povinelli's framing. In perlemoen poaching and in rhino poaching, the figure of the poacher serves as a lightning rod for public ire. Both are linked to a notorious "dark web"—an underworld linking them to aphrodisiac sellers in the Far East: arms dealers, drug traders, gangsters, and desperately poor Southern Africans who are willing to risk death by lions or sharks to earn some hard currency. In other words, the driving force in South Africa in both poaching scenarios is economic exclusion. The formal economy can easily, and apparently "morally," mobilise the military and police to defend dominant interests. Yet the "informal economy"—in this case, that of the excluded—is an ideal target for the underworld of poaching syndicates. All the militarised conservation in the world can deal with a symptom only if that core driver—economic exclusion—is addressed.

Ecological extinctions are tied up with economic exclusions. And the bulk of public attention to "illegal" environmental damage is focused on poachers: not the big-money game hunters, or the canned lion-hunting farms, or the mining companies who put community livelihoods at risk, or the municipalities who discharge untreated sewage into rivers or the sea.

In the Cape, abalone poaching is real; it is a problem that needs to be addressed. But to blame the figure of the abalone poacher for the extinction risks of abalone is at best a partial argument. It is not hard to see why the abalone poacher is a lightning rod for blame in the public, and in the scientific and military sectors: it is easier to blame individuals or even a whole population than to reckon with economic injustices and economic exclusions that began under apartheid and now thrive in the age of state capture, driven by a neoliberal economics that has deregulated finance to the degree that it has fostered offshore accounts in tax havens where corporates and the uber-wealthy have created their own illegal economy, and are poaching on taxpayers elsewhere. It is easier to bring public ire to bear on a small group with diving gear and a garage full of freezers than it is to address the causes of the current dysfunctional global economy that has created poverty by hiding billions away from government funds. And it is easier to attack independent scientists who research marine contamination than to take responsibility for the feedback loops among marine sewer outfalls, sanitation injustice, over-loaded waste-water treatment works, the selling of harmful chemicals, and persistent organic pollutants.

In one of her seminars at UCT, Isabelle Stengers delighted in telling the fable of the "twelfth camel," which goes like this: a father died and left half his camels to his eldest son, a quarter to his middle son, and a sixth to his youngest.[38] The problem was that the man had eleven camels. Not knowing what to do, the three approached a poor wise man for help. He said he had no knowledge to offer them, but if it would help, they could have his old, lame and blind camel. They accepted his generosity reluctantly—no one would want such a camel—but with twelve camels, they found they could give half to the eldest: six camels. A quarter of twelve meant that three camels went to the middle son. A sixth was two camels, which went to the youngest. That made eleven—and the blind and lame geriatric camel could go back to the old man.

Stengers's delight in telling this story is in the way that the old man's camel—useless to all—changed the staging of the problem, making the impossible resolution possible. Scholarship, she argues, can pose dilemmas in ways that remain stuck in the groove of intractability, or it can broaden the scope of the problems addressed in order to find what links them—what makes them capable of being thought through differently.

In South Africa at present, small-scale fishers and coastal communities bear the brunt of the blame for the commercial extinction risks of lobster and abalone. Scientists cannot understand why lobsters have moved. Harmful algal blooms proliferate with little explanation in Cape waters. Kayakers get ill. Small sessile organisms struggle with pharmaceuticals and household chemicals. It's a plethora of problems: a perfect storm of apparently unrelated difficulties that need studies and committees and court battles—or do they? Is there a twelfth camel?

What's missing from this picture is two struggles for justice in the Cape, both of which link to the ocean, and to dignity—the struggle to be recognised as persons and citizens. The first is the struggle for just, equitable access to fishing activity by coastal communities in which many are accused of poaching because they have not been allocated fishing quotas. The second is the struggle for just, equitable access to sanitation in shack settlements of the city of Cape Town, a struggle inseparable from broader questions about the city's discharge of sewage into the sea.

———————

The struggle for just, equitable access to fishing activity has been a long one. After the shift to democracy in 1994, the Marine Living Resources Act of 1998 sought to address historical inequities in the right to fish.

That law, signed into being by Nelson Mandela, affirmed the continued scientific management of the ocean via calculations of the "Total Allowable Catch" (TAC) and "Total Allowable Effort" (TAE) and it affirmed in law the methods of fortress conservation: the definition of marine protected areas, the establishment of a network of marine enforcement officials, a system of permits and quotas and methods of fishing. It criminalised non-compliance with the quota system as poaching, and defined categories of fishers in a manner that excluded small-scale fishers from economic participation. As neither "commercial" fishers involved in large-scale industry, nor "recreational fishers" who fished for leisure, nor "subsistence fishers" who fished only for their own households, small-scale fishers who wanted to fish for a living were not recognised in law.[39] The category "subsistence fishers" bound coastal fishers to a culturalist folk-biology model that was as limited in its theoretical use as it was limiting in everyday life: you can't pay a child's school fees with fish.

From a fisheries management point of view, the logic was that the three categories—commercial, recreational, subsistence—could both allow coastal fishers to fish (presumably for supper) and make scientific management possible via the allocation of sector-based fishing quotas. Broadly, the policy claimed to ensure that commercial fishing companies were no longer solely benefiting white owners and employees, in keeping with neoliberal post-apartheid policies that relied on businesses and big capital to drive redistribution.[40] However, "redistributive policies do not belong to the palette of neo-liberalism," observed Stefano Ponte, Simon Roberts, and Lance van Sittert in 2007.[41] In another article, Van Sittert notes that the state "has historically sought to advance and/or protect the symbiotic interests of big capital and science against popular demands for the redistribution and co-management of resources."[42]

Far from supporting the rise of a new conservation-minded fisheries sector, the arrests and convictions of subsistence fishers who were not compliant with those terms of the new law with which they didn't agree led to a loss of confidence in conservation and enforcement[43] as well as decreased compliance with legislation; increased hunger and the loss of food security;

increased poverty and dependence on the state; increased poaching; and diminished support for South Africa's ruling party among fishers.

A nine-year struggle to mobilise fishers to oppose the law, in the Western Cape, resulted in a successful challenge to the act in South Africa's Equality Court in 2007.[44] The judgment forced state fisheries managers to provide an "interim relief measure" quota to a new category of "artisanal fishers,"[45] a new umbrella category for fishers who had been excluded from the formal economy.[46]

Five years later, in 2012, a nationwide programme of activism and consultation among fishers headed by the non-governmental organisation Masifundise and the community-based organisation Coastal Links, along with several academic researchers working on environment and justice, led to the drafting of the Policy for the Small Scale Fisheries Sector in South Africa in 2012.[47]

Another year later, draft legislation in the form of the Marine Living Resources Amendment Bill of 2013 was released, based largely on the proposals in the 2012 policy. The Parliamentary Portfolio Committee for the Department of Agriculture, Forestry and Fisheries (DAFF) invited public comment to be presented in Parliament. Thus it was that an assembly sporting suits and T-shirts, chiffon and denim, balding heads, braids, and dreadlocks found themselves seated on the green leather benches of Committee Room V454.[48] Some of the same fishers who would speak here had chained themselves to the gates of Parliament a few years before.[49]

On the first day of the hearings, veteran fisheries activist Andy Johnston summed up the fishers' struggles:

> It has taken us the fishing communities six decades of struggle to get to this point of actually now being recognised as people who can take their rightful place in society. In the pursuit for fishing quotas there was a general tendency to tell a lie, live a lie and behave like a wolf in sheep's clothing, and this had and still will become the winning recipe for many opportunists, who were extremely successful in obtaining fishing quotas in order to enrich themselves . . . through this immoral individual allocation system that makes a mockery of ethics and righteous behaviour.[50]

What was remarkable about the hearings was the national response. An organiser of an Eastern Cape branch of the community-based organisation Coastal Links spoke in isiXhosa:[51]

> Honorable Chairperson and portfolio committee members, fishers of our beloved country, officials of DAFF, I greet you all. . . . I am Lulamile

Ponono from the clan of Tshawe, Togu, Ngconde, Madange kaTshiwo whose umbilical cord is buried in Cebe. My nearest town is Centane, under the Mnquma municipality in the Eastern Cape. Here I am going to speak as a fisher, whose parents were fishers, grandparents were fishers, and great-grandparents were fishers. . . . We fish as individuals but in a group and we share common rules not individual rules. It is always safe and less boring to fish as a group. Although I am from Cebe, I fish in the waters between Gqungqe on the north and Nxaxo on the south. I know fishers from Gqungqe and Nxaxo and they also know me. As fishers we have common experiences, aspirations and needs. We face the same conditions and challenges. Most importantly we have families to feed and communities to build.

From KwaZulu-Natal, Lindani Ngubane from Coastal Links made the case in isiZulu that oppressive laws such as the Marine Living Resources Act needed to be swiftly changed, and his community was happy that the bill would make changes that would liberate affected fisher communities. "Mr Ngubane," runs the official record of the meeting, "said that in Kosi Bay traditional fish traps were used . . . which allowed small fish to grow without being harvested. . . . Traditional techniques were also . . . conserving, managing and controlling the harvesting of the fish along that coastline [via these] 'fish kraals.'"[52]

Selene Smith from the town of Langebaan on the Western Cape coast spoke in Afrikaans:[53] "Our forefathers were born there [in Langebaan]. . . . In the old apartheid years we were taken from our own places where we lived. And now the sea has also been taken from us, where we fished for traditional fish for all the years."

She defended fishers' capacity to self-regulate: "We looked after our own fish. Our people had their own rules on how to catch fish. They can make their own rules among themselves: 'After Friday evening you don't go to sea any more, you can go to sea Sunday evening again.'"

She challenged the law enforcement: "Most of our people become criminals [because of these laws], . . . even if they get lost in the mist [and drift out of the demarcated zone] they get locked up."

She challenged the legal framework of environmentalism: "And I must say to you, people . . . we must have a hard look at the MPA [the Marine Protected Area] . . . because at the moment they are only interested in looking at conservation, that's all, but not at the livelihood and the social responsibility concerning the fishers."

She challenged science: "And I am sorry to say this to the Department: I sat at a round table . . . and I am sorry, but I just said to one of the researchers,

'I don't like your attitude at all, Mister, you must listen, you can't tell me [i.e., shut me up with the question] when last was the research done?'"

She challenged the idea that the tourist industry would benefit local communities: "People wring the tourist industry around the fishers' necks . . . and we are told 'We pay, we say.'"

She challenged the conservationists:

We, in Langebaan, to catch fish, we have to share it with thousands, millions of seals. And we cannot make a living. . . . But at the moment we respect the laws that are in place. But people, I need to say to you: what is valid for one must be valid for another. We are not allowed to fish in the MPA [Marine Protected Area]. But there are three whites who are allowed to catch fish there. And what are we supposed to call that? We must call it apartheid. . . . I cannot give it another name. . . . I grew up on a farm, and I must tell you this. When . . . the Nooi [Madam—reference to the farmer's wife] . . . asked me to come and sleep in her house when the baas and the kleinbaas [Boss and Little Boss—hurtful terms for "Master" under apartheid] were not there, I was not allowed to take my own mattress. I had to sleep on a hard mat under the kitchen table. I know apartheid. I know what it is to be hurt. I know tears. The same for all my community.

When fishers had heard that the Equality Court had ruled that they would be getting interim relief permits, she said:

We said [to ourselves], what we get, we will put in a pool and we will also help those who don't have. It is not all fishers who can make a decision like that. Because each must weigh for himself, that is why people are still fighting. But those fishers who decided that night [to work together] gave us a mandate to act on their behalf. We decided that the little that we had, we would share with other fishers . . . because they have experienced what goes on inside a house. They have experienced that what leads to hunger [in a policy] was affecting them too [*wat lei na honger toe het hulle honger ly*].

In a country whose racism has been framed by an obsession with defining personhood, humanity, and citizenship, and in a context where fishers had had to fight for the right to take their rightful place as people in society (in Andy Johnston's words), it was ironic that the focus of the debate in that Parliament of Fishers came to bear on the form of legal personhood that small-scale fishers should have. It was a tiny moment of infinite significance: what Selene Smith had described as the goal of setting up a co-operative, or an association of sharing, was now to be required to constitute itself in terms of established principles of corporate legal personhood, and be trained in the techniques and strategies of capitalist extractivism[54]—because the fishers' co-operatives needed legal personhood in order to be able to release capital investments and loans from the Department of Trade and Industry.

The principle of freedom from want that the small-scale fishers' activism had intended was being translated into shareholding. A fight for a commons translated into a right to corporate personhood. Their struggle to take their place as persons in the South African post-apartheid neoliberal economy now required them to reinvent themselves in the terms of *Homo oeconomicus*: as neoliberal economic actors, far from the co-operatives that Selene Smith had imagined with fishers in Langebaan. Decolonial theorist Sylvia Wynter's words ring clear:

> The human has ... been redefined, since the nineteenth century, on the *natural scientific model* of a *natural* organism. ... All the peoples of the world, whatever their religions/cultures, are drawn into the homogenizing global structures that are based on the-model-of-a-natural-organism world-systemic order. This is the enacting of a uniquely secular liberal monohumanist *conception* of the human—Man-as-*homo oeconomicus* ... (as if it is the *class of classes* of being human itself). ... We fell into the trap of modeling ourselves on the mimetic model of the Western bourgeoisie's liberal monohumanist Man. But ... what other model was there? Except, of course, for the hitherto neocolonially neglected yet uniquely ecumenically human model put forward by Frantz Fanon from what had been his activist "gaze from below" antibourgeois, anticolonial, anti-imperial perspective.[55]

A corporate structure, as a mode of organisation,[56] would translate fishers' proposed co-operatives into companies that would not level the playing fields with established commercial fishing companies, but would conform with the preferred model of the neoliberal state. This represented a lost opportunity for environmentalists and government to deepen the kinds of ecological concerns in the PSSFS, since a corporate structure offers very little in the way of useful scripts for values other than extracting resources. Integrated with the economy, but unintegrated with the ecumene or ecology, the proposed solution seemed to me, at the time, a betrayal of what had been fought for. Would care for one another's families flow from upgrading fishers' co-operatives to corporate structures? In what possible sense was the playing field level between fishing corporates and the new proposed corporate structures of small-scale fishers? How would the creation of companies alter the PSSFS values—addressing hunger in the community through catch-sharing—into a transactional relation? Why could relationships outside capital not be imagined? Was this perhaps in some way a replay of the way that the struggle of the enslaved for the right to freedom, in the early 1800s, was translated into the registration of ownership? Could a justice-based marine ecology only be managed in terms prescribed by capital? Where was the commons, in collective thinking? Could fisheries ecology be a matter not only of *rights over* other species, but also of *relationships with* and *accountabilities to* them?

That failure of imagination at the level of governance, however, was much discussed within fisher networks, and led to the development of an extraordinary partnership of scientists and justice scholars with fishers. Abalobi—the isiXhosa term for a small-scale fisher—is a mobile phone app that networks fishers, enabling information sharing about catches, commerce, and safety in real time. Built with open-source software after extensive consultations and workshops and partnerships with fishers, Abalobi takes fishers' concerns as the primary research problems the app must set out to resolve. Abalobi is described as "a co-designed and fisher-driven mobile-app suite to transform small-scale fisheries governance from hook to cook,"[57] and provides means by which fishers can link directly with markets and buyers and provide traceable seafood—"seafood with stories."[58]

Initiated by marine conservationist and justice-focused geographer Serge Raemaekers in the Department of Environmental and Geographical Science at the University of Cape Town, and supported by the ICT4D (Information and Communications Technology for Development) centre founded by the late Gary Marsden in the University of Cape Town Department of Computer Science, the project has leveraged the capacity of networked technology to challenge the hierarchies and gatekeeping associated with corporate structures, and make the supply chains that cater to big industrial fishing rather than small-scale fishers fairer.[59] Via Abalobi, small-scale fishers are able to sell their catches directly to restaurant chefs, consequently achieving prices that are higher up in the value chain.

It was "the realisation that local ecological knowledge can be a catalyst for ocean stewardship, for better fisheries management, [that] it's an asset for fisheries management, that spurred me on, together with a whole team, together with these fishers," says Raemaekers. "We never realised that providing fishers with a simple mobile log book on their smartphones could be such a powerful tool."[60]

Abalobi provides a means for fishers to log their catches in real time and on-site, and manage the sharing of that data. It includes a phone-based "wallet" for logging transactions and expenses, which, integrated with the co-operative, becomes a collective accounting tool. Fishers' knowledge and observations, logged and aggregated, give fishers for the first time a voice in scientific research and marine conservation: on the Abalobi suite of applications are tools for noting the appearance of unusual species, as well as ocean conditions, including sea temperatures and winds or signs of algal blooms. Signs of poaching or illegal fishing can be noted. A new "safety at sea" app has just been launched, which will enable fishers to share their locations and call for immediate assistance when in trouble. In the South African context, where a variety of seafood is available that is not commercially managed and therefore underutilised, the tool enables the marketing and management of different species—again building on fishers' knowledge. Abalobi, as a marketing tool, disrupts the existing hierarchical value chains that undercut fishers' earnings.[61]

Abalobi offers a view of the kinds of partnerships that are possible between management sciences and publics when the sciences slow down, think alongside, and take local concerns and values seriously. Most conservation scientists who are in pursuit of data, for example, would not consider including something like "safety at sea" as part of their research mandates. But, unmoored from traditional scholarly disciplines, and committed to providing a networking app that was what fishing collectives needed, Abalobi has done exactly that. Care for the fishers' lives, and care for their concerns, became the twelfth camel that Stengers looks for.

If this kind of partnership based on care, instead of quota, could be rolled out to include the entire South African coastline, it might well seed the kind of network that could enable care for ocean ecologies to flourish via those who are out on the sea in all seasons, instead of relying on a few teams of marine inspectors each expected to police hundreds of kilometres of coast.[62]

The Abalobi project offers the potential for the sort of paradigm shift that could seed the reframing of environmentalism in South Africa, outside of the current paradigm in which "nature" exists in a "reserve" that is set apart from all but the wealthy—and needing the kind of command-and-control relations described in chapter 5.

By contrast, Abalobi's use of science to pay attention to what people care about, they who live in and with and from an environment, is like the twelfth camel in a marine research science sector that is in crisis. There is no way that small and under-resourced, semi-militarised marine compliance offices, without any kind of adequate coast-guard-style navy, nor any sense of the kinds of relationships that are meaningful, can manage to police the coastline, not only because the scale and extent of the South African coast make it impossible to police, but because the kinds of command-and-control relationships that inhere in policing make it impossible for inspectors and their families to be part of the small coastal towns they live in. Marieke Norton's brilliant anthropological study of the experiences of marine compliance officers in a range of Western Cape coastal towns provides ample evidence of this point.[63]

Including fishers in ecological management, by contrast, shifts the relationships of compliance from a command-and-control orientation to a partnership. I'm not suggesting that this is a perfect intervention that would solve everything. But because it shifts relationships from control to co-management, the potential is there for citizens to take a primary role in reducing poaching.

The app situates fishers as producers of knowledge, logging their observations (sea temperatures, marine species' well-being, algal blooms, or lobster beaching) in the co-ordinates of space and time used by scientists. As such, Abalobi serves to bring fishers' knowledge into dialogue with that of scientists.

Abalobi also creates an economic network with entry points into the food value chain that have until now been controlled by corporate fisheries. It doesn't level the playing field; it makes its own. It makes it possible for fishers to supply buyers with species that are not part of the two dozen managed fisheries, potentially taking pressure off those species. And it minimises the likelihood that fishers will operate "under the radar" as "poachers."

The coming risks that fishers' co-operatives will face are many, however. They include higher ocean temperatures and consequent changes in both marine ecology and ocean currents. Increasing chemical and pharmaceutical contamination of the two oceans around the city of Cape Town, and other towns, will continue to affect fish and the small sessile creatures that inhabit rock pools.

Of particular concern is the effect of the post-2012 ocean regime shift that gave coastal countries the ownership of an "Exclusive Economic Zone" that pushes two hundred kilometres out to sea. Designed with the idea that countries would better be able to steward marine ecologies if they had ownership of them, what the Exclusive Economic Zone regime seems not to have foreseen was the headlong rush to deep-sea mining. Within a very short period of time, the entire South African Exclusive Economic Zone was parcelled out to corporate extractivists. Deep-sea mining, including deep-sea fracking, puts the South African ocean at greater risk of major chemical accidents and other potential damage. Such events would be catastrophic for the hard-won rights and emerging networks of small-scale fisheries. In this situation, small-scale fishers' care for the ocean and its ecologies is among the most important resources for building sciences of ocean observation that can bring challenge and change.

The Abalobi partnership is vital because it brings small-scale fishers into the process of knowledge production about the ocean. It offers a roadmap for an alliance of marine scientists with fishers who care about the ocean and its algal blooms and chemical pollution, and could be a vital antidote to officialdom that continues to declare that the marine sewer outfalls are sufficiently diluted by the ocean, and therefore are not a problem.

As the research work led by Leslie Petrik on ocean chemical pollution from sewage outfalls progressed, the findings of doctoral researcher Cecilia Ojemaye showed that pharmaceuticals were bioaccumulating in the food chain all the way up to larger fish that, historically, fed early Cape Town and continue to be an extremely important part of the city's food supply.[64] From the abstract:

> A comprehensive analysis of 15 target chemical compounds (pharmaceuticals and personal care products, perfluoroalkyl compounds and industrial chemicals) were carried out to determine their concentrations in selected commercially exploited, wild caught small and medium sized pelagic fish species and their organs (*Thyrsites atun* (snoek), *Sarda orientalis* (bonito), *Pachymetopon blochii* (panga) and *Pterogymnus laniarius* (hottentot)) obtained from Kalk Bay harbour, Cape Town.... The results revealed that perfluorodecanoic acid, perfluorononanoic acid and perfluoroheptanoic acid were the most predominant among the perfluorinated compounds.... Diclofenac had the highest concentration in these edible fish species out of all the pharmaceuticals detected.... The risk assessment values were above 0.5 and 1.0 for acute and chronic risk respectively which shows that these chemicals have a high health risk to the pelagic fish, aquatic organisms and to humans who consume them. Therefore, there is an urgent need for a precautionary approach and the adequate regulation of the use and disposal of synthetic chemicals that persist in aquatic/marine environment in this province and other parts of South Africa, to prevent impacts on the sustainability of our marine environment, livelihood and lives.

The article was widely reported in South African media,[65] but the response from the responsible official, Xanthea Limberg, was misleading. The study shows, she said on radio, "that the concentration of pollution in our marine environment is what we term at trace levels, meaning you would need to consume about 147 000 kgs of snoek or 60 000 whole snoek at a single sitting in order to ingest the equivalent of one Dicloflam tablet as an example." Her analysis, however, confused wet weight of fish with the dry weights used in laboratory science. Moreover, it had nothing to do with what the therapeutic dose is of one pharmaceutical: what ought to have been of concern to the responsible official was the finding in relation to acute and chronic risk levels. Further, Limberg's response ignores the big picture which is that the presence of diclofenac and the other chemicals demonstrates that chemicals and pharmaceuticals were not being adequately removed from sewage by Cape Town's water treatment facilities. Limberg's claim, in the same interview, that "No conventional wastewater treatment plant in the world can remove all those compounds from effluent," is deeply misleading, since the issue is that Cape Town does not yet offer comprehensive tertiary treatment, even though tertiary treatment facilities, including in the Western Cape town of Beaufort West, have been shown to significantly reduce these compounds.[66]

Much as US environmental scientists face dealing with official recalcitrance, reputational harm, obfuscation, distraction, "whataboutery," and outright lies under the Trump "Make America Great Again" presidency, Cape scientists tracking ocean pollution similarly have to deal with under a city administration determined to claim that their party, the Democratic Alliance, runs local government better than any other municipality in the country. Taken together with her distortion of the findings of the CSIR report on marine sewer outfalls cited earlier in this chapter, Limberg's denials and distractions prevent the problem from being addressed. Yet the continued discharge to sea of millions of litres of sewage in Cape Town's urban oceans is a slow violence that affects fishers directly, not only in the pharmaceutical and chemical bioaccumulations in locally caught fish, but in damage to the marine environment, including the marine reserve.

Cape Town can no longer afford to be a city that does not care about the sewage it sends to sea, has huge shack settlements without the dignity of sewage management, and has peri-urban farms that use the kinds of fertilisers that foster harmful algal blooms. Each one of those situations is premised on the absence of care.

Restoring care to science, as Abalobi has done, holds out the possibility for reimagining environmentalism in a way that restores the relationships that were broken by the advent of what Césaire described as "thingification" under colonialism and modernist thought.[67] "Thingification," in South Africa, has done enormous damage to people, species, oceans, and landforms. Restoring relations of care as central to the projects and questions of scientific research, as Abalobi has done, is essential to transforming science into a means for addressing the devastations that have yielded the Anthropocene: devastations that have resulted from treating atmosphere and oceans as trashcans, and people and creatures as objects. Attending to justice and histories of injustice, and attending to the challenges of restoring relationships between people, species, oceans, and landforms, will yield a transformative environmentalism capable of being supported by a broad-based public.

Urban sanitation affects the health of marine species, as well as fishers' well-being. The flows of water from city to sea are part of marine ecology. And an alliance of those to whom this matters holds the potential to make an ocean regime shift possible. Perhaps in time the story will be told of how Abalobi's alliance of fishers and scientists provided the leadership that enabled South African environmental science to reimagine itself at a critical juncture in the country's history.

COMPOSING ECOPOLITICS

The concept of "natural science" makes an unnatural claim: that it exists outside society.

That claim has been taken up enthusiastically by the environmental governance system adopted by post-apartheid policy-makers in South Africa eager to avoid the accusation of racism and associated injustices with which the country has wrestled. "Expert scientific consultancies" designed or hired by environmental managers have all too frequently presented their findings in a context that claims to be context-free. Even worse, many environmental managers and scientific consultants who aim to be funded or hired by them collude in their agreement that publics know little about science and are therefore incapable of playing a role in decision-making.[1]

While many academic researchers have sought to address both science and society in their work, the yawning gap between social science approaches and the conceptualisation of "social ecological systems" that are preferred within the environmental sciences has rendered unworkable many a potential transdisciplinary research partnership. Frustrated, anxious, and discouraged, the majority of environmental research scientists can only shake their heads at a social science that they see as chasing research questions that would not exist without a functioning ecology. And on the other side, social scientists find it impossible to work within a research paradigm that declines to challenge the powerful.

Trying to intervene in policy decisions, then, research scientists who want to do "socially relevant science" have uncritically accepted the terms of economic growth proposed by post-apartheid political leadership, and offer "ecosystem service" assessments in dollar values, hoping that this will lead to social change.

The chapters in this volume have sought to highlight the intractability of these approaches by digging wide and deep in specific fields in order to try to illuminate possibilities for transformative, transdisciplinary research questions.

The work of the pure sciences has been focused on designing ever more specific directional microphones that will do better and better at recording specific sounds in order to score them for a proverbial piano solo: all too often, reduced by specific methodological insistences to scoring them in C-major (all ivory; no ebony), and, at least in militarised approaches to conservation, setting them to a 4/4 marching time.

My goal as an anthropologist (perhaps, rather, an "anthropocenologist") has been to record every instrument and every sound I could find in each field, including those at a distance, and those that I could find in forgotten archives, on instruments barely remembered. I sought to understand proverbial rhythms and improvisations, modes of expression, and relationships—not always harmonic—between the musicians. The goal was to compose each chapter as a text that could strengthen the capacity of both natural and social scientists to grasp the polyphonics of each situation.

Five proposals for viable, transdisciplinary, and transformative scholarship are set out below. They sum up what I currently see as the issues that will set out the terms for a different environmental conversation, within the public sphere, in consultancies, and in research. The five are not claimed to be an exhaustive list, nor is my account of them offered as definitive. My hope is that they might serve as conversation-openers for the innovative scholarship that is needed in universities where the geohistory of the global South encounters colonial roots.

I | RESITUATING SCIENTIFIC AUTHORITY

Embarking on in-depth, detailed, big-picture research in six fields of environmental management in 2012, I had no idea what I would find. What I did know was that the environmental debates around me were intensely conflicted. Table Mountain's water, Karoo fracking, South African struggles over scientific authority in teaching and research, the lack of connection between land restitution and environmentalism, urban baboon management, urban ocean management, and fishers' struggles—each of these was interwoven with struggles over the right of scientists to speak with authority in matters of environmental governance. It seemed to me urgent to address the issues of why and how scientists and environmental officials were so often accused of racism, because without addressing that, South Africa would have difficulty in transitioning into an environmental democracy.[2]

However, much as I wanted to support the work of environmental governance, officials' claims to have used only "the best available science" were all too often accompanied, in the name of political neutrality, by a combination of moralism, authoritarianism, polemic, and dismissal. These are not attitudes that I associate with either excellent science or leading scientists, for whom the best science is always available for questions, discussion, and improvement.

Most disconcerting was to see so little attention, in environmental managerialism, to its entanglement with historical injustice—by definition, racist and patriarchal.[3] While extraordinary work was being done by the South African division of World Wide Fund for Nature with Saliem Fakir as its head of policy, and by geographers like Rachel Wynberg, on seed sovereignty; Merle Sowman, Serge Raemaekers, and Jackie Sunde on fisheries justice; chemist David Gammon and botanist Timm Hoffmann on plant medicine; process engineer Harro von Blotnitz on urban metabolism; Jane Battersby on urban food systems; and Maano Ramutsindela and Frank Matose on race, capitalism, nature, and the militarisation of conservation, similar commitments to addressing justice and inequity from the ground up were evident in very few of the environmental consultancies that advise government.

While the majority of my South African colleagues in the pure life and earth sciences are doing extraordinary research on extinctions risks and environmental change, I remain puzzled at the collective silence of formal science on intensely contested fields of environmental management. The response from five of the top City of Cape Town water management officials to my research team's challenge to city water quality data taught me how vulnerable scientific researchers may be to environmental managers who, amid party political struggles, may choose to attack researchers' reputations than countenance contrary data.[4] Where party political survivalism dominates, "Don't rock the boat!" becomes for many professional scientists a career survival strategy. The attempts to silence scientific study of sewage pollution of Cape Town's rivers and oceans is not far from the struggles over science depicted in the hit series *Chernobyl*, which dramatised the intense struggle between scientists and Soviet officials who were determined that the regime was incapable of imperfection.[5]

The shutting down of evidence of imperfection is apparent also wherever research scientists are reluctant to challenge the neoliberal claim that financialised economics provides a neutral meeting place for science and society. Dollar values on "ecosystem services" are an expression of fact, not value.[6]

Even though there is a real sense among many researchers that this is so, there is a general, if reluctant, acceptance that "this is the way things are" and that there is no other game in town. In the country with the greatest Gini coefficient in the world,[7] the charging of high access fees by SANParks in Cape Town (see the introduction) undermines the emergence of a broad-based environmental public. It is not a politically neutral decision, and it may well, in the long term, undermine the cause of environmentalism. Why is it not questioned by scientists who are desperate to defend Cape biodiversity?

Ideas about how game theory, an unproven approach drawn from behavioural economics, should be used to give effect to scientific findings was equally troubling. "People have got to be forced" was a line I heard often. "How can you have an environmental democracy if people don't want change?" Pointing out that trying to compel behaviour change through fines and arrests would cause a backlash against environmentalism, would generally yield a resigned shrug. "We don't have other options," they would reply. Few of my colleagues, so horrified by the stark data on climate change and extinctions risks, could see that command-and-control relations would entrench an authoritarian relationship between science and society (chapter 1) and score one own-goal after another. Arresting and fining fishers, for example, had built a culture of noncompliance, and, once criminalised, few had much stopping them from joining up with the poaching gangs (chapter 6).[8]

Where the vast majority of professional environmentalists are white, there is a risk that the modes of "whiteliness"[9]—the expert, the judge, the martyr—render environmentalism particularly vulnerable to what the poet A. R. Ammons called "the darkest swervings of the deepest heart."[10] Marilyn Frye explains "whiteliness" as follows:

> Minnie Bruce Pratt, a feminist and a white southerner, has spelled out some of what I would call the whitely way of dealing with issues of morality and change.[13] She said she had been taught to be a *judge*—a judge of responsibility and of punishment, according to an ethical system which countenances no rival; she had been taught to be a *preacher*—to point out wrongs and tell others what to do; she had been taught to be a *martyr*—to take all responsibility and all glory; she had been taught to be a *peacemaker*—because she could see all sides and see how it all ought to be. I too was taught something like this, growing up in a small town south of the Mason-Dixon line, in a self-consciously christian and white family. I learned that I, and "we," knew right from wrong and had the responsibility to see to it right was done; that there were others who did not know what is right and wrong and should be advised, instructed, helped and directed by us. I was

taught that *because* one knows what is right, it is morally appropriate to have and exercise what I now would call race privilege and class privilege.[11]

The observation that green policing was also offering a space for some white authoritarians to reassert themselves politically in the name of their claim to an unquestionable knowledge of nature was not open for discussion. Scientists seemed to fear that opening scientific authority to discussions of its occasional appropriation by racialised power might unravel democracy itself—without recognising that a "green white" environmental authoritarianism would achieve that anyway.

"I can't help that I'm white" was an inevitable response to my attempt to open a discussion; "the science I do has nothing to with my race. I am not a racist."[12] Thus individualised, the conversation would shut down.

Apologising for having the best advice—"I'm sorry to speak, I know I'm white, but actually the science is solid"—was an alternative response. It too was a conversation-stopper, preventing scientists from hearing what concerned those who opposed them.

What alternatives were there that might open the necessary conversation?

I had to struggle deeply with my own white shame and an ingrained habit, as a professional academic, of speaking before I had listened. What I learned amid student protest was that none of the students in my classes wanted my shame. What they wanted was the dignity of being heard in regard to present struggles, pasts that are present, and futures that could portend no better if knowledge formations were unable to change. I had to learn to listen differently.

Whiteness, as I now understand it, does not necessarily inhere in a white body; rather, it inheres in the fiction that being white accords one the authoritative voice in a situation. The political affects of expert/judge/martyr/world-saver foster a way of being in the world that is so certain of its political neutrality in bearing absolute universal truths to the world that its practitioners are blind to the struggles of black environmentalists who, on account of their opposition to injustice, were defined as "political" and therefore not embraced as allies in the struggle for an ecological future. If Whiteness is a club that limits the world to its knowledge alone, the work that is needed is how to get oneself removed from the club.[13]

In the life sciences, climate sciences, and engineering, very few academics recognised in black political struggles for land restitution a struggle to live from ecology as well as economy (chapter 4).

You can't cut generations of people off from the land on the basis of their race and then judge them for not being environmentalists in the way that you define environmentalism.

It was alarming to think that perhaps what I *was* seeing was that the framing paradigm accepted by environmental managers and contract researchers, reluctantly or not, was undermining the possibility that an environmental democracy would emerge to address climate crisis, pollution, biodiversity loss, and extinctions risks. Instead, consultancy advice on environmental governance sailed close to the winds of injustice—winds that, precisely because they weakened democratic processes by proposing command-and-control relations, put at risk the good ship *Ecology*.

The problem is not solely South African. Similar dynamics pertain in every "environmental mission" in former colonial countries where the assumptions of what it means to be an environmentalist exclude the voices of those who struggle for an ecological relation with the earth that was broken generations before. My choice of the word "mission" is deliberate, for the languages of saving souls and saving the earth tend to be very similar, except where one discourse describes saving the world with Jesus, the other offers GIS (Geographical Information System). Whether technological or spiritual, when the narrative of saviours becomes a performative script, one represents one's work as transcendent, neutral, and universal.

Rethinking the necessity for environmentalism to be politically neutral is vital.

Black environmental movements are many in South Africa, but few are recognised as such by predominantly white South African environmentalists (academic and otherwise) because they are struggles for land; struggles for sanitation and water; struggles for health; struggles for an affordable and nutritious food system. They may not directly address the extinctions risks that concern white conservationists, but the issues that they are fighting are local expressions of the damage to earth systems that define our era.

Alliances are everything.

Environmental publics and research scientists concerned about extinctions risks and loss of biodiversity could offer vital support to environmental activists like Nonhle Mbuthuma and Sinegugu Zukulu and Davine Cloete, who struggle against destructive titanium sand mining at Xolobeni and Lutzville (chapter 4) along with a supportive network of lawyers.

Activists fighting for agro-ecological reformation such as Nazeer Sonday and Susanna Coleman with the Philippi Horticultural Area Campaign (chapter 4), are working to protect Cape Town's farmland and its underlying aquifer: a cause that is in the interests of every Capetonian.

Struggles for healthy urban rivers and beaches (chapters 1 and 6) challenge water officials and the environmental managers who propagate the falsehood that sending sewage to sea has no effects on marine ecology.

So, too, baboon managers and scientists consulting to the City of Cape Town may find it useful to spend less energy dismissing those who care deeply about baboons, and instead develop research partnerships with residents that address questions of how baboon and human social lives might generate a multispecies neighbourliness in this "anthrozoological metropolis" (chapter 5).

Anti-frackers may find their environmentalism far more potent if they abandon the false claim that the Karoo is pristine, acknowledge the genocidal wars perpetrated on the !Xam and Khoena ancestors of contemporary farmworkers, and instead work for a socially transformative, justice-based ecology of land, water, energy, and rewilding (chapter 2).

Engaging with student objections to the authoritarianism experienced in faculties of science can offer environmental scientists a route into understanding how and why scientific advice is distrusted in so many government fora (chapter 3). So, too, working with the concepts and relationalities within traditional knowledges may offer a route to innovative research questions (chapter 3).

The first principle that I have learned from researching the aforegoing chapters, then, is this: to build an environmental public, encourage debate on matters of scientific authority. "Science for hire" has for too long been allowed to provide "science says so" lines in corporate, city, and government press statements, without peer accountability to the institution of science. Second, the embrace of behavioural economics by ecologists is not social science, and its assumptions about human motivation have little empirical grounding. *Homo sapiens* is not *Homo oeconomicus*. Third, command-and-control relations have yielded antipathy if not downright opposition to the sciences. Based on a misdiagnosis of the problem, the treatment has all too often exacerbated, not solved, ecological problems. Fourth, the conviction by environmental scientists and managers that their work is apolitical is not empirically valid. Acts of omission are as important politically as acts of commission. To ignore a historical context, or to turn a blind eye to the use of science to perpetuate injustice by those who join the dots that scientists decline to join, is to make a political choice. Perhaps it is time to abandon the idea of "environmentalism" and instead name and claim the field for what it is: ecopolitics. Then its terms, methods, and approaches will not claim transcendence, and will be less open to abuse by unscrupulous officials in cities, corporates, or national governance.

South African "state capture" by corrupt politicians, officials, and corporates was aided and abetted by accountants who abused their professional neutrality. In the process, many institutions of democracy have been compromised, as they have been under neoliberal governance principles in many other countries. One of those institutions vulnerable to capture is science. Environmental scientists must stand up and defend the abuse of scientific integrity by corporates and officials. For an ecological society is all but impossible as long as some spaces are, by virtue of their physical distance from the homes of the powerful, represented as *terra nullius* for extractions, expulsions, extinctions, or excreta.

II | TO DISASSEMBLE THE NATURE/SOCIETY DIVIDE, REPLACE "SOCIAL-ECOLOGICAL SYSTEMS"

Extractivism is made possible by the embrace of the binaries in modernist thought that conceptually separate society from ecology, life from non-life, and subjects from objects. There is no space on this planet that is outside a living system. The nature/society binary must be recognised for what it is: an untenable fiction.

Dominant in environmental management consultancy work is the approach known as social-ecological systems thinking, absented from which are questions about whether its core conceptualisation of "social" and "ecological" are actually the separate spheres that it claims it will connect.

The crucial problem is that the lens of social-ecological systems enables scientists to imagine that there are social wholes that operate as systems which impact on nature. At that point all manner of troubles enter the picture. Most damaging of all is the conflation of "the social" with one side of the modernist binary: that of "the mind." Social-ecological systems projects are notorious for requesting of social sciences endless studies of perceptions of reality. The research I would rather do, as a social scientist, is to work with peoples' concerns and insights to reframe the research questions, so that the research may have a possibility of generating a transformative social process. I've been told more times than I care to remember that such an approach is unable to respond to the urgency of the situation.

With data generated by compliant social science that will not question whether what is being counted is what counts, environmental managers allow themselves to proceed, believing it sufficient to have commissioned a "stakeholder perception" study. Such studies neither change the questions the hiring scientist asks, nor do they problematise the presumption that environmental science is inevitably in a command-and-control relation with society.

For environmental managers and conservationists who do the hiring of such consultancies, social science may be presumed solely to assist with science communication.[14] As one fisheries specialist put it in a social science planning workshop that I attended, "What is needed was social science that will assist with education, regulation, and enforcement." In other words, the function of a social scientist was to teach fishing communities what "the science says," design strategies to achieve compliance with that thinking, and advise on how to arrest and prosecute those who disagreed. Besides asking social scientists to be complicit in reproducing a violent relation between science and fishing communities, this approach was also doing violence to the sciences: presenting them as if the research being generated by marine ecologists and the fisheries mathematicians and the environmental justice geographers is of one accord, which it is not.

Since environmental managers seldom have any training in conceptualising social justice, their lack of social science training and reluctance to engage social science questions may lead to the sort of elementary mistakes that risk catastrophic consequences for their own data. Such a situation occurred when for five years the lobster catch take by recreational fishers was provided by a social science research consultancy that (a) relied on telephone surveys to ask how many lobsters recreational fishers had caught (a basic methodological error: a telephone conversation and a confessional have different truth conditions), and (b) only telephoned people who had in the previous two weeks received a catch license. These errors took five years to detect in an annually peer-reviewed science programme that not once included peer review of the contracted social science, evidently assuming that social science data required so little skill that no scholarly oversight was needed. When discovered, the catch data required a major backdating correction. Correcting the Excel spreadsheets, however, would not put back into the oceans the excess catch of lobsters that are at risk of commercial extinction.[15] Good social science matters.

The presence of an "Anthropocene" as a geological category demonstrates that the division of society from nature is an empirical failure. Why then do so many social and natural science research project methods and questions remain framed as if this were not so? The urgency of climate crisis and extinctions risks necessitates that universities resource the emergence of new forms of research practice.

What integrative research approaches will contribute to unmaking this "Anthropocene"? A number of approaches have arisen in the social sciences and humanities that are more useful than social-ecological systems thinking.

Beginning in the 1990s, multispecies approaches have enabled conceptualisation of the entanglement of human lives with those of other species. The human body depends on relationships with multiple species beyond the obvious need for food species of plants, fungi, and animals. Humans cannot live without a healthy gut biome made of diverse microbes. Humans could not have evolved without oxygen-producing ocean microbiota which changed the atmospheric concentration of oxygen. Companion species studies—horses, dogs, cats, cows, plants—challenge the conceptualisation of society as comprised solely of human relations, and feedlot studies have testified to the rise of capitalism's brutal relations with other species. Studies of the simultaneous rise of plantations with slavery, and the concomitant changes in landscapes and their impact on wild species, have led Anna Tsing and Donna Haraway to explore whether the Anthropocene may be better described as the "Plantationocene."[16] Other key studies focus on the impact of biogeochemical flows on non-human species, such as Thom van Dooren's study of the impact of the commonly used anti-inflammatory drug diclofenac on vulture extinctions in India.[17] Taken as a whole, multi species research focuses attention on the ways the nature/society divide makes it difficult to see or comprehend relations with non-human species, and their relations to each other. Contesting human exceptionalism, the field challenges both natural and social sciences to rethink their core analytical framings.

A second gap in social-ecological systems thinking is that it implicitly overlooks infrastructure, technology, and engineering, and therefore offers little with which to conceptualise the effects of fuels, machines, and built environments. A number of approaches have emerged in social sciences and environmental humanities to address this. Reflecting on the challenge of climate crisis to generate research at a planetary scale, Tim Morton offers the idea of the *hyperobject*.[18] The concept encompasses effects at scales that exceed the perceptual limits of individuals, disciplines, or even regions. At a planetary scale, Peter Haff's concept of the *technosphere* proposes that contemporary human technologies operate on the planet at the scale of atmosphere, hydrosphere, cryosphere, and lithosphere.[19] McKenzie Wark's *Molecular Red* traces matter, molecular flows, and collective labour in the work of writers addressing capitalism and socialism.[20]

A third gap in social-ecological systems approaches is that in its conviction that ecological and social systems are actors and agents whose intersecting links are to be understood, it is not well equipped to conceptualise constant, slow accretions of material flows. This places out of sight the geological accretions effected by processes of commodity extraction, production, distribution, and detritus: a field that is brilliantly drawn into view in social science approaches that focus on matter and materials.

There are many valuable approaches that have emerged and almost all have informed the preceding chapters. A brief overview of the field of *new materialism* in social sciences links well to emerging fields in geology, such as the study of anthropogenic rock like cement and tar and plastics in new volcanic magmas. In soil science, the study of anthropogenic soils similarly explores the physical accretions of urban society,[21] as does environmental chemistry in studies of persistent organic pollutants and other toxins in soils, water, and living creatures.

The relatively new field of *biogeochemistry* similarly explores the material flows between living creatures and the earth's geology. The relation described in chapter 2 between the termites that ended the carboniferous era by raising atmospheric carbon with their posterior methane emissions is a good example. So, too, the ocean microbiota that fill the earth's atmosphere with sufficient oxygen for life to thrive.

Following molecular and species flows in the social sciences in ways that are grounded in environmental chemistry and mindful of environmental justice and the necessity of thinking via the commons opens innovative ways to improve the empirical observations of natural sciences and the basis of environmental regulation.

One of the generators of the geological Anthropocene has been in allowing the conceptual inventions of territory and property law to supersede the empirical reality that life subsists within a commons that is shared by all people and all species. There is no Planet B, as the saying goes: Earth is a *commons*. The current planetary emergency requires environmental regulators everywhere to prioritise the commons over private land ownership. Property and territory law, for example, has no impact on fluid flows or atmospheres or the movements of certain species, which flow, dissolve, are ingested or inhaled, and move or migrate regardless of legal regimes. Following molecular flows through bodies and across legal boundaries makes it possible to see the irrationality of the ways in which the idea of private property and territory have taken form in fracking regulation (chapter 2), baboon management (chapter 5), or a sewage management policy that assumes that an urban ocean or an urban river can absorb city-scale effluent without any effects on its algal blooms or living creatures (chapter 6).

Closely allied to bioprocess engineering (the treatment of waste) is the field of *industrial ecology*, which traces the flows of industrial products and other urban excreta. Similarly, the field of *urban metabolism* (chapter 6), recognising that nothing goes nowhere, specifically looks at the cycle of waste from human bodies and manufacturing and consumption within the water and soil and waste streams of built environments. Having developed within engineering and regional planning, it builds closely on Karl Marx's observations that capitalism, by expelling small-scale farmers and animals from land in the name of agro-industry, disrupted soil metabolism—and is therefore ultimately unsustainable.[22]

Working on waste, *The Anthrobscene* by Jussi Parikka addresses the geological effects of the media era, in particular e-waste, and similar scholarship follows the flows of plastics, concrete, and other commodified substances of late modernity.[23] Vital work is necessary on technopolitical injustice, in regard to the excreta of commodity society—such as chemically and pharmaceutically contaminated sewerage; plastics; e-waste and mine dumps in particular—from rich urban areas to poorer urban areas and from the global North to the global South, both of which are traversing routes that intersect with racism. For consumer society's excreta are far more likely to be situated alongside black communities than white. Pieter Hugo's *Permanent Error* portrays the Agbogbloshie e-waste dump in Ghana, West Africa. Agbogbloshie is the physical manifestation of the ideas in a leaked World Bank memo signed by Lawrence Summers, who was then the institution's chief economist: "I think the economic logic behind dumping a load of toxic waste in the lowest-wage country is impeccable and we should face up to that. . . . I've always thought that under-populated countries in Africa are vastly under-polluted."[24]

The *energy humanities* has also become a major player in environmental humanities. The notion of energopolitics, developed by Dominic Boyer, and the book *Carbon Democracy* by Tim Mitchell look at the entanglements of power-producing companies with political power in democratic societies.[25] Eduardo Gudynas's work linking extractivism with corruption, under neoliberalism, is a major contribution.[26] Much of it has yet to be translated from Spanish, but it is exceptionally pertinent to debates in the global North and the global South about understanding and reframing the era that geologists call the Anthropocene.

Arundhati Roy's essays on *extractivism* in India in *Capitalism: A Ghost Story* describes this era of neoliberalism—the convergence of market and state; the privatisation of commons—as an era that promises wealth distribution by trickle-down, but has instead created an elite paradise via the practice of "gush-up."[27] Recent work published in *Nature* on the relation between environmental destruction and tax havens has demonstrated the ways in which the practice of creating shell companies enables taxes and wages in the extractivist sector to stay low, and environmental looting to be hidden from the formal economy.[28]

The field of *critical zone studies*, a field that has grown from an initial base in soil sciences, addresses the effects of extractivism on the zone of life on Earth:

The term critical zone addresses the critical dependence of humans on the terrestrial surface layer, which is under critical pressure from human activity and is a critical interface of Earth planetary systems. The critical zone exhibits dominant material and energy fluxes governed by land-atmosphere, soil-vegetation, shallow geosphere-lithosphere, and land-ocean interactions. The role of soil in the processing of energy, material, and biodiversity is essential to the critical zone provisioning of life-sustaining resources.[29]

While critical zone studies, as articulated above, diagnoses all humans as the trouble, it has not yet acknowledged the argument that this era is not only an Anthropocene, as geologists describe it, but a Capitalocene (the phrase is that of Jason Moore).[30] Moore's book *Capitalism in the Web of Life* explores the ways in which financialised relations, aimed at profit extraction, have come to redefine nature, including human nature, placing profit-making (and therefore capitalism) in nature, not outside nature.[31] Linked to this is the work of anthropologist Anna Tsing, who traces wild-picked mushrooms through global markets to theorise contemporary relations of damaged soil and life under capital.[32] This work builds on her earlier theorisation of frontier forests in Indonesia in the 1980s: forest destruction that is an effect of the rise of neoliberal philosophy of unlimited growth at that time.[33] That policy was pursued notwithstanding the cautions by the Club of Rome in the 1970s that unlimited growth was a catastrophically flawed idea: an idea on which several scholars are writing, including Jason Hickel, Kate Raworth, and winners of the Nobel Prize in Economics.[34]

Unmaking the geologists' "Anthropocene" requires more than attention to financialisation, however. It also necessitates attention to the *mythopoetics of human mastery*, as it is expressed in so many political myths and affects of what it is to be human, and to have knowledge. Viewing it from the South, I understand the Anthropocene to be what happens when coloniality has exported modernist thought and capitalist extractivism to the ends of the earth. Kathryn Yusoff's extraordinary work *A Billion Black Anthropocenes or None* underscores this argument: black communities have experienced Anthropocene-like conditions for a very long time.[35]

If the Anthropocene is understood as an effect of modernity-coloniality in its division of nature from society, then the work of decolonising knowledge is specifically to address the problem of objectification (for Césaire, thingification),[36] and its attendant methodological assumptions that data alone can remedy earth systems that have been artificially carved up into human and natural worlds. Any retooling of knowledge must attend to the relationalities, flows, and movements that make the knowledge in and of the era geologists call "the Anthropocene," and social science, the Capitalocene.

For philosopher Bernard Stiegler, negating "the Anthropocene" as a concept and as an earth system requires unmaking the concepts that have made both possible. He argues forcefully for unmaking the ways in which data production has rendered the effects of capitalism invisible, and unaccountable, in the knowledge economy.[37] Fractured and fragmented knowledges, he contends, are unable to unmake the present relations between capitalism and Earth.

Recognising the conceptual limits imposed by the knowledge economy on the questions posed by scientists and social scientists alike is necessary if research questions are to address the conditions that have generated this Capitalocene that geologists have named "the Anthropocene." The time frames of the knowledge economy, for example, are typically financial, in respect of either an annual financial cycle or the expected economic life of a "natural resource" that is targeted for extraction. But these are artificial time frames that are at odds with the permanence of detritus such as nuclear waste and fracking waste fluids. The claim that effects can be managed by environmental law when those effects can last into the hundreds of thousands of years (in the case of nuclear waste) or permanently (in the case of fracking fluids) is a myth that has nothing to do with "nature." It ought to be identified as such and struck from the list of "accepted truths" in environmental science and law.

Attending to the logics and rationalities for knowledge production that make a world gone mad seem perfectly reasonable—Amitav Ghosh calls it *The Great Derangement*[38]—opens up a broader palette of questions than biogeochemistry or urban metabolism. In the conviction that formal academic disciplines must step into the ring from their particular corners of knowledge-making, the humourist in me proposes a thirty-eight-letter unhyphenated conglomerate: "capitalothermobiogeochemicomythopoesis" to address relations of capital with heat and biogeochemistry, and the myths and rhetorics that enable people to continue living via the illusion of human mastery of the earth. While the word itself is something of a quip, I am quite serious about the need to draw political economy and political ecology into dialogue with earth systems, flows of matter, and the fictions of mastery and human exceptionalism. This kind of integrative approach can shift the way we story and research the world. Social-ecological systems have yielded few spaces to think about the power of stories, although stories are the most powerful way of changing worlds. "*Homo fabula*," Nigerian novelist Ben Okri writes, "we are storytelling beings."[39]

My invitation to scientists is this: to build alliances in society, change the story you are currently telling. To do this will require challenging the cosmopolitical story of our time that comes from an unreasonable faith in the gods of reason in the knowledge economy.

III | DEPOSING THE NEOLIBERAL GODS OF REASON

Having grown up in apartheid South Africa without imagining alternatives, I remember feeling bewilderment, as a child, at the condemnation white South Africa received in the *Newsweek* magazines I would read at the dentist's office. Only much later, thanks to the reading my high school English teacher gave of *King Lear*, did I begin to understand how power and privilege could make people blind to the evidence of harm, injustice, and unreason.[40]

The struggle against apartheid was a struggle to depose its rationalities and imagine alternatives. The struggle for an ecopolitics, in our time, is a struggle against the rationality of permanent economic growth and unlimited source extraction.

The three gods of reason of the knowledge economy have made their appearance several times in this volume. They serve to authorise extractivism, and their names (after Bruno Latour) are Technical Efficiency, Economic Profitability, and Scientific Objectivity.[41] Like a game of rock-paper-scissors in which each has the capacity to trump the other, their wielding has the power to close and limit the available evidence of harm to the earth, so that no other concerns might be imaginable, or seem viable, or reasonable.

Arundhati Roy's *Capitalism: A Ghost Story* speaks of "the end of imagination"—a phrase that became the title of a later collection of essays.[42] The extraordinary hold that neoliberalism has over the public imagination is described by Belgian magistrate Manuela Cadelli:

Fascism may be defined as the subordination of every part of the State to a totalitarian and nihilistic ideology.

I argue that neoliberalism is a species of fascism because the economy has brought under subjection not only the government of democratic countries but also every aspect of our thought. . . .

The austerity that is demanded by the financial milieu has become a supreme value, replacing politics. . . . The nihilism that results from this makes possible the dismissal of universalism and the most evident humanistic values: solidarity, fraternity, integration and respect for all and for differences. . . .

The foundations of our culture are overturned: every humanist premise is disqualified or demonetized because neoliberalism has the monopoly of rationality and realism.

Margaret Thatcher said it in 1985: "There is no alternative." Everything else is utopianism, unreason and regression. The virtue of debate and conflicting perspectives are discredited because history is ruled by necessity. Trust is broken. Evaluation reigns.[43]

Contrary to the claims of neoliberalism that market self-regulation and privatisation would lead to the creation of unprecedented wealth that would trickle down to create equality, the approach has generated the greatest inequality in the history of humanity, argues economist Jason Hickel.[44] Inequality generates the conditions that make people vulnerable to ruthless exploitation, leading to extraordinary environmental damage ranging from deforestation by illegal timber logging; to overfishing; to government-endorsed mining of beaches, nature reserves, and traditional lands; and to the killing of megafauna for skins, horns, and tusks.

Neoliberalism is not an environmentalist's friend. The energy era and the knowledge economy present a time of patents and property lines and corporate personhood, an economy of growth based on an extractivism that has generated zones of sacrifice leading to expulsions, extinctions, and the deposition of excreta ranging from mine tailings and toxic waste to poorly processed sewage. The trust that so many environmental managers, officials, and scientists have in the neutrality of neoliberal precepts is not deserved. They are not, and have never been, neutral.

Much as South African constitutional court justices Yvonne Makgoro and Albie Sachs drew on the African philosophy of ubuntu (see chapter 4) to redefine a dignity-based jurisprudence, environmental management sciences may find it far more beneficial to work toward a dignity-based ecopolitics.[45]

What would it mean to recover dignity as a principle in environmental management? I think of the care fishers have for the ocean and its species, in which the ocean is a mother, and species like lobsters have likes and dislikes: was it really environmentally useful to strip that away and replace it with biomass quota, so that the ocean became an ATM? How might the desire for life and collective well-being frame an approach to ecopolitics that would facilitate a transition to environmental democracy?

Perhaps the most surprising "find," in researching these chapters, was the power of figures drawn from Southern African thought that have generally been outside scientific research, to compel thinking in a new register. In chapter 1, the figure of Tsūi‖Goab may be valuable in restoring many

Capetonians' appreciation for the waters of Table Mountain, and the need to care for the rivers and oceans around it. In chapter 2, the Rain Being offers ways of living that embody respect for rain. In chapter 3, the insights of Namaqualand "bossiedokter" healers on "krag" (energy, vitality, power) enabled our research team to understand the limits of the concepts that frame the questions posed by pharmaceutical medicine. The focus in chapter 4 on the idea of being kin ("sons and daughters") with soil, so widespread in Southern African thought, provokes insights on the need, and the possibility, for addressing the metabolic rift between people, soils, plants, and water. The power of traditional stories and rock paintings of baboons, in chapter 5, serves to remind that there have, throughout history, been street-smart (or bush-smart) ways of relating to baboons in respect of their propensity to cause damage. And in chapter 6, the power of fishers' understanding the behaviour of species like lobsters makes it possible to imagine a conservation based not on biomass allocations, but on care for marine creatures like lobsters as relational and sentient beings.

Each of these situations speaks to the possibility for an ecopolitics based on relatedness, on kinship, on connectedness. And in the forms of relationship between people and ecologies, attached to each of these figures, there are resources for composing an understanding of the world based not on the mapping of things but on caring for its relations.

To see, in chapter 6, fishers' struggles for collective personhood—the emphasis on care, and on ensuring no one goes hungry—being translated into corporate personhood felt to me like a loss. Effectively, fishers' struggles in the Equality courts were resolved by granting corporate personhood to local companies. But corporate personhood means becoming a shareholder or a board member, and having collective standing in law. Companies are inherently exclusionary: that is their business model. Is this the equality of personhood that fishers fought for? Does that model with its basket of multi species extractive rights foster the kinds of care for the sea to which fishers have historically attested, before quotas turned the ocean into an ATM? What kind of environmental management might restore local concepts of care for bays and oceans?

IV | FROM A RELATION OF MASTERY
TO A RELATION OF PRESENCE

A representation of all of science, encompassing everything that is known from the scale of quantum particles to the scale of the expanding universe, is not the sum of all things. Nowhere in that scalar model is it possible to see relationships. Missing in it is life, consciousness, the relationships that make those objects knowable, interactions, emotions and the political power of affect, generosity and gifting, survival, predation, responsiveness, communication, signalling. Missing too, in that model, is the relationship of science, technology, and the scientist with specific technologies, and specific research funding agendas. Absented are questions about the frame that renders invisible the critically important relational aspects of knowledge-making.

A world imagined as if it were composed only of objects is not "natural." It is a world that has arisen as part of a modernist conception of the universe, framed by scholars in the 1600s who kept their bodies and their books from the ash heap by declaring that they did not contest the Church's dualism of spirit and matter, but confessed instead that theology was separate from science. Over time, studies of belief and society would give rise to the social sciences, in the belief that the study of belief should form the basis of the social sciences of religion, sociology, politics, anthropology, philosophy, and so on. As value was separated from fact, and society from nature, capitalism was handed on a plate the cosmology of a world available for extractions, extinctions, and expulsions, all in the name of economic productivity, and justified by scientific objectivity.

No environmental scientist I know would seriously propose that extractivism is all the world is good for. Yet if the binary distinction of facts from values, objects from subjects, life from non-life is accepted by science as "natural," all that is left for environmentalism (after the extraction of any politics) is moralism: a space of "oughts" and "ought nots." No wonder regional environmental scientists feel so disempowered! For their passionate care for the environment is discouraged by a misguided collective requirement of one another to avoid politics or emotion. Their language for caring about the world may be expressed in the abstract terms of "the social," but involvement with activist struggles is discouraged.[46]

Recognising that that conceptual approach has arisen historically rather than via "nature itself" can bring out the best in science: a humility in respect of its own histories of error, and an ability to encompass the scholarly necessity of questions and doubts. For the problem is "cosmopolitical"— pertaining to the versions of the world, nature, society, fact, value.

Black studies, feminist studies, and critics of modernist philosophy of science consistently demonstrate that what is named and known is an effect of the relationships, including technological relationships, that make it known. "Relation ... defines the elements ... at stake, and at the same time it affects (changes) them," wrote Édouard Glissant.[47]

Caribbean thinkers like Glissant and Aimé Césaire struggled with knowledges that claimed to represent them and the world—in which both they and the world were unrecognisable to them. Césaire's evoking of the relation of mastery, in the figures of Prospero and Caliban in his *A Tempest*, dramatises his insight that a master who imagines himself a "rational man" might name a person and or an island or a world—yet not know any of them, because mastery as a relation offers an illusion of knowing the truth that is unable to predict the tempests it generates, in people or lands, *because within its mindset there is no mastery: there is only truth.*

The climate adaptation scientist who wants to advise on climate-smart agriculture without addressing land restitution, or the impact of genetically modified seeds on soils and seed-sharing-based societies, is not acknowledging those relations. Yet without knowledge of those relations, she will be unable to predict the tempest that will come to farming productivity when an economy of exchange is criminalised. That scientist makes no connection between climate crisis and the false distinction between nature and society which their own science debunks. There is no space for critique, in this conceptual landscape, between the rise of neoliberalism and the abuse of natural science by disaster capitalism. To that scientist, what exists are the beings that are named and described to science: there is seed that addresses their concern about drought; that is all. What to them has "being" is simply what has being in the version of nature promoted in the knowledge economy. Those who try to name what is unseen, in that limited paradigm, risk marginalisation as "irrelevant" to the urgency of the task at hand. This is the approach that Glissant names as "Being-as-Being": the idea that what is already named and known is all there is to name and know.[48]

When I read Audre Lorde's famous words "The master's tools will never dismantle the master's house,"[49] I hear her arguing that scholarly practice in the mode of mastery will not achieve the change that is needed in the university or in activist circles. Decolonial thinkers' attention to the relational, so similar to works in contemporary feminist science studies like those of Val Plumwood, Karen Barad, and Isabelle Stengers,[50] points toward the necessity to attend to the relation of mastery in South African environmental science. It is knowledge-making in the mode of "Being-as-Being"—presuming what one names and knows to come from nature itself—that haunts environmental management, making it impossible to grasp the ways in which settler-colonial history renders fictions and spectres as certainties.

For Glissant, bringing relation into the arts of knowing is essential to the decolonisation of knowledge. In an exquisite exegesis of the art of knowing by attending to relation, he crafted a series of aphorisms:

Someone who thinks Relation thinks by means of it . . .
Relation contaminates, sweetens, as a principle or as flower dust.

That which would pre-exist (Relation) is vacuity of Being-as-Being.
Being-as-Being is not opaque but self-important.
Relation struggles and states itself in opacity. It defers self-importance. . . .
Relation . . . scatters abroad from Being-as-Being and confronts
    presence. . . .
Relation also dismantles the thought of non-Being. . . .
Non-Being could not be except outside Relation. . . .
Beings risk the being of the world, or being-earth. . . .
Relation is the knowledge in motion of beings, which risks the being of the
    world.
Never conceive of the being of the world or the being of the universe as
    being of Being or fitting themselves to it.
It depends upon Relation that the knowledge in motion of the being of
    the universe be granted through osmosis, not through violence.[51]

"Presence-to" rather than "being" has a wider presence in recent scholarship: Jean-Luc Nancy's *The Birth to Presence* attempts to avert the abuse of social science to foster ethnic-nationalist accounts of culture in Europe by exploring how to "language" (verb) collectivity in terms other than social wholes (as if they are forms, like a Platonic form or Archimedean solid). The fiction of those social wholes fosters an illusion of collective identity as a bounded entity, a "tribal being" with insiders and outsiders.

In much the same vein, Emmanuel Levinas's notion of the face-to-face focused on the relational encounter that restores personhood where social-wholes-based thinking, like the anti-Semitism that had placed him in one of the camps of the Holocaust, had rendered Jews as less-than-human objects.[52]

The decolonial invitation to restoring presence-to the world is a life-giving antidote to the logics of objectification that make necropolitical ecologies—the distribution of extraction, expulsion, excreta, extinctions—appear rational, reasonable, logical. Crucially, for this discussion of unmaking the "great derangement" that has made planetary-scale necropolitics possible, presence-to the world is an invitation to attend to the relationalities through which knowledge is made—that is, how things come to be named and known, and different versions of "nature" come into being.

One aim of this volume is to demonstrate that within history, different ecopolitical arrangements have resulted in different versions of "the one true nature" (see especially chapters 1, 2, and 5). Understanding the relationship between a specific ecopolitics and the kinds of "nature" that that generates is crucial to finding a route through the impasse over different knowledges and ways of knowing in South Africa.

A useful example comes from Norway, where attention has been on the difference between a salmon that is farmed and one that is wild caught.[53] The difference is more than two contending perceptions of a single truth. Although their biological classification is the same, their flesh has come into being through very different muscle movements and combinations of materials. One has grown on a wild diet, the other in a crowded pen, fed on pellets and antibiotics. A "pure science" approach that addresses only the taxonomy of the fish does not address the evidence that matters to a consumer.

The situation of Norwegian salmon is analogous to the distinction made in Mpondoland between the free-range homestead chicken *mleqwa*—"the one that you chase," with its huge drumsticks and textured flesh—and the *lamthuthu* or "stupid chicken" that has been raised industrially.[54]

What the chicken and salmon examples demonstrate is that a species name alone locates an object against a background devoid of relationships. But those who eat the flesh recognise in its texture the importance of the relationships through which a life comes into being. Those relationships emerge via flows and movement and relations.

The concept of ubuntu teaches that "a person is a person through other people." As an approach to how one comes into life and wellness, it invites presence-to the relationships that accompany other living beings in the course of their lives. The issue is more than a matter of mind, or versions of reality or even different ways of making knowledge: they follow different empirical trajectories with one following form, and the other following relations. The implication is that how something is brought into being really matters. Definitions don't come from nowhere. They are as embedded in relations as mleqwa and lamthuthu. It is not possible for a taxonomist to declare that all salmon or chicken or seeds of the same species are the same because all species share some essential "thingness." The examples demonstrate that even a science as solid as taxonomy does not constitute an essential, universal truth. How life grows matters too. That does not mean that the science of taxonomy is wrong, but it does mean that its essentialist approach to truth is partial. Not all salmon or chickens or seeds are raised in the same way. They inhabit different ecopolitical relations, which change them. Knowing what they are requires understanding not only their species (that is, their form), but the flows and relations through which they become present in the world.

That realisation invites an ecopolitical approach to knowledge that acknowledges definitions and the relations that produce them. Moreover, it recognises that every object is entangled in relations that bring it into being.

Césaire's writings against the essentialism via which colonial ethnographers claimed their version of blackness to be ultimate truth has relevance here. The postcolonial project contested the colonial belief that cataloguing of objects (in this case, people) as if they were devoid of relations constituted an absolutely true knowledge of the world. As a poet, Césaire's surrealism contested the irrationality of everyday practices of certainty, authoritarianism, and mastery. Foregrounding the concerns of the colonised, he destabilised the logics of colonial world-making that rendered people as objects. Science needs poets, Césaire argued, not because science must be taught in poetry (although the ancient Roman scholar Lucretius managed it), but because science needs to allow its object-based certainties to be unsettled and replaced.

The idea of "objectivity" as an approach to knowledge means knowing a thing in a manner that excludes any relations that might alter its essential "thingness." That analytic simultaneously hides the framing concerns of the community of practice endorsing the science and renders others' concerns unknowable, because it is certain that there is only one universal definition of everything.[55] For Césaire, poetry and theatre could compel presence-to the relationalities of a situation in a way that could surface what it is that some concerns evident and therefore knowable, but occluded other concerns and relegated them to matters of subjectivity or superstition.

Decolonial student protests against university curricula were demands for recognition of their presence, and their concerns, in the knowledge made and taught on university campuses. Lis Lange, deputy vice chancellor at the University of Cape Town, noted in a keynote address on curriculum transformation: "Of all the imagery that accompanied the student protest in the period 2015–2017, a photograph of a black woman student during the Rhodes Must Fall (#RMF) moment holding a placard with the inscription 'Look at me, I am here,' constitutes for me one of the most revealing statements of the issues that the protest movement attempted to address."[56]

She continues:

The colonial university was built on "scientific" notions of disrespect for the colonial subject that were carried over and refined under the Apartheid regime, even at the English-medium so-called open universities. The colonial university was built—it was complacently assumed—on white genius and ingenuity; it conformed to a set pattern of development to modernity that started and culminated with Western history. A hundred years later, in the context of an unsatisfactory and incomplete transition to democracy, the question about black recognition and white privilege in the academic space is being reopened. But this question is asked simultaneously across the domain of knowledge and that of identity for the colonial edifice. This is so because the colonial edifice was not built only on the assumption of the inferiority and incompleteness of its subjects, but also on the mastery and superiority of colonial knowledge, which scientifically has demonstrated the inferiority of the colonised.[57]

Lange concludes by calling for

a pedagogy and a curriculum of presence. This represents an affirmation of the students and their blackness, of their selves, their bodies, their identities and in particular their direct and indirect (intergenerational knowledge) experiences of the world. This requires a counter movement: the acknowledgement of the identity and the position of those who teach, as well as of white students. . . . If the university is to play a truly transformative role in our society, a pedagogy of presence should help all of the university's inhabitants to recentre themselves away from "white mythologies," creating the possibility of the development of new intersubjective relationships, new forms of learning and new respect for different modes of knowing. . . . A pedagogy of presence makes possible two important intellectual movements: the resizing of European knowledge and its provincialisation in a global world, and the incorporation of other epistemological traditions—African, Chinese, Indian, Latin American—into the horizon of global knowledge.[58]

What are the implications of rethinking research practice in relation to presence? Two areas seem to me vital. The first is to pay attention to how a specific question arises, and what concerns it addresses. The second is to attend to the relations, ideas, and procedures that make a being tangible and measurable as a Being, that is, as a fact. In this respect, Isabelle Stengers's work in *Another Science Is Possible,* Didier Debaise's volume *Nature as Event,* and Marisol de la Cadena's work *Earth Beings* are formative, demonstrating the potential for knowledge production that is not limited to matters of epistemology, but productively engages the ontological: the question of how things come to matter enough to be named and known.[59]

In contemporary environmental politics, aligning with nature increasingly takes the form of legal and constitutional struggles to claim for nature a legal personhood, in order to correct the objectification of nature. For example, in Paris during COP21 in November 2015, the International Tribunal on Rights of Nature, led by South African environmental lawyer Cormac Cullinan, was supported by leading southern environmentalists like Vandana Shiva from India, and Alberto Acosta, who drafted Ecuador's constitution, which includes the rights of nature; and Nnimmo Bassey, whose founding of the Health of Mother Earth Foundation in Nigeria builds on the legacy of Ken Saro-Wiwa, who was executed for opposing oil extraction in the Niger Delta.[60]

The tribunal built on legal innovations in Ecuador, Bolivia, and Peru to establish earth rights in laws and constitutions. Since those laws were drafted, the United Nations has been investing in an earth rights project. A river in New Zealand has, after almost a century of struggle, achieved legal personhood, as have dolphins in India and a chimpanzee in New York, and, more recently, two rivers and the glaciers that feed them, in the state of Uttarakhand in India. That judgment declared that "rivers, streams, rivulets, lakes, air, meadows, dales, jungles, forests, wetlands, grasslands, springs and waterfalls" must be given "corresponding rights, duties and liabilities of a living person, in order to preserve and conserve them."[61]

Leveraging the existing legal category of personhood for the protection of landforms (glaciers, rivers, mountains) or species (dolphins, chimpanzees) is a first step to exit the conceptual madness of a world turned into objects devoid of relations. It turns on its head the claim that private property is the solution.

Legal personhood may be a step toward equalising corporate mastery over nature by equalising the relation in law between companies and creatures or landforms. Instead of subject-object relations, it offers subject-subject relations.

There is a need for caution, however: two hundred years after the same debate was held in white courts over whether slaves were legal persons or legal property, the United States still needs #BlackLivesMatter. Attaining legal personhood is a strategy of survival in a system of objects versus subjects established in modernist thought and science, and imposed by coloniality and its courts. For every creature or earth form that wins an "upgrade" to personhood, there are those beings that are not upgraded. Are the only relations worth caring about those that are no longer legally made "other" as non-human, but made "same"? The figures of care described above suggest that there is more we can draw from; that something does not need to be granted the status of personhood to enter into relationships of care.

The problem with conferring legal personhood is not the value of legal personhood per se, but the underlying notion that subjects matter, while objects don't. As an approach, granting legal personhood to living and non-living does not escape the core conceptual problem of modernity that identifies things devoid of relations, as objects on a blank background. To address that, I want to propose that one more conceptual intervention is necessary for remaking ecopolitical thought that will be capable of unmaking the geologists'

Anthropocene. That is the necessity to address questions in environmental research and policy to the challenge of comprehending movement and flow, more than property, borders, and territory.

V | FLOW AND MOVEMENT: RETHINKING SPACE AND TIME

The innovative approaches listed in section II above—urban metabolism, critical zone scholarship, biogeochemistry—all have in common an attention to flows of matter, reflected also in the fascinating mathematical modelling of complex flows and feedback loops as developed in climate science analysis. Together these fields comprise a fascinating emerging scholarship: addressing a world in motion. Working with flows set in motion by people, whether via the food system, consumerism, travel and transport, medicine, and so on, these approaches recognise that nothing goes nowhere.

These fields offer a line of flight from the binaries of modernist thought. Rather than studying nature and society, or subjects and objects, or space and time as discrete if interlinked entities, scholarship that attends to flows and movement offers the beginnings of a route to a knowledge of nature that is free of the fiction that nature is composed of singular objects that may be described in science against a blank background of space and time. By analogy, the difference is between knowing a butterfly through its movements and relations in an ecology, versus knowing it pinned under glass in a white drawer. While the knowledge of how to identify the butterfly is undeniably important, it is not the sum of the knowledge of the butterfly's life, only of its form. Moreover, devoid of any acknowledgment of the butterfly's relations, that knowledge cannot generate care for the butterfly's presence in the world. Care for the non-extinction of the butterfly-under-glass is necessarily relegated to the world of "oughts" and "shoulds" which will mean nothing to those in ministries of mining, forestry, or agriculture, whose view of economy is that its growth is more important than a few extinctions. Objectivist science of this kind, limited as it is to the description of objects, cannot yield much more than the moralisms and finger-pointing that have an uneasy overlap with the moralisms associated with white superiority (see section I, this chapter).

By contrast, environmentalism that attunes to the effects of flows and movements in metabolisms, accretions, sedimentations, and morphogenesis is free of the Faustian bargain struck when objects are claimed as known without attention to their relations.

The crucial issue here is that while leading fields of soil science, climate science, archaeology, process engineering, and biogeochemistry now attend to flows, similar orientations have barely begun to inform environmental management, social science, or law. It is vital that this be remedied if practitioners, and the universities that train them, are to address the challenges of this moment in geological history.

How, for example, might environmental laws change once it is recognised that effects—of fracking, or nuclear waste, or acid mine drainage—will outlast the political systems that claim to govern them, and will flow over property boundaries and territorial borders, species, rocks, and waters into the commons that is the earth system?

How might the social science research requested so often by natural scientists shift from "perceptions of reality" to anthropogenic effects and their implications for justice and inequality? How might environmental managers, many of whom began their training to care for ecologies but who now find themselves stuck ticking boxes to meet environmental government departmental mandates, be empowered to speak to the wider "derangements" introduced in the knowledge economy?[62] How might psychologists begin to recognise the geological effects of psychopathic CEOs and narcissistic political administrations?

How might taxonomists in botany, marine ecology, and zoology change their approach to species once it is understood that flows of harmful chemicals in soils or seawater or plants may transform the species they study? The sciences of epigenetics speak to this, as do consumers concerned about their own health. Environmental justice that attends to pollution whether of carbon or industrial chemicals is more than environmentalism plus politics: it speaks to the emergence of a transformative scholarship, and a paradigm shift, in which it is possible to grasp that political systems have geological effects.

Research that attends to flows and movements and relations offers an opportunity to rethink the grounding assumptions—that is, the ontologies—that are in error in the modernist conceptual division of space from time. Indeed, modernist thought may be unique in this history of humanity to presume this distinction. In *Knowing the Day, Knowing the World*, I made the case that Palikur thought in the northern Amazon is based on movement as a theory of world-making.[63] As an anthropologist in an archaeological project, my task was to try to link Palikur narratives with archaeological sites. A simple task this was not! It required a systematic unmaking of the ontologies of modernist thought, as expressed in theories of personhood and what it is to know; rethinking history as set in chronology and geography as space and ethnoastronomy as the knowledge of constellations. In each field, what I found instead was movement, not territory or bodies, as the basis of making knowledge of the world.

Knowledges on every continent have been described that take form in relation to movement, including Aboriginal songlines in Australia, and some Amerindian taxonomies of creatures order them with via their type of movement (flying, swimming, walking, etc.).[64] In North America, the Din'é (Navajo) language integrates movement and place. That the Khoe root "au-b" refers to stream, blood, or snake (chapter 1) suggests that attention to movement and flow may have been an important concern in the formation of Khoe knowledge. Contemporary social science conversations initiated by anthropologist Tim Ingold have begun to pay attention to movement in and through places.[65] David Turnbull's work on traditional knowledge in Australia, too, notes careful attention to the organisation of knowledge around movements in and through places.[66] When and why did modernist knowledge that arose in the Enlightenment era begin to focus on forms in empty space?

The question gripped thinkers including Jacques Derrida, whose first book was on the problems accruing to scholarship in the early modernists' focus on static Euclidean geometries, and their conversion from the properties of shapes to algebras of space in the work of Descartes and his peers.[67] As a technology of knowing space that allowed it to be represented as empty and devoid of relations, the new technology of mapping worked perfectly in Europe at the time of the rise of capital, private property ownership, the extraction of raw materials, and the closure of the commons. Exported to the colonies in the age of European expansion (chapter 1), it provided the conceptual infrastructure for identifying objects of value (phenomena) in space that could be extracted for the benefit of companies and empires, again, overriding the relationships that formed them. Where state knowledge became focused on inventorying and quantifying assets, scientific questions too were caught up in the task of assessing extractability. To the extent that marine environmental management in the knowledge economy allows itself to be led by total allowable catch estimates (chapter 6), it continues that inheritance.

Michel Serres, in *The Birth of Physics* and in *On Geometry*, notes the limits of a physics based on the conceptualisation of matter, in equilibria, set in empty space.[68] Returning to the forbidden work of the Roman scholar Lucretius, whose works were banned in Europe for centuries, he proposes a knowledge that returns to movement: "For Lucretius, as for us, the universe is a global vortex of local vortices. . . . Nature is river and whirlwind. . . . What nature teaches is the streaming of the inexhaustible flow, the atomic cascade, and its turbulence."[69] So too the recent work of philosopher Thomas Nail, in *Lucretius: An Ontology of Motion*, describes Lucretius's work not as a study of natural *phenomena* but as a study of natural *kinomena*—the way movement generates the accretions, sediments, and folds that become forms and life-forms.[70] There is no difference for Lucretius, argues Nail, between living and non-living, for solid and liquid and life and air share matter that circulates between them, and all are products of morphogenesis:

> The flow of the material conditions produces order out of itself in the form of a bend, curve, or fold in the flow.... The flows create, expand, and continuously nourish the folds they sustain, like the leaves of a plant. The forms of the curve, the spiral, the circle, and all others emerge from the primacy and activity of the material flows themselves that generate and sustain them. The curve occurs in the flow, and the circle or fold occurs when the curve intersects with itself.
>
> In short, the existence of what we call discrete things ... with discrete forms are products of a more primary kinetic condition of distribution and folding that produces things by a process of self-ordering. Form emerges from matter....
>
> Just as form is created from these continuous flows folded into loops and sustained in cycles, the folds can also be untied, unbound, loosened, or opened.... Nature is both the process by which being is created, increased, and maintained, but also the process by which it is dissolved or unfolded. When the folds of being are unfolded, it does not destroy the flows, because flows can neither be created nor destroyed.... Form, for Lucretius, therefore has no existence independent from the kinetic activity of its material conditions....
>
> The form of things comes from the creativity of matter itself. Thus, for Lucretius, the division between organic and non-organic life is a false one. All matter is active, creative, self-organising, morphogenetic.[71]

Similar concepts emerge in the study of Chinese Han dynasty thought. Scholars Yuk Hui, Brook Ziporyn, and David Wade each reflect on the Chinese concepts of *qi* and *li* as alternative approaches to the production of knowledge.[72] For Ziporyn, *li* refers to flow, coherence, and ceremony, a concept that at once records, honours, and celebrates an understanding of the minimal energies that achieve desired consequences. Yuk Hui's *The Question Concerning Technology in China: An Essay in Cosmotechnics* addresses the way Chinese approaches to technology attended to the minimising of energy (*qi*) expenditure, and sought to "skilfully bring something together [in a way] that resonates with the cosmic order."[73] For David Wade, *li* may be seen in patterned flows of all kinds, from the patterns wind makes on sand dunes, to the raindrops on a windscreen. Learning to work with the patterns and flows and propensities of situations is to reconnect with the earth and to achieve desired outcomes with minimal energy use. *Li* offers a suite of principles for knowledge that follows matter and materials. It seems to me the opposite of a knowledge system that has grown up through modernity-coloniality that is based on mastery, manipulation, and control of objects, and the belief that there is somewhere that is a nowhere to which waste can go.

Modernist scholarship whose core conceptualisation of knowledge is that it focuses on objects set in space and time, without attention to their movements and relations, is the effect of an extraordinary rollout of science that is deeply invested in the logics of capitalism, taking as given truths the cosmopolitical relations of ownership and mastery (chapter 1). Rethinking environmental governance as a space for the facilitation of relations of care and attention, interactions and flows, is critically important if contemporary environmental management is to be able to escape the territorialism that leads to militarised conservation, and address instead research questions generated via attention to life, ecology, justice, equity, and well-being. Such an alteration of the way environmental problems are conceptualised bears within it the potential to generate an ecopolitics of well-being that can begin to address the metabolic rift (chapter 4)—the cutoff from land—that is the heart of South Africa's current political challenges, and in a similar yet different way, also at the heart of the Anthropocene, globally. For environmental questions that presume property, territory, ownership, and the mastery of nature to be the cosmic order, generate an unsustainable cosmopolitics that can only command and control; fence and wall; impose fees and fines. Conservation science deserves a better repertoire of possibilities: one that is guided by the question "How do we compose a common world?"[74]

In this time of displaced molecules and mass migrations, and struggles over extinctions, expulsions, extractions, and excreta, let us tell, and love, the stories of learning about flows, li, and kinomena to learn together how to unmake the Anthropocene by reconnecting with soil and with oceans. I look forward to the day when I turn on the radio and hear, in the five-minute segment at the end of the news, not overly simplified accounts of growing economies, but the stories of restorative ecopolitics that address urban metabolism, biogeochemistry, climate models, and environmental justice in the critical zone.

"Vision sets out / journeying somewhere, / walking the dreamwaters," wrote poet Denise Levertov, "arrives / not on the far shore but upriver / a place not evoked, discovered."[75]

7.1 | Malawian chair. Photo: Vanessa Cowling. Used with permission.

The situations explored in this book, which began with a sense that "something is wrong with this picture" as I cycled the Cape Peninsula, engender an invitation to finding and forming an ecopolitics that gives life. May our living and knowing become attuned to the vital flows and delicate relations between rock, water, and life, in our collective histories, presence, and futures.

NOTES

Book epigraphs: Aimé Césaire, cited in Edmond Moukala, *Rabindrânâth Tagore, Pablo Neruda, Aimé Césaire: For a Universal Reconciliation* (Paris: UNESCO, 2011), 108.

FOREWORD

1   John Dewey, "Pragmatic America," *New Republic*, 12 April 1922, 186.

INTRODUCTION

1   Barbara Maregele, "Redhill's Ruins: Cape Town's Forgotten District Six," GroundUp, 28 July 2014, https://www.groundup.org.za/media/features /redhill/redhill_2043.html, accessed 9 July 2019.
2   See section 15 of the closing submissions to the Arms Procurement Commission in respect of hours at sea, particularly for the period 2012–14. "Closing Submissions Addressing Term of Reference 1.2," accessed 9 July 2019, http://www .justice.gov.za/comm-sdpp/docs/20150610-ClosingSubmissions-ToR1-2.pdf.
3   Rosaleen Duffy, "Waging a War to Save Biodiversity: The Rise of Militarized Conservation," *International Affairs* 90, no. 4 (2014): 819–34; quote, 819.
4   Joel Netshitenze, "The Black Man's Burden," *Sunday Times*, 21 February 2016, https://www.timeslive.co.za/sunday-times/opinion-and-analysis/2016-02-21 -the-black-mans-burden-/.
5   Brett Stetka, "By Land or by Sea: How Did Early Humans Access Key Brain-Building Nutrients?," *Scientific American*, 1 March 2016, https://www .scientificamerican.com/article/by-land-or-by-sea-how-did-early-humans -access-key-brain-building-nutrients/.
6   Marilyn Frye, "White Woman Feminist, 1983–1992," in *Willful Virgin: Essays in Feminism 1976–1992*, 147–69 (Freedom, CA: Crossing Press, 1992).
7   Aimé Césaire, *Discourse on Colonialism*, trans. Joan Pinkham (New York: Monthly Review Press, 1972), 42.

8 See Bruno Latour, *Politics of Nature* (Cambridge, MA: Harvard University Press, 2004).

9 Jason Moore, *Capitalism in the Web of Life: Ecology and the Accumulation of Capital* (London: Verso, 2015).

10 Jean-Luc Nancy, *The Birth to Presence*, trans. Brian Holmes et al. (Stanford, CA: Stanford University Press, 1993).

11 Elspeth Probyn, *Eating the Ocean* (Durham, NC: Duke University Press, 2017).

12 Aimé Césaire, "Poetry and Knowledge," in *Refusal of the Shadow: Surrealism and the Caribbean*, ed. Michael Richardson, trans. Michael Richardson and Krzysztof Fijałkowski (London: Verso, 1996), 134–46; Édouard Glissant, *Poetics of Relation*, trans. Betsy Wing (Ann Arbor: University of Michigan Press, 1997).

13 Nancy, *Birth to Presence*, 97ff.

PART I | PASTS PRESENT

Epigraph: Jacques Derrida, *Specters of Marx: The State of the Debt, the Work of Mourning, and the New International*, trans. Peggy Kamuf (New York: Routledge, 1994), xix.

ONE | ROCK | CAPE TOWN'S NATURES

This chapter owes a debt to Caron von Zeil for reintroducing the forgotten springs to the city of Cape Town.

1 Theophilus Hahn, *Tsuni-||Goam: The Supreme Being of the Khoikhoi* (London: Trübner, 1881), 34–35, https://ia800500.us.archive.org/0/items /tsunillgoamsupreoohahnuoft/tsunillgoamsupreoohahnuoft.pdf. Hahn explains that while Tsūi||Goab was in everyday use, he used the older form of the name for the title of his book.

2 For my thinking about Khoena beings, I am indebted to Marisol de la Cadena, for her work in her book *Earth Beings: Ecologies of Practice across Andean Worlds* (Durham, NC: Duke University Press, 2015).

3 Hahn, *Tsuni-||Goam*, 30.

4 See Hahn, *Tsuni-||Goam*, 38.

5 Hahn, *Tsuni-||Goam*, 38.

6 Hahn, *Tsuni-||Goam*, 128.

7 Hahn, *Tsuni-||Goam*, 127–29.

8 Hahn, *Tsuni-||Goam*, 130.

9 Hahn, *Tsuni-||Goam*, 139.

10 Harold Scheub, *A Dictionary of African Mythology: The Mythmaker as Storyteller*, Kindle ed. (Oxford: Oxford University Press, 2000), Kindle locations 3891–93.

11 Charles Darwin, *Geological Observations on the Volcanic Islands Visited during the Voyage of H.M.S. Beagle, Together with Some Brief Notices of the Geology of Australia and the Cape of Good Hope* (London: Smith Elder, 1844), http://darwin-online.org.uk/content/frameset?itemID=F272&viewtype =side&pageseq=1.

12 Sharad Master, "Darwin as a Geologist in Africa—Dispelling the Myths and Unravelling a Confused Knot," *South African Journal of Science* 108, nos. 9–10 (2012): 1–5.

13 Department of Geological Sciences, University of Cape Town, "Cape Town Geology," accessed 29 December 2015, http://www.geology.uct.ac.za/cape /town/geology.

14 Patric Tariq Mellet, "Cape Indigenes," Camissa People, accessed 14 August 2019, https://camissapeople.wordpress.com/cape-indigenes-2/.

15 Patric Tariq Mellet, "Camissa," Camissa People, accessed 30 December 2015, https://camissapeople.wordpress.com/camissa/.

16 Nigel Penn, *The Forgotten Frontier: Colonist and Khoisan on the Cape's Northern Frontier in the 18th Century* (Athens: Ohio University Press, 2005), 32.

17 R. Raven-Hart, *Cape of Good Hope 1652–1702: The First Fifty Years of Dutch Colonisation as Seen by Callers*, vol. 1 (Cape Town: A. A. Balkema, 1971), http://www.dbnl.org/tekst/rave028cape01_01/rave028cape01_01.

18 The date of the reference for this citation is 26 June 1652, cited by Raven-Hart, *Cape of Good Hope.*

19 Nigel Worden, *Slavery in Dutch South Africa* (Cambridge: Cambridge University Press, 1985), 4.

20 See Andrew Smith in A. B. Smith and R. Pheiffer, *The Khoikhoi at the Cape of Good Hope: Seventeenth-Century Drawings in the South African Library* (Cape Town: South African Library, 1993), 17.

21 See Richard Elphick, *Kraal and Castle: Khoikhoi and the Founding of White South Africa* (New Haven, CT: Yale University Press, 1977), 100. See also A. B. Smith, "The Disruption of Khoi Society in the Seventeenth Century," in *African Seminar Collected Papers* (Cape Town: Centre for African Studies, University of Cape Town, 1983), 3:1–16.

22 See Pascal Dubourg Glatigny, Estelle Alma Maré, and Russel Stafford Viljoen, "*Inter se nulli fines*: Representations of the Presence of the Khoikhoi in Early Colonial Maps of the Cape of Good Hope," *South African Journal of Art History* 23, no. 1 (2008): 301–17. Also Elphick, *Kraal and Castle*; and I. Schapera, *The Khoisan Peoples of South Africa* (London: Routledge, 1930).

23 Olfert Dapper, "Kaffraria or Land of the Kafirs, Otherwise Named Hottentots," in *The Early Cape Hottentots*, by Olfert Dapper, Willem ten Rhyne, and Johannes Gulielmus de Grevenbroek, ed. and trans. I. Schapera and B. Farrington (Cape Town: The Van Riebeeck Society, 1933), 13–15, http://www .dbnl.org/tekst/dapp001earl01_01/dapp001earl01_01.pdf. Note that this account accords with a letter by Van Riebeeck of 29 July 1659: "The prisoner, one of the Capemen, who could speak Dutch fairly well, having been asked the reason why they caused us this trouble, declared for no other reason than that they saw that we kept in possession the best lands, and grazed our cattle where theirs used to do so, and that everywhere with houses and plantations we endeavoured to establish ourselves so permanently as if we intended never to leave again, but take permanent possession of this Cape land (which had belonged to them during all the centuries) for our sole use; yea! to such an extent that their cattle could not come and drink at the fresh water without

going over the corn lands, which we did not like them to do." Quoted in Dapper, ten Rhyne, and de Grevenbroek, *Early Cape Hottentots*, 15n19.

24   Huigen, *Knowledge and Colonialism: Eighteenth-Century Travellers in South Africa* (Leiden: Brill, 2009), 33ff.

25   Aimé Césaire, "Notebook of a Return to the Native Land," in *Aimé Césaire: The Collected Poetry*, trans. Clayton Eshleman and Annette Smith (1939; Berkeley: University of California Press, 1983).

26   Glatigny, Maré, and Viljoen, "*Inter se nulli fines*," 302.

27   Isabelle Stengers, "The Cosmopolitical Proposal," in *Making Things Public*, ed. Bruno Latour and Peter Weibel (Cambridge, MA: MIT Press, 2005), 994–1003.

28   Glatigny, Maré, and Viljoen, "*Inter se nulli fines*," 305 (emphasis mine).

29   Peter Kolbe, *Capvt Bonae Spei hodiernvm, das ist: Vollständige Beschreibung des africanischen Vorgebürges der Guten Hofnung: Worinnen in dreyen Theilen abgehandelt wird, wie es heut zu Tage, nach seiner situation und Eigenschaft aussiehet; ingleichen was ein Natur-Forscher in den dreyen Reichen der Natur daselbst findet und antrifft: wie nicht weniger, was die eigenen Einwohner die Hottentotten, vor seltsame Sitten und Gebräuche haben: und endlich alles, was die europaeischen daselbst gestifteten Colonien anbetrift. Mit angefügter genugsamer Nachricht, wie es auf des Auctoris Hinein* (Nuremberg: Peter Conrad Monath, 1719).

30   Huigen, *Knowledge and Colonialism*, 33ff.

31   In the Dutch edition published eight years after the German edition, the drawing of settlement at the Cape is inserted between pages 58 and 59. Peter Kolbe, *Naaukeurige en Uitvoerige Beschryving van de Kaap de Goede Hoop* (Amsterdam: Balthazar Lakeman, 1727).

32   Enrique Dussel, *Philosophy of Liberation*, trans. Aquilina Martinez and Christine Morkovsky (Eugene, OR: Wipf & Stock, 1985), 33. Dussel's commentary on scientism is prescient in neoliberal post-apartheid South Africa: "When a science prescinds from its social, economic, political (dialectical) conditioning, when it forgets that its mathematical formulas can help the soldier hit the target in Vietnam with incendiary bombs . . . that science becomes scientism. It is a science that believes that, just as it is, it has absolute autonomy; it is valid everywhere; its themes have originated from the internal exigencies of scientific discourse, which can be imitated by all the countries of the world as pure, uncontaminated, neutral mediations" (168).

33   Bruno Latour, "The Recall of Modernity: Anthropological Approaches," trans. Stephen Muecke, *Cultural Studies Review* 13, no. 1 (March 2007): 14.

34   William Shakespeare, *A Midsummer Night's Dream*, act 3, scene 1.

35   René Descartes, *Discourse on Method, Optics, Geometry and Meteorology* (1637; Indianapolis: Hackett Classics, 2001).

36   *Discourse on the Method of Rightly Conducting One's Reason and Seeking Truth in the Sciences* was one of the parts of Descartes's volume on method, the original full title of which was *Discourse on the Method of Rightly Conducting One's Reason and Seeking Truth in the Sciences, Together with the Dioptrics, the Meteorology, and the Geometry, Which Are Essays in This Method*. For a

discussion of Descartes's claim that his "practical philosophy" could render us "masters and possessors of nature," see Neil M. Ribe, "Cartesian Optics and the Mastery of Nature," *Isis* 88 (1997): 42–61.

37 Hahn, *Tsuni-||Goam*, 78–79.

38 See Lesley Green and David Green, *Knowing the Day, Knowing the World: Engaging Amerindian Thought in Public Archaeology* (Tucson: University of Arizona Press, 2013), ch. 5.

39 Hahn, *Tsuni-||Goam*, 64.

40 Raven-Hart, *Cape of Good Hope*, 21 (emphasis mine).

41 Raven-Hart, *Cape of Good Hope*, 27.

42 François Valentyn, *Oud en Nieuwe Oost-Indien* (Dordrecht: Johannes van Braam and Gerard Onder de Linden, 1726), 5(2)51–52, cited by Elphick, *Kraal and Castle*, 232.

43 Candice Haskins, "Cape Town's Sustainable Approach to Stormwater Management" (unpublished paper, Catchment, Stormwater and River Management Branch, City of Cape Town, January 2012), 2.

44 See the account of sailor Johan Jacob Saar, in Raven-Hart, *Cape of Good Hope*, 64.

45 André van Rensburg offers one account of the secret journey from the Cape to Guinea to buy slaves in 1658 in "The Secret Modus Operandi in Obtaining Slaves for the Cape: The Ship *Hasselt*, 1658," South Africa's Stamouers, accessed 21 May 2018, https://www.stamouers.com/people-of-south-africa /slaves/678-obtaining-slaves. For a detailed overview of slaving networks in West Africa in the 1700s and 1800s, see Mbaye Guèye, "The Slave Trade within the African Continent," in UNESCO, *The African Slave Trade from the Fifteenth to the Nineteenth Century*, General History of Africa: Studies and Documents 2 (Paris: UNESCO, 1979), 150–63.

46 H. C. V. Leibbrandt, *Precis of the Archives of the Cape of Good Hope, January 1656–December 1658, Part II: Riebeeck's Journal &c* (Cape Town: W. A. Richards & Sons, 1897), 165.

47 Karel Schoeman, *Portrait of a Slave Society: The Cape of Good Hope, 1717–1795* (Pretoria: Protea Book House, 2012), 23. See also Gerald Groenewald, "Slaves and Free Blacks in VOC Cape Town, 1652–1795," *History Compass* 8, no. 9 (2010): 964–83; Nigel Worden, "Cape Slaves in the Paper Empire of the VOC," *Kronos* 40, no. 1 (2014): 23–44; and Pumla Dineo Gqola, "'Like Having Three Tongues in One Mouth': Tracing the Elusive Lives of Slave Women in (Slavocratic) South Africa," in *Basus 'Timbokado, Bawel 'Imilambo/ They Remove Boulders and Cross Rivers: Women in South African History*, ed. Nomboniso Gasa (Cape Town: Human Sciences Research Council, 2007), 21–41.

48 Nick Shepherd, "Archaeology Dreaming: Post-apartheid Urban Imaginaries and the Bones of the Prestwich Street Dead," *Journal of Social Archaeology* 7, no. 1 (2007): 3–28.

49 N. Jade Gibson, "Skeletal (In-)Visibilities in the City—Rootless: A Video Sculptural Response to the Disconnected in Cape Town," *South African Theatre Journal* 26, no. 2 (2012): 151–71.

50 Nadine Gordimer, *The Conservationist* (London: Jonathan Cape, 1974).

51 Kevin Wall, "Water Supply: Reshaper of Cape Town's Local Government a Century Ago," in *Proceedings of the Water Distribution Systems Analysis Conference WDSA 2008, August 17–20, 2008, Kruger National Park, South Africa*, ed. J. E. van Zyl, A. A. Ilemobade, and H. E. Jacobs (Reston, VA: American Society of Civil Engineers, 2009).

52 Janie Swanepoel, "Custodians of the Cape Peninsula: A Historical and Contemporary Ethnography of Urban Conservation in Cape Town" (master's thesis, Stellenbosch University, 2013).

53 Bruno Latour, *We Have Never Been Modern*, trans. Catherine Porter (Cambridge, MA: Harvard University Press, 1993).

54 Italo Calvino, *Invisible Cities* (San Diego, CA: Harcourt Brace Jovanovich, 1974), 165.

55 City of Cape Town, Department of Water and Sanitation, "Vision, Mission Statement, Values, Strategic Objectives, 2013," accessed 9 March 2013, http://www.capetown.gov.za/en/Water/Documents/Vision_mission_statement_value_and_strategy.pdf.

56 Lesley Green, Nikiwe Solomon, Leslie Petrik, and Jo Barnes, "Environmental Management Needs to Be Democratised," *Daily Maverick*, 6 March 2019, https://www.dailymaverick.co.za/article/2019-03-06-environmental-management-needs-to-be-democratised/.

57 See Saskia Sassen, *Expulsions: Brutality and Complexity in the Global Economy* (Cambridge, MA: Harvard University Press, 2014).

58 City of Cape Town, Department of Water and Sanitation, "Consumer Service Charter for the Supply of Potable Water and Sanitation Services, 2014/2015," accessed 22 February 2018, http://resource.capetown.gov.za/documentcentre/Documents/Agreements%20and%20contracts/Consumer%20Service%20Charter%202014-15_a.pdf.

59 City of Cape Town, "Integrated Metropolitan Environmental Policy (IMEP)," approved 31 October 2001.

60 Candice Haskins, "Cape Town's Sustainable Approach to Stormwater Management," accessed 10 June 2012, http://www.capetown.gov.za/en/CSRM/Pages/Reportsandscientificpapers.asx.

61 In terms of the City of Cape Town's study of 2015, the lowest measured flow for the five largest springs in the City Bowl was 1,900 cubic metres per day, or 1,900,000 litres. City of Cape Town, "Hydrogeological Investigation of Existing Water Springs in the City of Cape Town and Environs: Strategy Document, 25 June 2015" (GEOSS Report No. 2014/10–08, Geohydrological and Spatial Solutions International, Stellenbosch, South Africa), 38.

62 Siddique Motala, "Higher Education Well-Being: Transcending Boundaries, Reframing Excellence" (paper presented at the conference of the Higher Education Learning and Teaching Association of South Africa, Durban, 21–24 November 2017).

63 Calculation based on a figure of 36.8 litres per second as supplied by the City of Cape Town, "Hydrogeological Investigation," 66.

64 City of Cape Town, "2010 FIFA World Cup™: Host City Cape Town. Green Goal Legacy Report," 46, accessed 12 March 2013, https://www.westerncape .gov.za/text/2011/7/green_goal2010_part1.pdf.

65 Woolworths Holdings, "Annual Report 2012," accessed 9 March 2013, http://www.woolworthsholdings.co.za/investor/annual_reports/ar2012 /sustainability/its_our_world/water.asp.

66 See Swanepoel, "Custodians."

67 "Slaves and Negro Apprentices: April 26 1816, Proclamation: Offices for the Enregisterment of Slaves Established in Cape Town and the Districts," in *Proclamations, Advertisements and Other Official Notices Published by the Government of the Cape of Good Hope from January 10th 1806 to May 21st 1825,* compiled by Richard Plaskett and T. Miller (Cape Town: Cape of Good Hope Government Press, 1827), 22.

68 Aileen Moreton-Robinson, *The White Possessive: Property, Power and Indigenous Sovereignty* (Minneapolis: University of Minnesota Press, 2015), xii.

69 While slavery at the Cape was initiated by European settlers, slave ownership was not limited to Europeans. See the "Inventory of Free Black Slaveowners Identifiable by Name, Cape Town and Cape District, 1816–1834," in Andrew Bank, "Slavery at the Cape from 1806 to 1834" (master's thesis, Department of History, University of Cape Town, 1991), 236.

70 Césaire, quoted in in Edmund Moukala, ed., *Tagore, Neruda, Césaire: For a Reconciled Universal* (Paris: UNESCO, 2011), 108, http://unesdoc.unesco.org /images/0021/002116/211645e.pdf.

71 City of Cape Town, "Hydrogeological Investigation."

72 See also Michel Serres, *The Five Senses: A Philosophy of Mingled Bodies,* trans. Margaret Sankey and Peter Cowley (London: Continuum Books, 2008), 160.

TWO | WATER | FRACKING THE KAROO

1 Michael Marshall, "The History of Ice on the Earth," *New Scientist Environment,* 24 May 2010, http://www.newscientist.com/article/dn18949-the -history-of-ice-on-earth.html. See also "Karoo Ice Age," Wikipedia, accessed 4 January 2015, then at http://en.wikipedia.org/wiki/Karoo_Ice_Age. The page was updated on 28 July 2018 to reflect the more current scientific name of the era: the Late Paleozoic.

2 Christian N. Jardine, Brenda Boardman, Ayub Osman, Julia Vowles, and Jane Palmer, "Climate Science of Methane," in *Methane UK,* 14–23 (Oxford: University of Oxford, Environmental Change Institute, n.d.), http://www.eci .ox.ac.uk/research/energy/downloads/methaneuk/chapter02.pdf.

3 "Plate Tectonics," Wikipedia, last modified 22 March 2015, http://en .wikipedia.org/wiki/Plate_tectonics; "Pangea," Wikipedia, last modified 7 July 2019, http://en.wikipedia.org/wiki/Pangaea.

4 National Museum, "Paleontology," National Museum (Bloemfontein, South Africa), accessed 5 August 2019, http://www.nasmus.co.za/departments /palaeontology/introduction.

5   Hillel J. Hoffman, "The Permian Extinction: When Life Nearly Came
    to an End," *National Geographic*, accessed 16 April 2015, https://www
    .nationalgeographic.org/news/permian-extinction-when-life-nearly-came
    -end/7th-grade/.

6   Philip V. Tobias, "Homo erectus," in *Encyclopaedia Britannica*, article pub-
    lished in 2015, http://www.britannica.com/EBchecked/topic/270386/Homo
    -erectus.

7   Shula E. Marks, "Southern Africa," in *Encyclopaedia Britannica*, article
    published 1 July 2014, http://www.britannica.com/EBchecked/topic/556618
    /Southern-Africa#toc234030.

8   Robert Turrell, *Capital and Labour on the Kimberley Diamond Fields,
    1871–1890* (Cambridge: Cambridge University Press, 1987).

9   Penn, *Forgotten Frontier*.

10 Sean Archer, "Technology and Ecology in the Karoo: A Century of Wind-
    mills, Wire and Changing Farming Practice," in *South Africa's Environmental
    History: Cases and Comparisons*, ed. Stephen Dovers, Ruth Edgecombe, and
    Bill Guest (Athens: Ohio University Press; Cape Town: New Africa Books,
    2003), 112–38.

11 Lance van Sittert, "The Supernatural State: Water Divining and the Cape
    Underground Water Rush 1891–1910," *Journal of Social History* 37, no. 4
    (2004): 915–37.

12 "Crude extrapolation of current rates of erosion, in conjunction with
    depths of incision into the badlands, suggests that badland development
    started around 200 years ago, probably as a response to the introduction of
    European-style stock farming which resulted in overgrazing." John Boardman,
    David Favis-Mortlock, and Ian Foster, "A 13-Year Record of Erosion on Bad-
    land Sites in the Karoo, South Africa," *Earth Surface Processes and Landforms*
    40, no. 14 (November 2015): 1964–81.

13 Colleen L. Seymour et al., "Twenty Years of Rest Returns Grazing Potential,
    but Not Palatable Plant Diversity, to Karoo Rangeland, South Africa," *Journal
    of Applied Ecology* 47, no. 4 (August 2010): 859–67.

14 Thorsten Wiegand et al., "Live Fast, Die Young: Estimating Size-Age Rela-
    tions and Mortality Pattern of Shrubs Species in the Semi-arid Karoo, South
    Africa," *Plant Ecology* 150 (2000): 115–31.

15 The colloquial names for *Brownanthus ciliatus, Ruschia spinosa, Tripteris
    sinuata*, and *Pteronia pallens* are from W. v. D. Botha and P. C. V. du Toit,
    "Grazing Index Values (GIV) for Karoo Plant Species" (unpublished hand-
    book, Grootfontein Agricultural Development Institute, Middelburg,
    Eastern Cape, South Africa, 2001). The colloquial name for *Galenia fruticosa*
    is from P. A. B. van Breda and S. A. Barnard, *100 Veld Plants of the Winter
    Rainfall Region*, Bulletin No. 422 (Pretoria: Department of Agricultural
    Development, 1991). Thanks to Timm Hoffman for assistance with colloquial
    names; he notes that "the main authority on common names is probably:
    C. A. Smith, *Common Names of South African Plants*, Botanical Survey Mem-
    oir No. 35, 1966, Department of Agricultural Technical Services, Pretoria. I
    have also consulted him on the above, but have included those names that

I hear most commonly during my wanderings in the Karoo" (pers. comm., 19 January 2016). The age ranges for the different species are from Wiegand et al., "Live Fast, Die Young," 128.

16 Nomalanga Mkhize, "Game Farm Conversions and the Land Question: Unpacking Present Contradictions and Historical Continuities in Farm Dwellers' Tenure Insecurity in Cradock," *Journal of Contemporary African Studies* 32, no. 2 (2014): 207–19.

17 Mkhize, "Game Farm Conversions," 211.

18 A presentation to government noted that South Africa had three private game ranches in 1960; by 2015 there were estimated to be between ten thousand and eleven thousand, with a significant increase in populations of species that were extinction risks. Gert Dry, "The Case for Game Ranching: A Biodiversity Economy Imperative" (paper presented to the Department of Environmental Affairs Workshop on Game Ranching, Pretoria, 2 December 2015), https://www.environment.gov.za/sites/default/files/docs/wildliferanching _contributionto_southafrica_economy.pdf.

19 Mkhize, "Game Farm Conversions," 215.

20 Wildlife Ranching South Africa has argued for ranching to be under a national industrial mandate and not a conservation mandate, catering to a global and national hunting sector that would allow breeding for hunters after unusual colour pelts, or even glow-in-the-dark horns—rejecting the conservationists' interests in pristine wildlife. See Gert Dry, "Game Ranching Is a Biodiversity Economy Asset Class and Not a Government Conservation Estate" (paper presented to the 49th South African Society for Animal Studies Congress, Stellenbosch, Western Cape, 4 July 2016), https://www.wrsa.co .za/wp-content/uploads/2016/07/49th-SASAS-GAME-RANCHING-IS -A-BIODIVERSITY-ECONOMY-ASSET-CLASS.pdf.

21 Dominic Boyer, "Energopolitics," in *Energy Humanities: An Anthology*, ed. Imre Szeman and Dominic Boyer (Baltimore: Johns Hopkins University Press, 2017), 184–204.

22 Mike Loewe, "Fracking Task Team Denied Opportunity to Address Meeting," *TimesLive*, 10 November 2015, https://web.archive.org/web /20160224204405/http://www.timeslive.co.za/scitech/2015/11/10/Fracking -task-team-denied-opportunity-to-address-meeting.

23 Michael H. Finewood and Laura J. Stroup, "Fracking and the Neoliberalization of the Hydro-social Cycle in Pennsylvania's Marcellus Shale," *Journal of Contemporary Water Research and Education* 147 (March 2012): 72–79.

24 Ken Saro-Wiwa, *Genocide in Nigeria: The Ogoni Tragedy* (Lagos, Nigeria: Saros International, 1992).

25 For a comparison of Nigerian and Ecuadorian petro-struggles, see Nnimmo Bassey, *To Cook a Continent: Destructive Extraction and the Climate Crisis in Africa* (Dakar, Senegal: Pambazuka Press, 2012).

26 *AfrikaBurn 2015 WTF? Guide*, 2015, accessed 23 May 2018, https://www .afrikaburn.com/wp-content/uploads/2012/04/AB_WTF-Guide-2015.pdf.

27 Created by artists Nathan Victor Honey and Isa Marques in the Karoo in 2014, the *SubTerraFuge* constellates a range of arguments contra fracking as

part of the AfrikaBurn event of 2014. AfrikaBurn, modelled on the Burning Man festival in Nevada, attempts to offer participants the experience of a gift economy for a week. Not without its own struggles over capital, commodity, and race, AfrikaBurn is an important site in which to read emergent forms of environmentalism in South Africa.

28   Kader Asmal, "The Perils of Fracking," *fin24*, 25 April 2011, http://www .fin24.com/Opinion/Columnists/Guest-Columnist/The-perils-of-fracking -20110425.

29   Precisely because the logic of reductionism has contributed to the scale of planetary damage that we now speak of as the Anthropocene, some have of-fered an alternate name for our geological era: the "Metricene"—a term that locates culpability not in all humans per se as the term "Anthropocene" does, nor even in capital per se, but in an approach to the earth based on measur-able relations, without recognition of the other forms of relation that enable planetary ecologies to subsist and evolve.

30   Lesley Green, "What Does It Mean to Require 'Evidence-Based Research' in Decision-Making on Hydraulic Fracturing?," in *Hydraulic Fracturing in the Karoo: Critical Legal and Environmental Perspectives*, ed. Jan Glazewski and Surina Esterhuyse (Cape Town: Juta, 2016), 366–79.

31   Avner Vengosh, Robert B. Jackson, Nathaniel Warner, Thomas H. Darrah, and Andrew Kondash, "A Critical Review of the Risks to Water Resources from Unconventional Shale Gas Development and Hydraulic Fracturing in the United States," *Environmental Science and Technology* 48, no. 15 (2014): 8334–48.

32   Vengosh et al., "A Critical Review of the Risks to Water Resources," 8334.

33   Jolynn Minnaar, dir., *Unearthed* (Cape Town, South Africa: Stage 5 Films, 2014), at 18:21.

34   Bruno Latour, *Politics of Nature: How to Bring the Sciences into Democracy*, trans. Catherine Porter (Cambridge, MA: Harvard University Press, 2004).

35   Achille Mbembe, "Necropolitics," trans. Libby Meintjies, *Public Culture* 15, no. 1 (Winter 2003): 11–40.

36   Conor Joseph Cavanagh and David Himmelfarb, "'Much in Blood and Money': Necropolitical Ecology on the Margins of the Uganda Protectorate," *Antipode* 47, no. 1 (2015): 55–73.

37   See Michel Serres, *The Natural Contract*, trans. Elizabeth MacArthur and William Paulson (Ann Arbor: University of Michigan Press, 1995).

38   Hermann Scheer, cited in Dominic Boyer, "Energopolitics and the Anthro-pology of Energy," *Anthropology News*, May 2011, 5, https://anthrosource .onlinelibrary.wiley.com/doi/abs/10.1111/j.1556-3502.2011.52505.x.

39   For a detailed discussion on the Renewable Energy Independent Power Producer Procurement Programme (REIPPPP), see Anton Eberhard, Joel Kolker, and James Leigland, *South Africa's Renewable Energy IPP Procure-ment Program: Success Factors and Lessons* (Washington, DC: Public-Private Infrastructure Advisory Facility, World Bank, 2014).

40   See Eberhard, Kolker, and Leigland, *Renewable Energy IPP Procurement Program*; and David Fig, "Hydraulic Fracturing in South Africa: Correcting

the Democratic Deficits," in *New South African Review 3*, ed. J. Daniel, P. Naidoo, D. Pillay, and R. Southall (Johannesburg: Wits University Press, 2013), 173–94.

41 A report titled *The Localisation Potential of Photovoltaics and a Strategy to Support the Large-Scale Roll-Out in South Africa* notes, "The development opportunity in the utility-scale PV market segment is significant in South Africa considering the high irradiance observed in the country. Other drivers that create the potential in the market (and to other market segments) include, inter alia: South Africa (SA) has good solar resources and this is predominant in areas that have low population densities; SA is not faced with space restrictions that affect some other countries, specifically Europe; There is an increasing demand for power in the country; Good business opportunity, as South Africa is seen as the gateway to Africa with a potentially emerging market; However, the growth of the utility-scale PV market is entirely reliant on the government's policy regarding REIPPPP, which limits the total installed capacities that can be established in South Africa within this market segment." Chris Ahfeldt, *The Localisation Potential of Photovoltaics and a Strategy to Support the Large-Scale Roll-Out in South Africa,* Report Prepared for the South African Photovoltaic Industry Association, the World Wildlife Fund, and the Department for Trade and Industry, 2013, http://awsassets.wwf.org .za/downloads/ the_localisation_potential_of_pv_and_a_strategy_to_support_large_ scale_roll_out_in_sa.pdf.

42 See, for example, Lucy Baker and Holle Linnea Wlokas, "South Africa's Renewable Energy Procurement: A New Frontier" (Working Paper 159, Norwich, UK: Tyndall Centre for Climate Change Research, 2014).

43 |han≠kass'o, "Rain Changes People into Frogs," 9 September 1878, in Digital Bleek and Lloyd Collection, University of Cape Town, accessed 21 January 2015, http://lloydbleekcollection.cs.uct.ac.za/stories/737/.

44 Summary of |han≠kass'o, "Rain Changes People into Frogs."

45 Digital Bleek and Lloyd Collection, University of Cape Town, accessed 21 January 2015, http://lloydbleekcollection.cs.uct.ac.za/.

46 Isabelle Stengers, "The Intrusion of Gaia," trans. Andrew Goffey, Modes of Existence, 29 December 2015, http://modesofexistence.org/isabelle-stengers -the-intrusion-of-gaia/.

PART II | PRESENT FUTURES

1 Aimé Césaire, *A Tempest*, trans. Richard Miller (New York: Ubu Repertory Theater Publications, 1969).

2 Nelson Maldonado-Torres, *Against War: Views from the Underside of Modernity* (Durham, NC: Duke University Press, 2008).

3 Emmanuel Levinas, "Ethics and Infinity," *CrossCurrents* 34, no. 2 (1984): 191–203.

4 Jacques Derrida, *Edmund Husserl's "Origin of Geometry,"* trans. John P. Leavey (Lincoln: University of Nebraska Press, 1989).

5 Derrida, *Specters of Marx.*

6   Jacques Derrida, *On Touching—Jean-Luc Nancy* (Stanford, CA: Stanford University Press, 2005).

7   Quoted in translation into English by Gary Wilder, *Freedom Time: Negritude, Decolonization and the Future of the World* (Durham, NC: Duke University Press, 2015), 51.

8   Isabelle Stengers, *Thinking with Whitehead: A Free and Wild Creation of Concepts*, trans. Michael Chase (Cambridge, MA: Harvard University Press, 2011).

9   Green and Green, *Knowing the Day*, ch. 5.

10   Serge Gutwirth and Isabelle Stengers, "The Law and the Commons" (paper presented at the third Global Thematic International Association for the Study of the Commons Conference on the Knowledge Commons, Paris, 20–22 October 2016).

11   Jean-Paul Sartre, "Black Orpheus," trans. John MacCombie, *Massachusetts Review* 6, no. 1 (autumn 1964–winter 1965): 13–52.

THREE | LIFE | #SCIENCEMUSTFALL

Epigraph: Thomas Sankara, "Imperialism Is the Arsonist of Our Forests and Savannahs," in *Thomas Sankara Speaks* (Atlanta: Pathfinder, 1988).

1   The latter two threats were so serious that had this been a formally declared international war, and had the threats been acted upon, the burning of the chemical engineering building and the trashing of the infectious diseases lab may have had consequences serious enough to be declared crimes of war.

2   Jacques Pauw, *The President's Keepers: Those Keeping Zuma in Power and out of Prison* (Cape Town: Tafelberg, 2017).

3   Transcribed from the video "Science Must Fall?," YouTube, video posted by UCT Scientist, 16 October 2016, https://youtu.be/C9SiRNibD14.

4   "Newton's Law of Universal Gravitation," Wikipedia, accessed 21 April 2018, https://en.wikipedia.org/wiki/Newton%27s_law_of_universal_gravitation.

5   Niccolo Guicciardini, *Isaac Newton and Natural Philosophy* (London: Reaktion Books, 2018), 205.

6   Karen Barad, *Meeting the Universe Halfway: Quantum Physics and the Entanglement of Matter and Meaning* (Durham, NC: Duke University Press, 2007).

7   Lesley Green, "Beyond South Africa's 'Indigenous Knowledge–Science' Wars," *South African Journal of Science* 108, nos. 7–8 (2012): 44–54 (article 631), http://dx.doi.org/10.4102/sajs.v108i7/8.631.

8   Bruno Latour, *Pandora's Hope: Essays on the Reality of Science Studies* (Cambridge, MA: Harvard University Press, 1999); Latour, "Recall of Modernity"; Stengers, *Thinking with Whitehead*; Barad, *Meeting the Universe Halfway*; and Didier Debaise, *Nature as Event: The Lure of the Possible*, trans. Michael Halewood (Durham, NC: Duke University Press, 2017).

9   Lesley Green, "Retheorizing the Indigenous Knowledge Debate," in *Africa-Centred Knowledges: Crossing Fields and Worlds*, ed. Brenda Cooper and Robert Morrell (London: James Currey, 2014), 36–50.

10   Latour, "Recall of Modernity."

11   Quoted in Guicciardini, *Newton and Natural Philosophy*, 121.

12   Michel Serres, *Geometry: The Third Book of Foundations*, trans. Randolph Burks (New York: Bloomsbury, 2017), Kindle.

13   The late Paulus Gerdes published many exquisite studies on African mathematics. Among many examples in a lifetime of research, see his *Otthava: Making Baskets and Doing Geometry in the Makhuwa Culture in the Northeast of Mozambique* (self-pub., Lulu, 2010); and *Lunda Geometry: Mirror Curves, Designs, Knots, Polyominoes, Patterns, Symmetries* (self-pub., Lulu, 2008).

14   Guicciardini, *Newton and Natural Philosophy*, 17.

15   Guicciardini, *Newton and Natural Philosophy*, 16.

16   "The quantum disrupts this tidy affair," continues Barad. "A bit of a hitch, a tiny disjuncture in the underlying continuum, and causality becomes another matter entirely. Strict determinism is stopped in its tracks, but the quantum does not leave us with free will either; rather, it reworks the entire set of possibilities made available. Agency and causality are not on-off affairs." *Meeting the Universe Halfway*, 233.

17   Quoted in Guicciardini, *Newton and Natural Philosophy*, 205.

18   In this regard, see Guillhermo Vega Sanabria, "Science, Stigmatization and Afro-pessimism in the South African Debate on AIDS," *Vibrant* 13, no. 1 (2016): 22–51.

19   Nicoli Nattrass, *The AIDS Conspiracy: Science Fights Back* (New York: Columbia University Press, 2012).

20   Lesley Green, "Fisheries Science, Parliament and Fishers' Knowledge in South Africa: An Attempt at Scholarly Diplomacy," *Marine Policy* 60 (2015): 345–52.

21   Mhlaba Memela, "MEC's Call for Probe into Lightning," *Sowetan*, 4 January 2011, https://www.sowetanlive.co.za/news/2011-01-07-mec-dube-calls-for-lightning-probe/.

22   For a more thoughtful response, see Philip de Wet, "How to Deal with a Million Bolts a Month," *Mail and Guardian*, 28 January 2011, https://mg.co.za/article/2011-01-28-how-to-deal-with-a-million-bolts-month/.

23   South African Weather Service figures are cited in "Tragedy as Lightning Strikes More Than Twice in KZN," *Injobo* (KwaZulu-Natal Cooperative Governance and Traditional Affairs internal newsletter), February 2015, accessed 15 March 2018, http://www.kzncogta.gov.za/Portals/0/Documents/newsletters/2015/injobo%20Newsletter%20February%20Edition%202015.pdf (link no longer active).

24   Letitia Watson, "Insurance Claims Flood In," *Independent Online*, 27 July 2014, https://www.iol.co.za/personal-finance/insurance-claims-flood-in-1726042.

25   Department of Environmental Affairs, South Africa, "Minister Edna Molewa Officially Opens Lightning Conference," press release, 12 November 2015, https://www.environment.gov.za/mediarelease/molewa_opens_lightningconference.

26   Arundhati Roy, *Capitalism: A Ghost Story* (Chicago: Haymarket Books, 2014), 7.

27 Arundhati Roy, *Field Notes on Democracy: Listening to Grasshoppers* (Chicago: Haymarket Books, 2009), 169.

28 For a discussion of Stengers's "infernal questions," see Didier Debaise, "The Minoritarian Powers of Thought: Thinking beyond Stupidity with Isabelle Stengers," *SubStance* 145, vol. 47, no. 1 (2018): 17–28; Isabelle Stengers, *In Catastrophic Times: Resisting the Coming Barbarism*, trans. Andrew Goffey (London: Open Humanities Press, 2015).

29 Nattrass, *AIDS Conspiracy*.

30 Joshua Cohen, "Kruiedokters [Herb Doctors], Plants and Molecules: Relations of Power, Wind, and Matter in Namaqualand" (PhD diss., University of Cape Town, 2015).

31 Amelia Hilgart, "Determination of a Robust Metabolic Barcoding Model for Chemotaxonomy in Aizoaceae Species: Expanding Morphological and Genetic Understanding" (PhD diss., University of Cape Town, 2015); and Nicola Wheat, "An Ethnobotanical, Phytochemical and Metabolomics Investigation of Plants from the Paulshoek Communal Area, Namaqualand" (PhD diss., University of Cape Town, 2014).

32 In Lesley Green et al., "Plants, People and Health: Three Disciplines at Work in Namaqualand," *South African Journal of Science* 111, nos. 9–10 (2015): 3a. See also N. L. Etkin and E. Elisabetsky, "Seeking a Transdisciplinary and Culturally Germane Science: The Future of Ethnopharmacology," *Journal of Ethnopharmacology* 100, nos. 1–2 (2005): 23–26.

33 Green et al., "Plants, People and Health," 7.

34 Eduardo Viveiros de Castro, "Perspectival Anthropology and the Method of Controlled Equivocation," *Tipití: Journal of the Society for the Anthropology of Lowland South America* 2, no. 1 (2004): 3–22.

35 Green et al, "Plants, People and Health," 9.

36 Isabelle Stengers and Vinciane Despret, *Women Who Make a Fuss: The Unfaithful Daughters of Virginia Woolf* (Minneapolis: University of Minnesota Press, 2014), 85.

37 For readers unfamiliar with actor-network theory, the key text is Bruno Latour, *Reassembling the Social: An Introduction to Actor-Network Theory* (Oxford: Oxford University Press, 2005). Readers interested in understanding different modes of knowledge-making should see Bruno Latour, *An Inquiry into Modes of Existence*, trans. Catherine Porter (Cambridge, MA: Harvard University Press, 2013).

38 "The meaning of Being is not existence and the timeless preservation of essence, but event, the opening up of the horizon, and the spawning of temporary orders." Peter Sloterdijk, *Philosophical Temperaments: From Plato to Foucault* (New York: Columbia University Press, 2013).

39 Latour, "Recall of Modernity," 11.

40 See Isabelle Stengers, "Experimenting with Refrains: Subjectivity and the Challenge of Escaping Modern Dualism," *Subjectivity* 22, no. 1 (2008): 38–59.

41 See Val Plumwood, *Feminism and the Mastery of Nature* (London: Routledge, 1993).

42 Christopher Munyaradzi Mabeza, *Water and Soil in Holy Matrimony? A Smallholder Farmer's Innovative Agricultural Practices for Adapting to Climate Change* (Bamenda, Cameroon: Langaa RPCIG, 2017).

43 Daniel Eloff, "The Objective Absurdity of Decolonising Science," Rational Standard, 14 October 2016, https://rationalstandard.com/objective-absurdity-decolonizing-science/.

44 In addition to the ABC project, see also Green, "Retheorizing."

45 See Isabelle Stengers, "Book VII: The Curse of Tolerance," in *Cosmopolitics II*, trans. Robert Bononno (Minneapolis: University of Minnesota Press, 2011).

46 Guicciardini, *Newton and Natural Philosophy*.

47 Fukagawa Hidetosi and Tony Rothman, *Sacred Mathematics: Japanese Temple Geometry* (Princeton, NJ: Princeton University Press, 2008).

48 Paulus Gerdes, *Lunda Geometry: Mirror Curves, Designs, Knots, Polyominoes, Patterns, Symmetries* (self-published, Lulu, 2008), and *Otthava: Making Baskets and Doing Geometry in the Makhuwa Culture in the Northeast of Mozambique* (self-published, Lulu, 2010).

FOUR | ROCK | RESISTANCE IS FERTILE

Epigraph: Cited in Elizabeth DeLoughrey and George B. Handley, *Postcolonial Ecologies: Literatures of the Environment* (Oxford: Oxford University Press, 2011), Kindle locations 221–22.

1 Bobby Jordan, "Vengeance Is a Mine, Sayeth the Chairman," *Sunday Times*, 29 November 2015, https://www.pressreader.com/south-africa/sunday-times/20151129/281732678403889.

2 For a discussion of the two terms, see Jason Moore, ed., *Anthropocene or Capitalocene?* (Oakland, CA: PM Press, 2016).

3 CDP Worldwide, *The Carbon Majors Database: CDP Carbon Majors Report 2017* (London: CDP, 2017).

4 Deborah Danowski and Eduardo Viveiros de Castro, *The Ends of the World* (Cambridge, UK: Polity, 2017).

5 Ryley Grunenwald, dir., *The Shore Break* (documentary film), DVD, Frank Films/Marie-Vérité Films, 2014, 90 min., https://www.theshorebreakmovie.com.

6 This was confirmed in a letter written by Palala to anti-hunting activists posing as prospective hunting clients. HTTH Ban Trophy Hunting Facebook page, 1 October 2016, accessed 21 March 2018. An account of their communications, posted on their Facebook page: "We are grateful that you are affording us the opportunity to host you in our upmarket Mama Tau Lodge situated on 7000 acres of prime bushveld in the Limpopo Province. A lion hunt, as with other dangerous game, is a 10-day hunt and in that period you can hunt the sable/s and warthog as well as any other of the big variety of animals on our farm/s (42 000 acres). Lion prices start at U$10,000 and according to the mane, which determines the size of the trophy, can go up to 30/35K." When they were asked if Mama Tau Lodge (the lion-breeding farm) has ties to Palala Boutique Game Lodge and Spa, they said the following: "Yes, Palala Boutique

Hotel is owned by us and is our eco facility. Our hunting lodge Mama Tau is an upmarket Lodge where most of our hunting is done. We have changed the name of Palala Safaris to Hunting Safari S.A. and Palala Game Lodge to Palala Boutique Hotel, because it caused confusion in the market place."

7   Achille Mbembe, *Critique of Black Reason* (Durham, NC: Duke University Press, 2017), 32.

8   Mbembe, *Critique of Black Reason*, 49–50.

9   Mbembe, *Critique of Black Reason*, 42.

10 Marilyn Frye, "White Woman Feminist, 1983–1992," in *Willful Virgin: Essays in Feminism 1976–1992* (Berkeley: Crossing Press, 1992), 147–69.

11 David Graeber, *Debt: The First 5000 Years* (New York: Melville House, 2011).

12 Karl Marx, "Debates on the Law on Thefts of Wood," *Rheinische Zeitung*, 25 October–3 November 1842, trans. Clemens Dutt, https://www.marxists.org /archive/marx/works/download/Marx_Rheinishe_Zeitung.pdf.

13 Karl Marx, *Capital*, vols. 1–3, 1867–94, trans. Ben Fowkes (London: Penguin, 1990–92).

14 For an extended discussion of Marx's concept of metabolic rift, see John Bellamy Foster, *Marx's Ecology: Materialism and Nature* (New York: Monthly Review Press, 2000), 141–77.

15 David Montgomery, *Dirt: The Erosion of Civilizations* (Berkeley: University of California Press, 2008), 110.

16 Colin Turing Campbell, "Cawood, David," in *British South Africa, a History of the Colony of the Cape of Good Hope from Its Conquest 1795 to the Settlement of Albany by the Emigration of 1819 [A.D. 1795–A.D. 1825] with Notices of Some of the British Settlers of 1820* (London: John Haddon, 1897), 218.

17 A. W. Beck (town clerk), "Names of Voters within the Electoral Division of Graham's Town, August 27 1853," *Grahamstown Journal*, July–September 1853.

18 See André Boshoff, Jack Skead, and Graham Kerley, "Elephants in the Broader Eastern Cape—An Historical Overview," in Graham Kerley, Sharon Wilson, and Ashley Massey, "Elephant Conservation and Management in the Eastern Cape: Workshop Proceedings" (Port Elizabeth: Terrestrial Ecology Research Unit, University of Port Elizabeth, 2002), n. 12. See also Ivan Mitford-Barberton and Violet White, *Some Frontier Families: Biographical Sketches of 100 Eastern Province Families before 1840* (Cape Town: Human and Rousseau, 1969).

19 See Timothy J. Stapleton, "Hintsa (c. 1790–1835)," in *Encyclopedia of African Conflicts*, vol. 1, ed. Timothy J. Stapleton (Santa Barbara, CA: ABC-CLIO, 2017). For a semi-fictionalised personal account of a woman caught up in this battle, see W. C. Scully, *The White Hecatomb and Other Stories* (London: Methuen, 1897), ch. 1.

20 Denver Arnold Webb, "Kraals of Guns and Redoubts of Authority: Military Conflict and Fortifications in the Wars of Dispossession and Resistance in South Africa's Eastern Cape, 1780–1894," PhD diss., University of Fort Hare, 2015.

21 See Premesh Lalu, *The Deaths of Hintsa: Post-apartheid South Africa and the Shape of Recurring Pasts* (Cape Town: HSRC Press, 2009).

22 Nathaniel Morgan, "An Account of the Amakosae," *South African Quarterly Journal* 1, no. 3 (October 1833): S.1–12, 33–48, 65–71.

23 Tim Keegan, *Colonial South Africa: The Origins of Racial Order* (Cape Town: David Philip, 1996), 141.

24 Morgan, "Account of the Amakosae."

25 Cecil Rhodes, "Confession of Faith," 2 June 1877, in *The British Sense of Mission as a Ruling People*, by Craig M. White (Australia: History Research Projects, 2001), appendix 5, 111–13.

26 See "Sketch Map of South Africa Showing British Possessions, July 1885," The British Empire (website of Stephen Luscombe), accessed 12 November 2016, http://www.britishempire.co.uk/images3/southafrica1885map.jpg.

27 Joanne Bloch, *The Tree Man: Robert Mazibuko's Story* (Pietermaritzburg: New Readers), accessed 5 February 2017, http://live.fundza.mobi/home/library/non-fiction-articles-profiles/the-tree-man-robert-mazibukos-story.

28 Sol Plaatje, *Native Life in South Africa: Before and since the European War and the Boer Rebellion* (London: P. S. King, 1916).

29 See "Sketch Map of South Africa."

30 Plaatje, *Native Life*, 59–60.

31 Bloch, *Tree Man*.

32 Mabeza, *Water and Soil*, 69.

33 See Francis B. Nyamnjoh, "Cameroonian Bushfalling: Negotiation of Identity and Belonging in Fiction and Ethnography," *American Ethnologist* 38, no. 4 (2011): 701–13; Francis B. Nyamnjoh, "Fiction and Reality of Mobility in Africa," *Citizenship Studies* 17, nos. 6–7 (2013): 653–80.

34 Sankara, "Imperialism Is the Arsonist."

35 Penine Uwimbabazi, "Analysis of Umuganda: The Policy and Practice of Community Work in Rwanda" (PhD diss., College of Humanities at the University of KwaZulu-Natal, Pietermaritzburg, South Africa, 2012). See also Dan Ngabonziza, "Rwanda to Plant 43,000 Hectares of Trees in Six Months," *KT Press*, 26 October 2018.

36 Wangari Maathai, *Unbowed: A Memoir* (New York: Alfred A. Knopf, 2006), 119–39.

37 Wangari Maathai, *Replenishing the Earth: Spiritual Values for Healing Ourselves and the World* (New York: Doubleday, 2010), 129.

38 Byron Caminero-Santangelo, *Different Shades of Green: African Literature, Environmental Justice and Political Ecology* (Charlottesville: University of Virginia Press, 2014).

39 Gabriel Setiloane, "How the Traditional World-View Persists in the Christianity of the Sotho-Tswana," 1978, Michigan State University Libraries African e-Journals Project typescript, 38, http://pdfproc.lib.msu.edu/?file=/DMC/African%20Journals/pdfs/PULA/pula001001/pula001001003.pdf, original at http://digital.lib.msu.edu/projects/africanjournals/html/issue.cfm?colid=256.

40 Hahn, *Tsuni-‖Goam*, 112.

41 Derrida, *On Touching*.

42 Césaire, "Return to the Native Land," in *Aimé Césaire: The Collected Poetry*, trans. Clayton Eshleman and Annette Smith (1939; Berkeley: University of California Press, 1983).

43 Nancy, *Birth to Presence*.

44 From the *Oxford English Dictionary*: "Anglo-Norman and Middle French culture (French culture) action of cultivating land, plants, etc., husbandry (12th cent. in Anglo-Norman), (piece of) cultivated land (12th cent. in Anglo-Norman), formation, training (13th cent. in Anglo-Norman), worship or cult of someone or something (14th cent. or earlier in Anglo-Norman), cultivation, development (of language, literature, etc.) (1549), mental development through education (1691), intellectual and artistic conditions of a society or the (perceived) state of development of those conditions, also the ideas, customs, etc. of a society or group (1796, after German Kultur) and its etymon classical Latin cultūra cultivation, tillage, piece of cultivated land, care bestowed on plants, mode of growing plants, training or improvement of the faculties, observance of religious rites (2nd cent. A.D. in this sense), in post-classical Latin also rites (Vetus Latina), veneration of a person (late 2nd or early 3rd cent. in Tertullian), training of the body (5th cent.) < cult-, past participial stem of colere to cultivate, to worship (see cult n.) + -ūra -ure suffix1. In branch III., and especially in senses 6 and 7, also influenced by German Kultur, both directly and via French. The German word is a 17th-cent. borrowing < French, but the transfer of the meaning 'state of intellectual development' from an individual to the whole of a society occurred in German in the mid-18th cent." OED *Online*, s.v. "culture, n.," accessed 22 January 2017, http://www.oed.com/viewdictionaryentry/Entry/45746.

45 For a comprehensive review of ubuntu in Constitutional Court rulings in South Africa, see Drucilla Cornell and Nyoko Muvangua, eds., *uBuntu and the Law: African Ideals and Postapartheid Jurisprudence* (New York: Fordham University Press, 2012).

46 Nkonko M. Kamwangamalu, "Ubuntu in South Africa: A Sociolinguistic Perspective to a Pan-African Concept," *Critical Arts: A South-North Journal of Cultural and Media Studies* 13, no. 2 (1999): 24–41.

47 Kamwangamalu, "Ubuntu in South Africa," 25.

48 Munyaradzi Murove, "Ubuntu," *Diogenes* 59, nos. 3–4 (2012): 10, https://doi .org/10.1177/0392192113493737.

49 Artwell Nhemachena, "Knowledge, Chivanhu and Struggles for Survival in Conflict-Torn Manicaland, Zimbabwe" (PhD diss., University of Cape Town, 2014).

50 Environmental Affairs and Development Planning, Western Cape Government, *Western Cape Climate Change Response Strategy*, February 2014, https://www.westerncape.gov.za/text/2015/march/western_cape_climate _change_response_strategy_2014.pdf.

51 Gina Ziervogel et al., "Inserting Rights and Justice into Urban Resilience: A Focus on Everyday Risk," *Environment and Urbanization* 29, no. 1 (2017): 134.

52 Quoted in Alister Doyle, "South Africa Compares World Climate Change Plan to 'Apartheid,'" *Mail and Guardian*, 20 October 2015, http://mg

.co.za/article/2015-10-20-south-africa-compares-global-climate-plan-to
-apartheid.

53 Isabelle Stengers, *Another Science Is Possible: A Manifesto for Slow Science*,
trans. Stephen Muecke (Cambridge, UK: Polity, 2018), 120.

54 Seth Denizen and Etienne Turpin, "The Stratophysics of Urban Soil Production," in *Land and Animal and Non-animal*, ed. Anna-Sophie Springer and
Etienne Turpin (Berlin: K-Verlag, 2015), 50–73.

PART III | FUTURES IMPERFECT

Epigraph: Didier Debaise, "The Minoritarian Powers of Thought: Thinking beyond Stupidity with Isabelle Stengers," *SubStance* 145, vol. 47, no. 1 (2018): 19.

1 Isabelle Stengers, "Thinking with Deleuze and Whitehead: A Double Test,"
in *Deleuze, Whitehead, Bergson: Rhizomatic Connections*, ed. Keith Robinson, 28–44 (London: Palgrave Macmillan, 2009).

2 Marisol de la Cadena, *Earth Beings: Ecologies of Practice across Andean Worlds*
(Durham, NC: Duke University Press, 2015).

3 Isabelle Stengers, *In Catastrophic Times: Resisting the Coming Barbarism*,
trans. Andrew Goffey (London: Open Humanities Press, 2015), 55.

4 Jennifer Ferguson performed at Monkey Valley, Noordhoek, Cape Town, 17
November 2016.

5 Val Plumwood, *Feminism and the Mastery of Nature* (London: Routledge,
1993), 3.

FIVE | LIFE | WHAT IS IT TO BE A BABOON . . . ?

1 ANC Western Cape, "The Real Cape Town Story—ANC WCape," Politicsweb,
10 May 2011, https://www.politicsweb.co.za/opinion/the-real-cape-town
-story-anc-wcape.

2 City of Cape Town, *Integrated Annual Report 2014/15*, n.d., accessed 23
May 2018, http://splshortcourses.co.za/available-courses/diploma-in
-public-accountability-i-2016/07.-public-financial-accounting-1/additional
-documents/Integrated%20annual%20report%20City%20of%20CT.pdf, 131.

3 Lucie A. Moller, *Of the Same Breath: Indigenous Animal and Place Names*
(Bloemfontein: Sun Press, 2017), 137.

4 Moller, *Of the Same Breath,* 139.

5 William Challis, "The Roots of Power: Baboons and the Medicines of Protection," in "The Impact of the Horse on the Amatola 'Bushmen'" (D.Phil. diss.,
University of Oxford, 2008), 144–76.

6 In "The Roots of Power" (20–21), Challis notes, "Black farmers from the
southern Drakensberg . . . who migrated to the eastern Cape during the
turmoil of the early nineteenth century, rose to prominence as diviners among
the Cape Nguni people who adopted them—drawing on their status as
people who had learned religious practices from the Bushmen. Some of these
diviners had personal sacred animals, among which the baboon was prominent.
Baboon fur hats were part of their religious regalia. One specific category of

diviners among the Cape Nguni came to prominence during times of war—the wardoctor—who administered special medicines to the army. Among the Xhosa of the eastern Cape these wardoctors were known as amatola, and the category of root medicines they administered was believed to protect one from harm—particularly projectiles—and also to confound and incapacitate one's enemies. These medicinal roots were also appropriated by ordinary people and used as protection while on livestock raiding expeditions. Importantly, Khoe-speakers also employed root medicines as war amulets, and Bushmen—from whom the practice may have stemmed—believed that the roots were used by baboons, who were considered invulnerable to harm and sickness."

7 Brian Hungwe, "Letter from Africa: Thomas Mapfumo," *BBC News*, 2 May 2018, http://www.bbc.com/news/world-africa-43963049.

8 See Dorothy L. Cheney and Robert M. Seyfarth, *Baboon Metaphysics: The Evolution of a Social Mind* (Chicago: University of Chicago Press, 2007), 31.

9 Pieter du Plessis, "Jack the Signalman," accessed 24 May 2018, https://railwayjade.wordpress.com/2015/09/30/railway-signalling-jack-the-baboon-signalman/.

10 F. W. Fitzsimons, *The Monkeyfolk of South Africa*, 2nd ed. (London: Longmans, Green, 1924), 38–39.

11 Cheney and Seyfarth, *Baboon Metaphysics*, 17.

12 Ancient Egyptians appear to have been so impressed by baboon sexuality that they added baboon faeces to aphrodisiacs. See Cheney and Seyfarth, *Baboon Metaphysics*, 16.

13 Cheney and Seyfarth, *Baboon Metaphysics*, 19.

14 Cheney and Seyfarth, *Baboon Metaphysics*, 21.

15 Dirk Klopper, "Boer, Bushman, and Baboon: Human and Animal in Nineteenth-Century and Early Twentieth-Century South African Writings," *Safundi: The Journal of South African and American Studies* 11, nos. 1–2 (2010): 3–18.

16 John P. Foley, "The 'Baboon Boy' of South Africa," *American Journal of Psychology* 53, no. 1 (January 1940): 128–33.

17 Robert Zingg, "More about the 'Baboon Boy' of South Africa," *American Journal of Psychology* 53, no. 3 (1940): 455–62.

18 Eckard Smuts, "The 'Baboon Boy' of the Eastern Cape and the Making of the Human in South Africa," *Social Dynamics* 44, no. 1 (2018): 146–57.

19 See Jeff Wicks, "'It's Just the Facts'—Penny Sparrow Breaks Her Silence," *News24*, 4 January 2016, https://www.news24.com/SouthAfrica/News/its-just-the-facts-penny-sparrow-breaks-her-silence-20160104.

20 Office of the ANC Chief Whip, Luthuli House, Johannesburg, "Specific Law to Criminalize Racism and Promotion of Apartheid Necessary," 5 January 2016, accessed 27 January 2016, http://www.anc.org.za/caucus/docs/pr/2016/pr0105.html.

21 Richard Dukelow, *The Alpha Males: An Early History of the Regional Primate Centers* (Lanham, MD: University Press of America, 1995), 184.

22 For a detailed discussion of racism and "simianisation," see Wulf D. Hund, Charles W. Mills, and Silvia Sebastiani, eds., *Simianization: Apes, Gender,*

Class, and Race (Berlin: Lit, 2015). Some of the many simian insults of Barack Obama are described on p. 10.

23 Malegapuru Makgoba, "Wrath of Dethroned White Males," *Mail and Guardian*, 25 March 2005, http://mg.co.za/article/2005-03-25-wrath-of-dethroned-white-males.

24 Donna Haraway, *When Species Meet* (Minneapolis: University of Minnesota Press, 2008), 309.

25 By the end of 2014, Cape Town was assessed as the fastest-growing city in South Africa, according to a report commissioned by the City of Cape Town: *State of Cape Town 2014: Celebrating 20 Years of Democracy*, accessed 23 March 2015, http://www.capetown.gov.za/en/stats/CityReports/Documents/SOCT%2014%20report%20complete.pdf (no longer available online).

26 Esme Beamish, "Causes and Consequences of Mortality and Mutilation in the Cape Peninsula Baboon Population" (master's thesis, University of Cape Town, 2009).

27 Vinciane Despret, *What Would Animals Say If We Asked the Right Questions?*, trans. Brett Buchanan (Minneapolis: University of Minnesota Press, 2014).

28 In correspondence between O'Riain and Trethowan between 26–28 October 2016, shared with the author by Trethowan, O'Riain advised her that data collected by the UCT Baboon Research Unit was subject to UCT copyright law since BRU had funded and collected all baboon population census data from 2003 to 2013 and had never placed it in the public domain for public use. Some of that data, he noted, had been used by postgraduates for their dissertations, in terms of the university's intellectual property policy, and that in terms of that same policy, she was not permitted to distribute, adapt, perform or display the work in public. If his claim was correct, that would imply that no completed dissertation lodged in the UCT library was available for data citation. In a letter to the trustees of Baboon Matters on 29 October 2016, Trethowan noted that "The 1999–2012 data were supplied to me freely by Ruth Kansky who actually did the ground counts. Ruth has nothing to do with UCT or BRU. . . . The ground counts and research mentioned on that slide are to do with the fact that they (UCT, BRU, City of Cape Town) would NOT make any information available despite my many requests for the data. . . . [On] the slide information [was] credited to EB (Esme Beamish, a BRU graduate researcher) and HWS was sourced from reports paid for by the City of Cape Town and contained on their web sites. . . . In 2005 ground counts were done separately by both Kansky and Beamish. Beamish's count was identical to Kansky's count of 2005. In the written document I refer only to Kansky's data for the ground counts 1999–2012."

29 Primatology was at the core of the science wars in the United States in the 1990s. For an attempted mediation of these wars among primatologists, see Shirley C. Strum and Linda Marie Fedigan, *Primate Encounters: Models of Science, Gender and Society* (Chicago: University of Chicago Press, 2000). For an extraordinary overview of racism, patriarchy, and coloniality in primate sciences, see Donna Haraway, *Primate Visions: Gender, Race and Nature*

in the World of Modern Science (New York: Routledge, 1989). For a review of primatologies focused on male dominance, see Vinciane Despret, "H for Hierarchies: Might the Dominance of Males Be a Myth?," in *What Would Animals Say*, 53–59.

30 Despret, *What Would Animals Say.*

31 Thelma Rowell, "The Concept of Social Dominance," *Behavioral Biology* 11 (1974): 131–54; and Barbara Smuts, *Sex and Friendship in Baboons* (New Brunswick, NJ: Transaction, 1985).

32 Anselm Franke and Hila Peleg, *Ape Culture* (Berlin: Haus der Kulturen der Welt, 2015).

33 Beamish, "Mortality and Mutilation."

34 T. S. Hoffman and M. J. O'Riain, "The Spatial Ecology of Chacma Baboons (*Papio ursinus*) in a Human Modified Environment," *International Journal of Primatology* 32 (2010): 308–28; T. S. Hoffman and M. J. O'Riain, "Monkey Management: Using Spatial Ecology to Understand the Extent and Severity of Human-Baboon Conflict in the Cape Peninsula, South Africa," *Ecology and Society* 17, no. 3 (2012), http://dx.doi.org/10.5751/ES-04882-170313; and T. S. Hoffman and M. J. O'Riain, "Troop Size and Human-Modified Habitat Affect the Ranging Patterns of a Chacma Baboon Population in the Cape Peninsula, South Africa," *American Journal of Primatology* 74, no. 9 (September 2012): 853–63.

35 J. Drewe, M. J. O'Riain, E. Beamish, and H. Currie, "A Survey of Infections Transmissible between Baboons and Humans in Cape Town, South Africa," *Emerging Infectious Diseases* 18, no. 2 (2012): 298–301.

36 Bentley Kaplan, Justin O'Riain, Rowen van Eeden, and Andrew J. King, "A Low-Cost Manipulation of Food Resources Reduces Spatial Overlap between Baboons," *International Journal of Primatology* 32, no. 6 (2011): 1397–1412.

37 Quoted in Kimon de Greef, "On the Front Lines of South Africa's Baboon Wars," *Outside Online*, 5 September 2017, https://www.outsideonline.com /2231291/frontlines-south-africas-human-vs-baboon-war.

38 Donna Haraway, email to the author, 2 April 2015.

39 For summaries of the range of animals living in and around Cape Town, see Belinda Ashton, *Living with Our Wild Neighbours* (Cape Town: The Nature Connection, 2013); and Wally Petersen, *Cape of Good Hope Wildlife Guide* (Cape Town: Kommetjie Environmental Action Group, 2015).

40 Sarah Ommanney, *Lacuna: Groote Schuur Zoo* (Cape Town: Centre for Curating the Archive, Michaelis School of Fine Art, University of Cape Town, 2012), 27.

41 The figure of 350 was offered by the late Wally Petersen in his account of terrestrial animals on the Cape Peninsula for the Kommetjie Environmental Action Group. Were marine creatures to be included, the figure would be much higher.

42 Douglas Hey, "The Control of Vertebrate Problem Animals in the Province of the Cape of Good Hope, Republic of South Africa," in *Proceedings of the 2nd Vertebrate Pest Control Conference*, paper 11 (Lincoln: University of Nebraska Press, 1964), http://digitalcommons.unl.edu/vpc2/11.

43 According to Jerome Singh, "the TRC [Truth and Reconciliation Commission] also heard from Dr Jan Lourens, the head of Protechnik (another Project Coast front organization), who testified that before starting the company, he had designed equipment for animal experiments taking place at RRL. These included a 'restraint chair' into which baboons were strapped for experiments, a transparent 'gas chamber' into which the chair and baboon were fitted for tests and a 'stimulator and extractor' to obtain semen from baboons. Lourens named Dr Riana Borman as the scientist in charge of the baboon experiments 'to control virility and fertility' with a view to reducing the birth rate among blacks." Singh, "Project Coast: Eugenics in Apartheid South Africa," *Endeavour* 32, no. 1 (2008): 5–9, https://wikileaks.org/gifiles/attach/169/169033_ProjectCoastEugenics.pdf.

44 After the Truth Commission, Basson continued to practice as a specialist cardiologist in Cape Town, notwithstanding a lengthy battle with the Health Professions Council of South Africa.

45 Johan G. Brink and David K. C. Cooper, "Heart Transplantation: The Contributions of Christiaan Barnard and the University of Cape Town/Groote Schuur Hospital," *World Journal of Surgery* 29 (2005): 953–61.

46 Peta S. Cook, "Science Stories: Selecting the Source Animal for Xenotransplantation" (paper presented at the Social Change in the 21st Century Conference, Queensland University of Technology, 27 October 2006, Brisbane, Australia).

47 Michael de Vaal, "In Vivo Mechanical Loading Conditions of Pectorally Implanted Cardiac Pacemakers: Feasibility of a Force Measurement System and Concept of an Animal-Human Transfer Function" (master's thesis, University of Cape Town, 2012); and Laura A. Cox et al., "Baboons as a Model to Study Genetics and Epigenetics of Human Disease," *Institute for Laboratory Animal Research (ILAR) Journal* 54, no. 2 (2013): 106–21.

48 Beamish, "Mortality and Mutilation," 8.

49 Elisa Galgut, "Raising the Bar in the Justification of Animal Research," *Journal of Animal Ethics* 5, no. 1 (2015): 5–19.

50 Beamish, "Mortality and Mutilation," 8.

51 At that time, the Baboon Management Team consisted of representatives from all the authorities and KEAG, Baboon Matters, a primatologist (at that time, Kansky), representatives of ratepayers from the villages affected, and the SPCA.

52 One opponent described Trethowan, head of Baboon Matters, as the "Mother Theresa of Baboons." Jon Abbott, "Who's the Monkey If Baboons Are Free While People Live in Cages?," *Dearjon Letter* (blog), 30 August 2011, http://dearjon-letter.blogspot.com/2011/08/whos-monkey-if-baboons-are-free-while.html.

53 Shirley C. Strum, "Activists and Their Anthropomorphism Remain the Greatest Threat to Baboons," *Cape Times*, 23 July 2012.

54 Notes Kansky, "BM [Baboon Matters] was a convenient scapegoat at the time because they had been opposing the culling—actually not so much opposing culling per se but rather saying authorities needed to do more to manage

waste and people so that baboons would not be needed to be culled." Pers. comm., 4 April 2015.

55 Justin O'Riain of the BRU, University of Cape Town, as minuted in City of Cape Town, SANParks, and Cape Nature, "Proceedings of Baboon Expert Workshop," 2 July 2009, http://resource.capetown.gov.za/documentcentre /Documents/Procedures,%20guidelines%20and%20regulations/Baboon _Expert_Workshop_Proceedings_2009-07.pdf, 20 (no longer available online).

56 Among the methods O'Riain proposed researching were a baboon monitor system; GPS collars and virtual barriers; provisioning; electric fencing; the use of light prisms; baboon vocalisation playbacks; condition taste aversion; and the use of bear bangers and bullwhips.

57 Ruth Kansky, *Baboons on the Cape Peninsula: A Guide for Residents and Tourists* (Cape Town: International Fund for Animal Welfare, 2002).

58 Kansky, pers. comm., 4 April 2015.

59 In the discussion at the crisis meeting cited above, O'Riain is minuted as commenting, "People are scared to take ownership of their territory." City of Cape Town, SANParks, and Cape Nature, "Proceedings of Baboon Expert Workshop," 29.

60 Human Wildlife Solutions, *Monthly Report: August 2014* (Tokai, South Africa: Human Wildlife Solutions, 2014), 5.

61 Human Wildlife Solutions, *Monthly Report: August 2014*, 9.

62 Human Wildlife Solutions, *Monthly Report: August 2014*, 7.

63 Haraway, *When Species Meet*.

64 The phrasing is drawn from Judith Butler and developed in Kathryn A. Gillespie and Patricia J. Lopez, "Introducing Economies of Death," in *Economies of Death: Economic Logics of Killable Life and Grievable Death*, ed. Patricia J. Lopez and Kathryn A. Gillespie (London: Routledge, 2015).

65 De Greef, "South Africa's Baboon Wars."

66 Shanaaz Eggington, "'I Am Living in a War Zone,' Says Resident as Baboons Take Over," *TimesLive*, 19 January 2016, http://www.timeslive.co.za/thetimes /2016/01/19/I-am-living-in-a-war-zone-says-resident-as-baboons-take-over.

67 Ruth Kansky, interview with the author, 14 March 2015.

68 Kansky, interview (sentence reconstructed from notes).

69 Haraway, *Primate Visions*, 7–8.

70 Haraway, *Primate Visions*, 2.

71 Donna Haraway, "Animal Sociology and a Natural Economy of the Body Politic, Part II: The Past Is the Contested Zone: Human Nature and Theories of Production and Reproduction in Primate Behavior Studies," in "Women, Science, and Society," special issue, *Signs* 4, no. 1 (Autumn 1978): 37.

72 Franke and Peleg, *Ape Culture*, 13–14.

73 Dussel, *Philosophy of Liberation*, 33.

74 Maldonado-Torres, *Against War*.

75 Franke and Peleg, *Ape Culture*, 12.

76 Thom van Dooren and Deborah Bird Rose, "Storied-Places in a Multispecies City," *Humanimalia* 3, no. 2 (2012): 2.

77 From Scarborough resident Ushka Devi's Facebook page, 12 February 2016, accessed 16 February 2016, https://www.facebook.com/ushka.mrkusic.

78 Dia!kwain, "Baboons Should Not Be Spoken With," in Digital Bleek and Lloyd Collection, University of Cape Town, accessed 21 January 2015, http://lloydbleekcollection.cs.uct.ac.za/stories/544/index.html.

79 Dia!kwain, "Baboons Dance the ≠gebbi-ggu," in Digital Bleek and Lloyd Collection, University of Cape Town, accessed 21 January 2015, http://lloydbleekcollection.cs.uct.ac.za/stories/559/index.html.

80 Dia!kwain, "Baboons Know Our Names," in Digital Bleek and Lloyd Collection, University of Cape Town, accessed 21 January 2015, http://lloydbleekcollection.cs.uct.ac.za/stories/550/index.html; see also Dia!kwain, "1876," in Pippa Skotnes, *Claim to the Country: The Archive of Wilhelm Bleek and Lucy Lloyd* (Johannesburg: Jacana Media; Athens: Ohio University Press, 2007), 332.

81 For example, in correspondence from late 2017 between Julia Wood in her capacity as chair of the Baboon Technical Team for the city of Cape Town and Jenni Trethowan in her capacity as chair of the Baboon Matters Trust, Trethowan requests the revised protocols for sick and injured baboons; protocol for euthanasia of baboons; population census for 2016 as completed in February 2017; information regarding water tanks for specified troops of baboons where there is a risk of their running out of water; dates when baboons identified as suffering from mange (photographs supplied and dates given) would be treated (email, Trethowan to Wood, 27 November 2017).

None of the protocols was sent. On allegations of inhumane paint-balling, which had been detailed in prior correspondence, Wood advised that the rangers were trained and monitored and that experts had advised paintballing; further, the letter noted that "full details and supporting documentation of any incidents are required. It is difficult to use emotional tirade against the practice of paintballing as given by animal rights activists, conspiracy theories or unverified hearsay for the purpose of processing disciplinary action" (email, Wood to Trethowan, 28 November 2017). In other words, Trethowan's specified concerns giving names and exact situations were dismissed as emotional hysteria or conspiracy theories. Where Trethowan had supplied photographs from a few days before of baboons suffering mange, she was told the condition had been treated six months before. At no point in the correspondence were Trethowan's concerns for or specific observations of baboon well-being taken seriously. The message is that the experts know what is best; that monitors are monitored; that everything is under control; and that Trethowan's organisation is unreasonable and irrational. The letter is a form of managerial power play that, using neutered terms, renders a citizen powerless by gendering them female. That the Baboon Management Team is eyes and ears on the ground that are valuable with regard to baboon well-being clearly eludes the city's Baboon Technical Team here. This is not what one could or should reasonably expect from a technical team that is surely tasked with investigating evidence that citizens provide.

82 Isabelle Stengers, "An Ecology of Practices," *Cultural Studies Review* 11, no. 1 (2005): 193.

83 Stengers and Despret, *Women Who Make a Fuss*, 66.

SIX | WATER | OCEAN REGIME SHIFT

1 D. A. Horstman et al., "Red Tides in False Bay, 1959–1989, with Particular Reference to Recent Blooms of *Gymnodinium* sp.," *Transactions of the Royal Society of South Africa* 47, nos. 4–5 (1991): 611–28.

2 Oliver Schultz, "Belonging on the West Coast: An Ethnography of St Helena Bay in the Context of Marine Resource Scarcity" (master's thesis, University of Cape Town, 2010).

3 CSIR, Cape Town Outfalls Monitoring Programme: Surveys Made in 2015/2016. CSIR Report CSIR/NRE/ECOS/IR/2017/0035/B. 2017.

4 City of Cape Town, "CSIR Confirms Sewage Outfalls Pose No Significant Risks" (press release), 21 November 2017.

5 CSIR, Cape Town Outfalls Monitoring Programme, 224.

6 Jacques Coetzee, "Job Losses Threaten the CSIR's Reputation," *Mail and Guardian*, 22 March 2019. https://mg.co.za/article/2019-03-22-00-job-losses-threatens-the-csirs-reputation.

7 Coetzee, "Job Losses Threaten the CSIR's Reputation."

8 Sue Segar, "Raw Deal: Desalinator's Job Is to Remove Salt, Not Sewage," *Noseweek* 236, 1 June 2019. https://www.noseweek.co.za/articles/4277/Raw-deal-desalinators-job-is-to-remove-salt,-not-sewage.

9 Melanie Gosling, "Cape Town Council's Desalination Debacle: Seawater 400% More Polluted Than City of Cape Town's Tender Data Indicated," *News24*, 27 May 2019, https://www.news24.com/SouthAfrica/News/va-desalination-debacle-seawater-400-more-polluted-than-citys-tender-data-indicated-20190527.

10 "Councillor Xanthea Limberg, mayoral committee member for water and waste, maintained that QFS was responsible for 'taking cognisance of normal variations in water quality' and for building a plant 'robust enough to accommodate raw water quality during all seasonal variations.'" Aletta Harrison, "Exclusive: Desalination Company to Take City of Cape Town to Court for Breach of Contract," *News24*, 13 June 2019.

11 S. J. Johnston and D. S. Butterworth, "Rock Lobster Scientific Working Group Agreed Recreational Catch Estimates" (unpublished paper, Department of Mathematics and Applied Mathematics, University of Cape Town, 2010), https://open.uct.ac.za/handle/11427/18920.

12 See Nancy N. Chen, "Speaking Nearby: A Conversation with Trinh T. Minh-Ha," *Visual Anthropology Review* 8, no. 1 (1992): 82–91.

13 Elspeth Probyn, *Eating the Ocean* (Durham, NC: Duke University Press, 2016).

14 Despret, *What Would Animals Say?*

15 Césaire, "Poetry and Knowledge."

16 William R. Bascom, *African Dilemma Tales* (The Hague: Mouton, 1975).

17 De Castro, "Perspectival Anthropology."

18 For further discussion of this point, see Green and Green, *Knowing the Day*, chs. 1 and 4.

19 Francois Jullien, *The Propensity of Things: Toward a History of Efficacy in China* (Cambridge, MA: Zone Books, 1999).

20 Blue Flag home page, accessed 2 January 2017, http://www.blueflag.global.

21 Lesley Green, "The Changing of the Gods of Reason: Cecil John Rhodes, Karoo Fracking and the Decolonising of the Anthropocene," in "Supercommunity," special issue, *e-flux Journal*, no. 65 (May–August 2015), http://supercommunity.e-flux.com/texts/the-changing-of-the-gods-of-reason/.

22 Lance van Sittert, "The Fire and the Eye: Fishers' Knowledge, Echo-Sounding and the Invention of the Skipper in the St. Helena Bay Pelagic Fishery ca. 1930–1960," *Marine Policy* 60 (2015): 300–308, https://doi.org/10.1016/j.marpol.2014.07.028.

23 Lizeth Botes, "Taxonomy, Distribution and Toxicity of Dinoflagellate Species in the Southern Benguela Current, South Africa" (PhD diss., University of Cape Town, 2003), 60–61.

24 Carl David Rundgren, "Aspects of Pollution in False Bay, South Africa (with Special Reference to Subtidal Pollution)" (master's thesis, University of Cape Town, 1992).

25 Jason R. Westrich et al., "Saharan Dust Nutrients Promote *Vibrio* Bloom Formation in Marine Surface Waters," *Proceedings of the National Academy of Sciences* 113, no. 21 (2016): 5964–69.

26 An earlier version of this section was published in Lesley Green, "Calculemos *Jasus lalandii*: Accounting for South African Lobster," in *Reset Modernity!*, ed. Bruno Latour and Christophe Leclerq (Cambridge, MA: MIT Press, 2016), 139–51.

27 Oliver Sacks, "The Mental Life of Plants and Worms," *New York Review of Books*, 24 April 2014, http://www.nybooks.com/articles/archives/2014/apr/24/mental-life-plants-and-worms-among-others/.

28 Brendon M. H. Larson, "The War of the Roses: Demilitarizing Invasion Biology," *Frontiers in Ecology and Environment* 3, no. 9 (2005): 495–500.

29 Mike M. Webster et al., "Environmental Complexity Influences Association Network Structure and Network-Based Diffusion of Foraging Information in Fish Shoals," *American Naturalist* 181, no. 2 (February 2013): 235–44; see also M. M. Webster, A. J. W. Ward, and P. J. B. Hart, "Shoal and Prey Patch Choice by Co-occurring Fishes and Prawns: Inter-taxa Use of Socially Transmitted Cues," *Proceedings of the Royal Society B: Biological Sciences* 275 (2008): 203–8; and K. N. Laland, N. Atton, and M. M. Webster, "From Fish to Fashion: Experimental and Theoretical Insights into the Evolution of Culture," *Philosophical Transactions of the Royal Society B: Biological Sciences* 366 (2011): 958–68.

30 E. A. Kravitz, "Serotonin and Aggression: Insights Gained from a Lobster Model System and Speculations on the Role of Amine Neurons in a Complex Behavior," *Journal of Comparative Physiology* 18 (2000): 221–38.

31 See Daisuke Sato and Toshiki Nagayama, "Development of Agonistic Encounters in Dominance Hierarchy Formation in Juvenile Crayfish," *Journal of Experimental Biology* 215 (2012): 1210–17.

32 Larry C. Boles and Kenneth J. Lohmann, "True Navigation and Magnetic Maps in Spiny Lobsters," *Nature* 421 (2003): 60–63; R. D. Bertelsen et al., "Spiny Lobster Movement and Population Metrics at the Western Sambo Ecological Reserve" (paper presented at Linking Science to Management: A Conference and Workshop on the Florida Keys Marine Ecosystem, Duck Key, Florida, 19–22 October 2010).

33 L. J. Atkinson, S. Mayfield, and A. C. Cockroft, "The Potential for Using Acoustic Tracking to Monitor the Movement of the West Coast Rock Lobster *Jasus lalandii*," *African Journal of Marine Science* 27, no. 2 (2005): 401–8.

34 Latour, "Recall of Modernity."

35 Lucy Towers, "South African Fisheries Win Major Victory in New York Court," The Fish Site, 8 July 2013, https://thefishsite.com/articles/south-african-fisheries-win-major-victory-in-new-york-court.

36 S. J. Johnston and D. S. Butterworth, "Rock Lobster Scientific Working Group Agreed Recreational Catch Estimates" (unpublished paper, Department of Mathematics and Applied Mathematics, University of Cape Town, 2010), https://open.uct.ac.za/handle/11427/18920.

37 Elizabeth Povinelli, *Geontologies: A Requiem to Late Liberalism* (Durham, NC: Duke University Press, 2016). Povinelli prefers the term "late liberalism," but for readers' ease, in this short invocation of her argument, I am using the more familiar term "neoliberalism."

38 Stengers and Despret, *Women Who Make a Fuss*, 56–57.

39 For a summary of the argument in the mid-2000s, see Naseegh Jaffer and Jackie Sunde, "Fishing Rights or Human Rights?," *Samudra Report* 44 (July 2006): 20–24.

40 See Patrick Bond, *Elite Transition: From Apartheid to Neoliberalism in South Africa*, rev. ed. (London: Pluto, 2014).

41 Stefano Ponte, Simon Roberts, and Lance van Sittert, "Black Economic Empowerment, Business and the State in South Africa," *Development and Change* 38, no. 5 (2007): 933–55.

42 Lance van Sittert, "'Those Who Cannot Remember the Past Are Condemned to Repeat It': Comparing Fisheries Reforms in South Africa," *Marine Policy* 26 (2002): 302–3.

43 See Marieke Norton, "At the Interface: Marine Compliance Inspectors at Work in the Western Cape" (PhD diss., University of Cape Town, 2014); Marieke van Zyl, "Ocean, Time and Value: Speaking about the Sea in Kassiesbaai," *Anthropology Southern Africa* 32, nos. 1–2 (2009): 48–58; Oliver Schultz, "Belonging on the West Coast: An Ethnography of St Helena Bay in the Context of Marine Resource Scarcity" (master's thesis, University of Cape Town, 2010); and Maria Hauck, "Rethinking Small-Scale Fisheries Compliance," *Marine Policy* 32 (2008): 635–42.

44 For the judgment, see *Minister of Environmental Affairs and Tourism v George and Others*, 437/05, 437/05 (2006) ZASCA 57; 2007 (3) SA 62 (SCA) (18 May 2006), http://www.saflii.org/za/cases/ZASCA/2006/57.html.

45 See Moeniba Isaacs, "Small-Scale Fisheries Governance and Understanding the Snoek (*Thyrsites atun*) Supply Chain in the Ocean View Fishing Commu-

nity, Western Cape, South Africa," *Ecology and Society* 18, no. 4 (2013), http://www.ecologyandsociety.org/vol18/iss4/art17/.

46 Jackie Sunde notes, "The Court allocation of IR [Interim Relief quotas] was made to the fishers involved in the court action only and hence was restricted to the class of fishers represented by Masifundise Development Trust and Coastal Links in the Western and Northern Cape" (pers. comm., 16 March 2014).

47 Department of Agriculture, Forestry and Fisheries, Republic of South Africa, Policy for the Small Scale Fisheries Sector in South Africa, 2012, https://www.gov.za/af/documents/policy-small-scale-fisheries-sector-south-africa.

48 Venue V454 is familiar on South African television screens as the old Chamber of Parliament. It is the room in which the laws of apartheid had been defined via an alliance of race sciences and social sciences; it was the same room in which the architect of apartheid, H. F. Verwoerd, had been assassinated, and in which the last apartheid president, F. W. de Klerk, had announced that Nelson Mandela would be released from prison unconditionally. In the same speech, he prayed that the "Almighty Lord will guide and sustain us on our course through uncharetered waters." ("Uncharted" would have meant "unknown," but "uncharetered" could be read as those not yet governed by corporate privilege. Details, details.) F. W. de Klerk, "Address by the State President Mr F. W. de Klerk at the Opening of the Second Session of the Ninth Parliament of the Republic of South Africa," Cape Town, 2 February 1990, F. W. de Klerk Foundation, https://www.fwdeklerk.org/index.php/en/historically-significant-speeches.

49 "Fishermen Protest at Parly," *fin24*, 29 May 2005, http://www.fin24.com/Business/Fishermen-protest-at-parly-20050529-3.

50 Quoted from the report of the Parliamentary Monitoring Group, "Marine Living Resources Amendment Bill [B30-2013]: Public Hearings Day 2," 16 October 2013, http://www.pmg.org.za/report/20131016-marine-living-resources-amendment-bill-b30-2013-public-hearings-day-2.

51 Parliamentary Monitoring Group, "Marine Living Resources Amendment Bill."

52 Parliamentary Monitoring Group, "Marine Living Resources Amendment Bill."

53 I have translated Smith's words into English from the audio recording provided by the Parliamentary Monitoring Group at https://pmg.org.za/committee-meeting/16545/.

54 Paul Nadasdy, *Hunters and Bureaucrats: Power, Knowledge, and Aboriginal-State Relations in the Southwest Yukon* (Vancouver: University of British Columbia Press, 2003).

55 Sylvia Wynter and Katherine McKittrick, "Unparalleled Catastrophe for Our Species? Or, to Give Humanness a Different Future: Conversations," in *Sylvia Wynter: On Being Human as Praxis*, ed. Katherine McKittrick (Durham, NC: Duke University Press, 2015), 21–22.

56 Researchers will find it useful to work with Latour's rethinking of "economics" as modes of organisation, modes of attachment, and modes of morality. See the final section on economics in his *An Inquiry into Modes of Existence*.

57 "From Hook to Cook," Abalobi, accessed 28 April 2018, http://abalobi.info/hooktocook/.

58 Serge Raemaekers, "'Storied' Seafood and Community-Led Technology," TEDx Talks, published 23 January 2018, video, 11:28, https://www.youtube.com/watch?v=dArhoEfNbwY.

59 "From Hook to Cook."

60 Raemaekers, "'Storied' Seafood."

61 Raemaekers, "'Storied' Seafood." See also Serge Raemaekers and Jackie Sunde, "Extending the Ripples," *Samudra Report* 75 (January 2017), 11–16 https://www.icsf.net/images/samudra/pdf/english/issue_75/4243_art_Sam75_e_art04.pdf.

62 Norton, "At the Interface."

63 Norton, "At the Interface."

64 See Cecilia Y. Ojemaye and Leslie Petrik, "Occurrences, Levels and Risk Assessment Studies of Emerging Pollutants (Pharmaceuticals, Perfluoroalkyl and Endocrine Disrupting Compounds) in Fish Samples from Kalk Bay Harbour, South Africa," *Environmental Pollution* 252A (2019): 562–72, doi: https://doi.org/10.1016/j.envpol.2019.05.091; and Cecilia Y. Ojemaye and Leslie Petrik, "Pharmaceuticals in the Marine Environment: A Review," *Environmental Reviews* 27, no. 2 (2019): 151–65, https://doi.org/10.1139/er-2018-0054.

65 See the widely syndicated article: Steve Kretzmann, "Pharmaceuticals and Industrial Chemicals Found in Fish Caught off Cape Town's Coast," *GroundUp*, 25 June 2019, https://www.groundup.org.za/article/were-eating-our-waste/.

66 For an overview of tertiary treatment efficacy and innovation, see Kassim Olasunkanmi Badmus, Jimoh Oladejo Tijani, Emile Massima, and Leslie Petrik, "Treatment of Persistent Organic Pollutants in Wastewater Using Hydrodynamic Cavitation in Synergy with Advanced Oxidation Process," *Environmental Science and Pollution Research* 25 (2018): 7299–7314, https://doi.org/10.1007/s11356-017-1171-z; and C. D. Swartz, B. Genthe, J. Chamier, L. F. Petrik, J. O. Tijani, A. Adeleye, C. J. Coomans, A. Ohlin, D. Falk, and J. G. Menge, "Emerging Contaminants in Wastewater Treated for Direct Potable Re-use: The Human Health Risk Priorities in South Africa, Volume III: Occurrence, Fate, Removal and Health Risk Assessment of Chemicals of Emerging Concern in Reclaimed Water for Potable Reuse," Report to the Water Research Commission TT 742/3/18, March 2018, Gezina, South Africa.

67 Césaire, *Discourse on Colonialism.*

CODA

1 For further discussion, see Dianne Scott and Clive Barnett, "Something in the Air: Civic Science and Contentious Environmental Politics in Post-apartheid South Africa," *Geoforum* 40 (2009): 373–82, esp. 374.

2 See, for example, Lesley J. F. Green, "Fisheries Science, Parliament and Fishers' Knowledge in South Africa: An Attempt at Scholarly Diplomacy," *Marine Policy* 60 (2015): 345–52.

3 See Jacob Dlamini, "'To Know the African Wild Was to Know the African Subject'—What Training as a Field Guide Taught Jacob Dlamini about Culture, Nature, Power and Race," *Johannesburg Review of Books*, 1 July 2019.

4 See Lesley Green, Nikiwe Solomon, Leslie Petrik, and Jo Barnes, "Environmental Management Needs to Be Democratised," *Daily Maverick*, 6 March 2019, accessed 11 July 2019, https://www.dailymaverick.co.za/article /2019-03-06-environmental-management-needs-to-be-democratised/. This was followed by a response from the City of Cape Town, "Academics Bombshells Trigger Panic Free of Accountability," *Daily Maverick*, 22 March 2019. Our team responded: see Lesley Green, Nikiwe Solomon, Leslie Petrik, and Jo Barnes, "Distortions, Distractions and Falsehoods in the City of Cape Town's Riposte to Kuils River Effluent Article," *Daily Maverick*, 10 April 2019.

5 *Chernobyl,* HBO mini-series, dir. Johan Renck, 2019.

6 Sian Sullivan, "The Natural Capital Myth; or, Will Accounting Save the World? Preliminary Thoughts on Nature, Finance and Values," Leverhulme Centre for the Study of Value Working Papers #3, 2014, http://thestudyofvalue.org/wp- content/uploads/2013/11/WP3-Sullivan-2014-Natural-Capital-Myth.pdf.

7 The Gini coefficient is a measure of the wealth gap in a country. A recent report by the World Bank in partnership with the South African government "found that the country had a Gini coefficient of 0.63 in 2015, the highest in the world and an increase since 1994." World Bank, *Overcoming Poverty and Inequality in South Africa: An Assessment of Drivers, Constraints and Opportunities* (Washington, DC: World Bank Publications, 2018), xv.

8 Marieke Norton, "At the Interface: Marine Compliance Inspectors at Work in the Western Cape" (PhD diss., University of Cape Town, 2014).

9 Marilyn Frye, *Willful Virgin: Essays in Feminism, 1976–1992* (Freedom, CA: Crossing Press, 1992).

10 A. R. Ammons, "The City Limits," in *A. R. Ammons: The Selected Poems: 1951–1977*, expanded ed. (New York : W. W. Norton, 1987).

11 Frye, *Willful Virgin*. See also Minnie Bruce Pratt, "Identity: Skin Blood Heart," in *Yours in Struggle*, ed. Elly Bulkin, Minnie Bruce Pratt, and Barbara Smith (Brooklyn, NY: Long Haul Press, 1984).

12 For climate scientists working in a global arena, the same dynamic was apparent: the social systems that are required to implement "the best scientific advice" are assumed neutral. The carbon trading system translates, in practice, to the global South doing the hard work of adaptation (such as not using REDD+ forests) while polluters in the global North could carry on polluting as long as they paid more to do so. Shifting to "climate-resilient" GM seeds, too, is presented as if they are neutral, even though in their implementation they criminalise the non-monetised economy of seed-gifting and exchanging that is the mainstay of rural African farming society, and rely on herbicides that damage soils.

13 Frye, *Willful Virgin*.

14 Unlike the professional registration required for scientists and engineers in South Africa, there is no registration system for professional social science

researchers. That also means that there is no continuous training requirement; no peer review mechanism to assess the social science consultancies that are done; nor is there a required level of training—anyone who claims to have adequate social science training can set up a consultancy company and, if they are in the right networks, they will get government contracts, and play a major role in environmental policy planning and implementation.

15 S. J. Johnston and D. S. Butterworth, "Rock Lobster Scientific Working Group Agreed Recreational Catch Estimates," Marine Resource Assessment and Management Group, 2010, accessed 2 August 2019, https://open.uct. ac.za/bitstream/item/21646/Johnston_Rock_lobster_scientific_2010.pdf.

16 Donna Haraway, "Anthropocene, Capitalocene, Plantationocene, Chthulu-cene: Making Kin," *Environmental Humanities* 6 (2015): 159–65; Anna Tsing, "More-Than-Human Sociality: A Call for Critical Description," in *Anthropology and Nature*, ed. Kirsten Hastrup, 27–42 (London: Routledge, 2013).

17 Thom van Dooren, "Vultures and Their People in India: Equity and Entanglement in a Time of Extinctions," *Australian Humanities Review* 50 (2011): 45–61.

18 Tim Morton, *Hyperobjects* (Minneapolis: University of Minnesota Press, 2013).

19 Peter Haff, "Humans and Technology in the Anthropocene: Six Rules," *Anthropocene Review* 1, no. 2 (2014): 126–36.

20 McKenzie Wark, *Molecular Red: Theory for the Anthropocene* (London: Verso, 2015).

21 Seth Denizen and Etienne Turpin, "The Stratophysics of Urban Soil Production," in *Land and Animal and Non-animal*, ed. Anna-Sophie Springer and Etienne Turpin, 50–73 (Berlin: K-Verlag, 2015).

22 John Bellamy Foster, *Marx's Ecology* (New York: Monthly Review Press, 2000), 1–20.

23 Jussi Parikka, *The Anthrobscene* (Minneapolis: University of Minnesota Press, 2015).

24 See Basil Enwegbara, "Toxic Colonialism: Lawrence Summers and Let Africans Eat Pollution," *The Tech* 121, no. 16 (6 April 2001), http://tech.mit .edu/V121/N16/col16guest.16c.html; Virginia MacKenny and Lesley Green, "African Artists in Post-colonial Colonial Landscapes: Re-membering the Discarded" (unpublished ms.). Pieter Hugo's images of Agbogbloshie are published in Hugo, *Permanent Error* (New York: Prestel, 2011).

25 See Dominic Boyer, "Energopolitics," in *Energy Humanities: An Anthology*, ed. Imre Szeman and Dominic Boyer, 184–204 (Baltimore: Johns Hopkins University Press, 2017); Timothy Mitchell, *Carbon Democracy: Political Power in the Age of Oil* (New York: Verso, 2011).

26 See Eduardo Gudynas, "Value, Growth, Development: South American Lessons for a New Ecopolitics," *Capitalism Nature Socialism* 30, no. 2 (2019): 234–43. See also Eduardo Gudynas, "Extractivismos y corrupción en América del Sur: Estructuras, dinámicas y tendencias en una íntima relación," *RevIISE* (University of San Juan) 10, no. 10 (2017): 73–87.

27 Roy, *Capitalism*.

28 Victor Galaz, Beatrice Crona, Alice Dauriach, Jean-Baptiste Jouffray, Henrik Österblom, and Jan Fichtner, "Tax Havens and Global Environmental Degradation," *Nature Ecology and Evolution* 2 (2018): 1352–57.

29 Steven A. Banwart, Nikolaos P. Nikolaidis, Yong-Guan Zhu, Caroline L. Peacock, and Donald L. Sparks, "Soil Functions: Connecting Earth's Critical Zone," *Annual Review of Earth and Planetary Sciences* 47, no. 1 (2019): 333–59.

30 Jason Moore, *Anthropocene or Capitalocene? Nature, History and the Crisis of Capitalism*, ed. Jason Moore (Oakland, CA: Kairos PM Press, 2016).

31 Jason Moore, *Capitalism in the Web of Life* (New York: Verso, 2015).

32 Anna Tsing, *The Mushroom at the End of the World* (Princeton, NJ: Princeton University Press, 2015).

33 Anna Tsing, *Friction: An Ethnography of Global Connection* (Princeton, NJ: Princeton University Press, 2005).

34 Donnella H. Meadows, Dennis L. Meadows, Jørgen Randers, and William W. Behrens III, *The Limits to Growth: A Report for the Club of Rome's Project on the Predicament of Mankind* (New York: Universe Books, 1972); Jason Hickel, *The Divide* (London: Penguin, 2017); Kate Raworth, *Doughnut Economics: Seven Ways to Think Like a 21st-Century Economist* (London: Random House Business, 2017); Amartya Sen, Jean-Paul Fitoussi, and Joseph Stiglitz, *Mismeasuring Our Lives: Why GDP Doesn't Add Up* (New York: New Press, 2010).

35 Kathryn Yusoff, *A Billion Black Anthropocenes or None* (Minneapolis: University of Minnesota Press, 2019).

36 Césaire, *Discourse on Colonialism*.

37 Bernard Stiegler, *The Neganthropocene*, trans. Daniel Ross (London: Open Humanities Press, 2018).

38 Amitav Ghosh, *The Great Derangement: Climate Change and the Unthinkable* (Chicago: University of Chicago Press, 2016).

39 Ben Okri, "The Joys of Storytelling III: Aphorisms and Fragments," in *The Joys of Storytelling* (London: Phoenix House, 1997), 114.

40 William Shakespeare, *King Lear*, ed. G. K. Hunter (London: Penguin Books, 1972).

41 Latour, "Recall of Modernity."

42 Roy, *Capitalism*; Arundhati Roy, *The End of Imagination* (Chicago: Haymarket Books, 2016).

43 Manuela Cadelli, "Neoliberalism Is a Species of Fascism," *Off-Guardian*, 13 July 2016, https://off-guardian.org/2016/07/13/neoliberalism-is-a-species-of-fascism/.

44 Jason Hickel, *The Divide: Global Inequality from Conquest to Free Markets* (New York: W. W. Norton, 2018).

45 For a discussion and casebook of ubuntu jurisprudence, see Drucilla Cornell and Nyoko Muvanga, eds., *uBuntu and the Law: African Ideals and Postapartheid Jurisprudence* (New York: Fordham University Press, 2012).

46 Bruno Latour, "'Thou Shalt Not Freeze Frame,' or, How Not to Misunderstand the Science and Religion Debate," in *On the Modern Cult of the Factish Gods*, 99–124 (Durham, NC: Duke University Press, 2010).

47   Édouard Glissant, *Poetics of Relation*, trans. Betsy Wing (Ann Arbor: University of Michigan Press, 1997), 169.

48   Glissant, *Poetics of Relation*, 185–88.

49   Audre Lorde, "The Master's Tools Will Never Dismantle the Master's House," in *Sister Outsider: Speeches and Essays* (Berkeley: Crossing Press, 1984), 110–14.

50   Val Plumwood, *Feminism and the Mastery of Nature* (London: Routledge, 1993); Barad, *Meeting the Universe Halfway*; and Stengers, *Thinking with Whitehead*.

51   Glissant, "That Those Beings Be Not Being," in *Poetics of Relation*, 185–88, my emphasis. The excerpt here cannot convey the depth of Glissant's text, which deserves to be read in full.

52   Emmanuel Levinas, *Otherwise Than Being*, trans. A. Lingis (Dordrecht: Nijhoff, 1974).

53   Marianne Lien and John Law, "Emergent Aliens: On Salmon, Nature and Their Enactment," *Ethnos* 76, no. 1 (2010): 65–87.

54   Thank you, Sinegugu Zukulu, for the language, the chicken-chase, and the feast.

55   Césaire, *Poetry and Knowledge*.

56   Lis Lange, "The Institutional Curriculum, Pedagogy and the Decolonisation of the South African University," in *Decolonisation in Universities: The Politics of Knowledge*, ed. Jonathan Jansen (Johannesburg: Wits University Press, 2019), 67.

57   Lange, "The Institutional Curriculum," 76

58   Lange, "The Institutional Curriculum," 83.

59   See Didier Debaise, *Knowledge as Event: The Lure of the Possible*, trans. Michael Halewood (Durham, NC: Duke University Press, 2017); Isabelle Stengers, *Another Science Is Possible: A Manifesto for Slow Science*, trans. Stephen Muecke (Cambridge, UK: Polity, 2018); Marisol de la Cadena, *Earth Beings: Ecologies of Practice across Andean Worlds* (Durham, NC: Duke University Press, 2015).

60   See Cormac Cullinan, *Wild Law: A Manifesto for Earth Justice*, 2nd ed. (London: Chelsea Green, 2011).

61   Quoted in Kamal Jagati, "Uttarakhand High Court Declares Gangotri, Yamunotri Glaciers as Living Entities," *Hindustan Times*, 31 March 2017, http://www.hindustantimes.com/india-news/uttarakhand-high-court-declares-gangotri-yamunotri-glaciers-as-living-entities/story-q1e7sjBnAGefEKT5cpezkO.html.

62   Ghosh, *The Great Derangement*.

63   Lesley Green and David Green, *Knowing the Day, Knowing the World: Amerindian Thought and Public Archeology* (Tucson: University of Arizona Press, 2013).

64   See Gerardo Reichel-Dolmatoff, *Rainforest Shamans: Essays on the Tukano Indians of the Northwest Amazon* (Devon: Themis, 1997), 35.

65   See Tim Ingold, *Lines: A Brief History* (London: Routledge, 2007).

66 David Turnbull, "Maps, Narratives and Trails: Performativity, Hodology and Distributed Knowledges in Complex Adaptive Systems—An Approach to Emergent Mapping," *Geographical Research* 45 (2007): 140–49. See also David Turnbull, "Movement, Boundaries, Rationality and the State: The Ngaanyatyarra Land Claim, the Tordesillas Line and the West Australian Border," in *Moving Anthropology: Critical Indigenous Studies*, ed. T. Lea, E. Kowal, and G. Cowlishaw, 185–200 (Darwin: Charles Darwin University Press, 2006).

67 Jacques Derrida, *Edmund Husserl's Origin of Geometry: An Introduction*, trans. John P. Leavey (Lincoln: University of Nebraska Press, 1989).

68 Michel Serres, *The Birth of Physics*, trans. Jack Hawkes (Manchester, UK: Clinamen, 2000); Michel Serres, *Geometry: The Third Book of Foundations*, trans. Randolph Burks (London: Bloomsbury Academic, 2017).

69 Serres, *Birth of Physics*, 127.

70 Thomas Nail, *Lucretius: An Ontology of Motion* (Edinburgh: Edinburgh University Press, 2018), 68.

71 Nail, *Lucretius,* 55–56.

72 Brook Ziporyn, *Ironies of Oneness and Difference: Coherence in Chinese Thought; Prolegomena to the Study of Li* (Albany: SUNY Press, 2012); Brook Ziporyn, *Beyond Oneness and Difference: Li and Coherence in Chinese Thought and Its Antecedents* (Albany: SUNY Press, 2013); David Wade, *Li: Dynamic Form in Nature* (New York: Walker & Company/Wooden Books, 2003).

73 Yuk Hui, *The Question Concerning Technology in China: An Essay in Cosmotechnics* (Falmouth, UK: Urbanomic, 2016), 62.

74 Bruno Latour, "Some Advantages of the Notion of 'Critical Zone' for Geopolitics," *Procedia Earth and Planetary Science* 10 (2014): 3–6.

75 Denise Levertov, "Learning the Alphabet," in *Poems 1968–1972* (New York: New Directions, 1987), 99.

BIBLIOGRAPHY

|han≠kass'o. "Rain Changes People into Frogs." 9 September 1878. Digital Bleek and Lloyd Collection, University of Cape Town. Accessed 21 January 2015. http://lloydbleekcollection.cs.uct.ac.za/stories/737/.

AfrikaBurn 2015 WTF? Guide. Accessed 23 May 2018. https://www.afrikaburn.com /wp-content/uploads/2012/04/AB_WTF-Guide-2015.pdf.

Ahfeldt, Chris. *The Localisation Potential of Photovoltaics and a Strategy to Support the Large-Scale Roll-Out in South Africa.* Report prepared for the South African Photovoltaic Industry Association, the World Wildlife Fund, and the Department for Trade and Industry, 2013. http://awsassets.wwf.org.za /downloads/ the_localisation_potential_of_pv_and_a_strategy_to_sup-port_large_ scale_roll_out_in_sa.pdf.

Ammons, A. R. "The City Limits." In *A. R. Ammons: The Selected Poems: 1951– 1977.* Expanded ed. New York: W. W. Norton, 1987.

ANC Western Cape. "The Real Cape Town Story—ANC WCape." Politicsweb, 10 May 2011. https://www.politicsweb.co.za/opinion/the-real-cape-town -story-anc-wcape.

Archer, Sean. "Technology and Ecology in the Karoo: A Century of Windmills, Wire and Changing Farming Practice." In *South Africa's Environmental History: Cases and Comparisons,* edited by Stephen Dovers, Ruth Edgecombe, and Bill Guest, 112–38. Athens: Ohio University Press; Cape Town: New Africa Books, 2003.

Ashton, Belinda. *Living with Our Wild Neighbours.* Cape Town: Nature Connection, 2013.

Asmal, Kader. "The Perils of Fracking." *fin24,* 25 April 2011. http://www.fin24.com /Opinion/Columnists/Guest-Columnist/The-perils-of-fracking-20110425.

Atkinson, L. J., S. Mayfield, and A. C. Cockroft. "The Potential for Using Acoustic Tracking to Monitor the Movement of the West Coast Rock Lobster *Jasus lalandii.*" *African Journal of Marine Science* 27, no. 2 (2005): 401–8.

Badmus, Kassim Olasunkanmi, Jimoh Oladejo Tijani, Emile Massima, and Leslie Petrik. "Treatment of Persistent Organic Pollutants in Wastewater Using

Hydrodynamic Cavitation in Synergy with Advanced Oxidation Process." *Environmental Science and Pollution Research* 25 (2018): 7299–7314. https://doi.org/10.1007/s11356-017-1171-z.

Baker, Lucy, and Holle Linnea Wlokas. "South Africa's Renewable Energy Procurement: A New Frontier." Working Paper 159. Norwich, UK: Tyndall Centre for Climate Change Research, 2014.

Banwart, Steven A., Nikolaos P. Nikolaidis, Yong-Guan Zhu, Caroline L. Peacock, and Donald L. Sparks. "Soil Functions: Connecting Earth's Critical Zone." *Annual Review of Earth and Planetary Sciences* 47, no. 1 (2019): 333–59.

Barad, Karen. *Meeting the Universe Halfway: Quantum Physics and the Entanglement of Matter and Meaning.* Durham, NC: Duke University Press, 2007.

Bascom, William R. *African Dilemma Tales.* The Hague: Mouton, 1975.

Bassey, Nnimmo. *To Cook a Continent: Destructive Extraction and the Climate Crisis in Africa.* Dakar, Senegal: Pambazuka Press, 2012.

Beamish, Esme. "Causes and Consequences of Mortality and Mutilation in the Cape Peninsula Baboon Population." Master's thesis, University of Cape Town, 2009.

Beck, A. W. "Names of Voters within the Electoral Division of Graham's Town, August 27 1853." *Grahamstown Journal*, July–September 1853.

Bertelsen, R. D., Larry C. Boles, and Kenneth J. Lohmann. "True Navigation and Magnetic Maps in Spiny Lobsters." *Nature* 421 (2003): 60–63.

Bertelsen, R. D., et al. "Spiny Lobster Movement and Population Metrics at the Western Sambo Ecological Reserve." Paper presented at Linking Science to Management: A Conference and Workshop on the Florida Keys Marine Ecosystem, Duck Key, Florida, 19–22 October 2010.

Bloch, Joanne. *The Tree Man: Robert Mazibuko's Story.* Pietermaritzburg, South Africa: New Readers. http://live.fundza.mobi/home/library/non-fiction-articles-profiles/the-tree-man-robert-mazibukos-story/.

"Blue Flag Beach Criteria and Explanatory Notes." Accessed 2 January 2017. https://static1.squarespace.com/static/55371ebde4b0e49a1e2ee9f6/t/56cc2a59859fd03dbee43223/1456220762132/Beach+Criteria+and+Explanatory+Notes.pdf.

Boardman, John, David Favis-Mortlock, and Ian Foster. "A 13-Year Record of Erosion on Badland Sites in the Karoo, South Africa." *Earth Surface Processes and Landforms* 40, no. 14 (November 2015): 1964–81.

Boles, Larry C., and Kenneth J. Lohmann. "True Navigation and Magnetic Maps in Spiny Lobsters." *Nature* 421 (2003): 60–63.

Bond, Patrick. *Elite Transition: From Apartheid to Neoliberalism in South Africa.* Rev. ed. London: Pluto, 2014.

Botes, Lizeth. "Taxonomy, Distribution and Toxicity of Dinoflagellate Species in the Southern Benguela Current, South Africa." PhD diss., University of Cape Town, 2003.

Botha, W. v. D., and P. C. V. du Toit. "Grazing Index Values (GIV) for Karoo Plant Species." Unpublished handbook, Grootfontein Agricultural Development Institute, Middelburg, Eastern Cape, South Africa, 2001.

Bouzrara, Nancy, and Tom Conley. "Cartography and Literature in Early Modern France." In *The History of Cartography*, vol. 3, pt. 1, edited by David Woodward, 427–37. Chicago: University of Chicago Press, 2007.

Boyer, Dominic. "Energopolitics." In *Energy Humanities: An Anthology*, edited by Imre Szeman and Dominic Boyer, 184–204. Baltimore: Johns Hopkins University Press, 2017.

Boyer, Dominic. "Energopolitics and the Anthropology of Energy." *Anthropology News*, May 2011, 5–7.

Brink, Johan G., and David K. C. Cooper. "Heart Transplantation: The Contributions of Christiaan Barnard and the University of Cape Town/Groote Schuur Hospital." *World Journal of Surgery* 29 (2005): 953–61.

Cadelli, Manuela. "Neoliberalism Is a Species of Fascism." *Off-Guardian*, 13 July 2016. https://off-guardian.org/2016/07/13/neoliberalism-is-a-species-of-fascism/.

Calvino, Italo. *Invisible Cities*. San Diego, CA: Harcourt Brace Jovanovich, 1974.

Caminero-Santangelo, Byron. *Different Shades of Green: African Literature, Environmental Justice and Political Ecology*. Charlottesville: University of Virginia Press, 2014.

Campbell, Colin Turing. "Cawood, David." In *British South Africa, a History of the Colony of the Cape of Good Hope from Its Conquest 1795 to the Settlement of Albany by the Emigration of 1819 [A.D. 1795–A.D. 1825] with Notices of Some of the British Settlers of 1820*, 218–19. London: John Haddon, 1897.

Cavanagh, Conor Joseph, and David Himmelfarb. "'Much in Blood and Money': Necropolitical Ecology on the Margins of the Uganda Protectorate." *Antipode* 47, no. 1 (2015): 55–73.

CDP Worldwide. *The Carbon Majors Database: CDP Carbon Majors Report 2017*. London: CDP, 2017.

Césaire, Aimé. *Aimé Césaire: The Collected Poetry*. Translated by Clayton Eshleman and Annette Smith. 1939. Berkeley: University of California Press, 1983.

Césaire, Aimé. *Discourse on Colonialism*. Translated by Joan Pinkham. New York: Monthly Review Press, 1972.

Césaire, Aimé. "Poetry and Knowledge." In *Refusal of the Shadow: Surrealism and the Caribbean*, edited by Michael Richardson, translated by Michael Richardson and Krzysztof Fijałkowski, 134–46. London: Verso, 1996.

Césaire, Aimé. *A Tempest*. Translated by Richard Miller. New York: Ubu Repertory Theater Publications, 1969.

Challis, William. "The Roots of Power: Baboons and the Medicines of Protection." In "The Impact of the Horse on the Amatola 'Bushmen,'" 144–76. DPhil diss., University of Oxford, 2008.

Chen, Nancy N. "Speaking Nearby: A Conversation with Trinh T. Minh-Ha." *Visual Anthropology Review* 8, no. 1 (1992): 82–91.

Cheney, Dorothy L., and Robert M. Seyfarth. *Baboon Metaphysics: The Evolution of a Social Mind*. Chicago: University of Chicago Press, 2007.

City of Cape Town. "CSIR Confirms Sewage Outfalls Pose No Significant Risks." Press release, 21 November 2017.

City of Cape Town. "Hydrogeological Investigation of Existing Water Springs in the City of Cape Town and Environs: Strategy Document, 25 June 2015." GEOSS Report No. 2014/10-08, Geohydrological and Spatial Solutions International, Stellenbosch, South Africa.

City of Cape Town. *Integrated Annual Report 2014/15.*

City of Cape Town. "Integrated Metropolitan Environmental Policy (IMEP)." Approved 31 October 2001, reviewed 25 June 2008.

City of Cape Town. *State of Cape Town 2014: Celebrating 20 Years of Democracy.*

City of Cape Town. "2010 FIFA World Cup™: Host City Cape Town. Green Goal Legacy Report." City of Cape Town, Department of Water and Sanitation. "Consumer Service Charter for the Supply of Potable Water and Sanitation Services, 2014/2015."

City of Cape Town, Department of Water and Sanitation. "Vision, Mission Statement, Values, Strategic Objectives, 2013."

City of Cape Town, SANParks, and Cape Nature. "Proceedings of Baboon Expert Workshop." 2 July 2009.

Coetzee, Jacques. "Job Losses Threaten the CSIR's Reputation." *Mail and Guardian*, 22 March 2019. https://mg.co.za/article/2019-03-22-00-job-losses-threatens -the-csirs-reputation.

Cohen, Joshua. "Kruiedokters, Plants and Molecules: Relations of Power, Wind, and Matter in Namaqualand." PhD diss., University of Cape Town, 2015.

Cook, Peta S. "Science Stories: Selecting the Source Animal for Xenotransplantation." Paper presented at the Social Change in the 21st Century Conference, Centre for Social Change Research, Brisbane, QL, Australia: Queensland University of Technology, 27 October 2006.

Cornell, Drucilla, and Nyoko Muvangua, eds. *uBuntu and the Law: African Ideals and Postapartheid Jurisprudence.* New York: Fordham University Press, 2012.

Cox, Laura A., Anthony G. Comuzzie, Lorena M. Havill, Genesio M. Karere, Kimberly D. Spradling, Michael C. Mahaney, Peter W. Nathanielsz, et al. "Baboons as a Model to Study Genetics and Epigenetics of Human Disease." *Institute for Laboratory Animal Research (ILAR) Journal* 54, no. 2 (2013): 106–21.

CSIR. "Cape Town Outfalls Monitoring Programme: Surveys Made in 2015/2016." CSIR Report CSIR/NRE/ECOS/IR/2017/0035/B. 2017.

Cullinan, Cormac. *Wild Law: A Manifesto for Earth Justice.* 2nd ed. London: Chelsea Green, 2011.

Danowski, Deborah, and Eduardo Viveiros de Castro. *The Ends of the World.* Cambridge: Polity, 2017.

Dapper, Olfert. "Kaffraria or Land of the Kafirs, Otherwise Named Hottentots." In *The Early Cape Hottentots*, by Olfert Dapper, Willem ten Rhyne, and Johannes Gulielmus de Grevenbroek, edited and translated by I. Schapera and B. Farrington, 1–79. Cape Town: The Van Riebeeck Society, 1933. http:// www.dbnl.org/tekst/dapp001earlo1_01/dapp001earlo1_01.pdf.

Darwin, Charles. *Geological Observations on the Volcanic Islands Visited during the Voyage of H.M.S. Beagle, Together with Some Brief Notices of the Geology of Australia and the Cape of Good Hope.* London: Smith Elder, 1844. http://darwin -online.org.uk/content/frameset?itemID=F272&viewtype=side&pageseq=1.

Debaise, Didier. "The Minoritarian Powers of Thought: Thinking beyond Stupidity with Isabelle Stengers." *SubStance* 145, vol. 47, no. 1 (2018): 17–28.

Debaise, Didier. *Nature as Event: The Lure of the Possible*. Translated by Michael Halewood. Durham, NC: Duke University Press, 2017.

de Greef, Kimon. "On the Front Lines of South Africa's Baboon Wars." *Outside Online*, 5 September 2017. https://www.outsideonline.com/2231291/frontlines-south-africas-human-vs-baboon-war.

de Klerk, F. W. "Address by the State President Mr F. W. de Klerk at the Opening of the Second Session of the Ninth Parliament of the Republic of South Africa, Cape Town, 2 February 1990." F. W. de Klerk Foundation, n.d. https://www.fwdeklerk.org/index.php/en/historically-significant-speeches.

de la Cadena, Marisol. *Earth Beings: Ecologies of Practice across Andean Worlds*. Durham, NC: Duke University Press, 2015.

DeLoughrey, Elizabeth, and George B. Handley. *Postcolonial Ecologies: Literatures of the Environment*. Kindle ed. Oxford: Oxford University Press, 2011.

Denizen, Seth, and Etienne Turpin. "The Stratophysics of Urban Soil Production." In *Land and Animal and Non-animal*, edited by Anna-Sophie Springer and Etienne Turpin, 50–73. Berlin: K-Verlag, 2015.

Department of Environmental Affairs, South Africa. "Minister Edna Molewa Officially Opens Lightning Conference." Press release, 12 November 2015. https://www.environment.gov.za/mediarelease/molewa_opens_lightningconference.

Department of Geological Sciences, University of Cape Town. "Cape Town Geology." Accessed 29 December 2015. http://www.geology.uct.ac.za/cape/town/geology.

Derrida, Jacques. *Edmund Husserl's "Origin of Geometry."* Translated by John P. Leavey. Lincoln: University of Nebraska Press, 1989.

Derrida, Jacques. *On Touching—Jean-Luc Nancy*. Stanford, CA: Stanford University Press, 2005.

Derrida, Jacques. *Specters of Marx: The State of the Debt, the Work of Mourning, and the New International*. Translated by Peggy Kamuf. New York: Routledge, 1994.

Descartes, René. *Discourse on Method, Optics, Geometry and Meteorology*. 1637. Indianapolis: Hackett Classics, 2001.

Despret, Vinciane. *What Would Animals Say If We Asked the Right Questions?* Translated by Brett Buchanan. Minneapolis: University of Minnesota Press, 2014.

de Vaal, Michael. "In Vivo Mechanical Loading Conditions of Pectorally Implanted Cardiac Pacemakers: Feasibility of a Force Measurement System and Concept of an Animal-Human Transfer Function." Master's thesis, University of Cape Town, 2012.

de Wet, Philip. "How to Deal with a Million Bolts a Month." *Mail and Guardian*, 28 January 2011. https://mg.co.za/article/2011-01-28-how-to-deal-with-a-million-bolts-month/.

Dewey, John. "Pragmatic America." *New Republic*, 12 April 1922, 185–87.

Dlamini, Jacob. "'To Know the African Wild Was to Know the African Subject'—What Training as a Field Guide Taught Jacob Dlamini about Culture, Nature, Power and Race." *Johannesburg Review of Books*, 1 July 2019.

Doyle, Alister. "South Africa Compares World Climate Change Plan to 'Apartheid.'" *Mail and Guardian*, 20 October 2015. http://mg.co.za/article/2015-10-20-south-africa-compares-global-climate-plan-to-apartheid.

Drewe, J., M. J. O'Riain, E. Beamish, and H. Currie. "A Survey of Infections Transmissible between Baboons and Humans in Cape Town, South Africa." *Emerging Infectious Diseases* 18, no. 2 (2012): 298–301.

Dry, Gert. "The Case for Game Ranching: A Biodiversity Economy Imperative." Paper presented to the Department of Environmental Affairs (DEA) Workshop on Game Breeding, Pretoria, 2 December 2015. https://www.environment.gov.za/sites/default/files/docs/wildliferanching_contributionto_southafrica_economy.pdf.

Dry, Gert. "Game Ranching Is a Biodiversity Economy Asset Class and Not a Government Conservation Estate." Paper presented to the 49th South African Society for Animal Studies (SASAS) Congress, Stellenbosch, Western Cape, 4 July 2016. https://www.wrsa.co.za/wp-content/uploads/2016/07/49th-SASAS-GAME-RANCHING-IS-A-BIODIVERSITY-ECONOMY-ASSET-CLASS.pdf.

Duffy, Rosaleen. "Waging a War to Save Biodiversity: The Rise of Militarized Conservation." *International Affairs* 90, no. 4 (2014): 819–34.

Dukelow, Richard, *The Alpha Males: An Early History of the Regional Primate Centers*. Lanham, MD: University Press of America, 1995.

Dussel, Enrique. *Philosophy of Liberation*. Translated by Aquilina Martinez and Christine Morkovsky. Eugene, OR: Wipf and Stock, 1985.

Eberhard, Anton, Joel Kolker, and James Leigland. *South Africa's Renewable Energy IPP Procurement Program: Success Factors and Lessons*. Washington, DC: Public-Private Infrastructure Advisory Facility, World Bank, 2014.

Eggington, Shanaaz. "'I Am Living in a War Zone,' Says Resident as Baboons Take Over." *TimesLive*, 19 January 2016. http://www.timeslive.co.za/thetimes/2016/01/19/I-am-living-in-a-war-zone-says-resident-as-baboons-take-over.

Eloff, Daniel. "The Objective Absurdity of Decolonising Science." In *Fallism*. Cape Town: Rational Standard, 2016.

Elphick, R. *Kraal and Castle: Khoikhoi and the Founding of White South Africa*. New Haven, CT: Yale University Press, 1977.

Environmental Affairs and Development Planning, Western Cape Government. *Western Cape Climate Change Response Strategy*. February 2014. https://www.westerncape.gov.za/text/2015/march/western_cape_climate_change_response_strategy_2014.pdf.

Enwegbara, Basil. 2001. "Toxic Colonialism: Lawrence Summers and Let Africans Eat Pollution." *The Tech* 121, no. 16 (6 April 2001). http://tech.mit.edu/V121/N16/col16guest.16c.html.

Etkin, N. L., and E. Elisabetsky. "Seeking a Transdisciplinary and Culturally Germane Science: The Future of Ethnopharmacology." *Journal of Ethnopharmacology* 100, nos. 1–2 (2005): 23–26.

Fig, David. "Hydraulic Fracturing in South Africa: Correcting the Democratic Deficits." In *New South African Review 3*, edited by J. Daniel, P. Naidoo, D. Pillay, and R. Southall, 173–94. Johannesburg: Wits University Press, 2013.

Finewood, Michael H., and Laura J. Stroup. "Fracking and the Neoliberalization of the Hydro-social Cycle in Pennsylvania's Marcellus Shale." *Journal of Contemporary Water Research and Education* 147 (March 2012): 72–79.

"Fishermen Protest at Parly." *fin24*, 29 May 2005. http://www.fin24.com/Business /Fishermen-protest-at-parly-20050529-3.

Fitzsimons, F. W. *The Monkeyfolk of South Africa*. 2nd ed. London: Longmans, Green, 1924.

Foley, John P. "The 'Baboon Boy' of South Africa." *American Journal of Psychology* 53, no. 1 (January 1940): 128–33.

Foster, John Bellamy. *Marx's Ecology: Materialism and Nature*. New York: Monthly Review Press, 2000.

Franke, Anselm, and Hila Peleg. *Ape Culture*. Berlin: Haus der Kulturen der Welt, 2015.

Frye, Marilyn. "White Woman Feminist, 1983–1992." In *Willful Virgin: Essays in Feminism 1976–1992*, 147–69. Freedom, CA: Crossing Press, 1992.

Galaz, Victor, Beatrice Crona, Alice Dauriach, Jean-Baptiste Jouffray, Henrik Österblom, and Jan Fichtner. "Tax Havens and Global Environmental Degradation." *Nature Ecology and Evolution* 2 (2018): 1352–57.

Galgut, Elisa. "Raising the Bar in the Justification of Animal Research." *Journal of Animal Ethics* 5, no. 1 (2015): 5–19.

Geldenhuys, Odette, and Ryley Grunewald, dirs. *The Shore Break*. Cape Town: Frank Films/Randburg, South Africa: Marie-Vérité Films, 2014. Documentary film, 90 min. DVD. https://www.theshorebreakmovie.com.

Gerdes, Paulus. *Lunda Geometry: Mirror Curves, Designs, Knots, Polyominoes, Patterns, Symmetries*. Self-published, Lulu, 2008.

Gerdes, Paulus. *Otthava: Making Baskets and Doing Geometry in the Makhuwa Culture in the Northeast of Mozambique*. Self-published, Lulu, 2010.

Ghosh, Amitav. *The Great Derangement: Climate Change and the Unthinkable*. Chicago: University of Chicago Press, 2016.

Gibson, N. Jade. "Skeletal (In-)Visibilities in the city—*Rootless*: A Video Sculptural Response to the Disconnected in Cape Town." *South African Theatre Journal* 26, no. 2 (2012): 151–71.

Gillespie, Kathryn A., and Patricia J. Lopez. "Introducing Economies of Death." In *Economies of Death: Economic Logics of Killable Life and Grievable Death*, edited by Patricia J. Lopez and Kathryn A. Gillespie. London: Routledge, 2015.

Glatigny, Pascal Dubourg, Estelle Alma Maré, and Russel Stafford Viljoen. "*Inter se nulli fines*: Representations of the Presence of the Khoikhoi in Early Colonial Maps of the Cape of Good Hope." *South African Journal of Art History* 23, no. 1 (2008): 301–17.

Glissant, Édouard. *Poetics of Relation*. Translated by Betsy Wing. Ann Arbor: University of Michigan Press, 1997.

Gordimer, Nadine. *The Conservationist*. London: Jonathan Cape, 1974.

Gosling, Melanie. "Cape Town Council's Desalination Debacle: Seawater 400% More Polluted Than City of Cape Town's Tender Data Indicated." *News24*, 27 May 2019. https://www.news24.com/SouthAfrica/News/va-desalination -debacle-seawater-400-more-polluted-than-citys-tender-data-indicated -20190527.

Gqola, Pumla Dineo. "'Like Having Three Tongues in One Mouth': Tracing the Elusive Lives of Slave Women in (Slavocratic) South Africa." In *Basus 'Iimbokado, Bawel 'Imilambo/They Remove Boulders and Cross Rivers: Women in South African History*, edited by Nomboniso Gasa, 21–41. Cape Town: Human Sciences Research Council, 2007.

Gqola, Pumla Dineo. *What Is Slavery to Me? Postcolonial/Slave Memory in Post-apartheid South Africa*. Johannesburg: Wits University Press, 2010.

Graeber, David. *Debt: The First 5000 Years*. Brooklyn, NY: Melville House, 2011.

Green, Lesley. "Beyond South Africa's 'Indigenous Knowledge–Science' Wars." *South African Journal of Science* 108, nos. 7–8 (2012): 44–54 (article 631). http://dx.doi.org/10.4102/sajs.v108i7/8.631.

Green, Lesley. "Calculemos *Jasus lalandii*: Accounting for South African Lobster." In *Reset Modernity!*, edited by Bruno Latour and Christophe Leclerq, 139–51. Cambridge, MA: MIT Press, 2016.

Green, Lesley. "The Changing of the Gods of Reason: Cecil John Rhodes, Karoo Fracking and the Decolonising of the Anthropocene." In "Supercommunity," special issue, *e-flux Journal* 65 (May–August 2015).

Green, Lesley. "Fisheries Science, Parliament and Fishers' Knowledge in South Africa: An Attempt at Scholarly Diplomacy." *Marine Policy* 60 (2015): 345–52.

Green, Lesley. "Retheorizing the Indigenous Knowledge Debate." In *Africa-Centred Knowledges: Crossing Fields and Worlds*, edited by Brenda Cooper and Robert Morrell, 36–50. London: James Currey, 2014.

Green, Lesley. "What Does It Mean to Require 'Evidence-Based Research' in Decision-Making on Hydraulic Fracturing?" In *Hydraulic Fracturing in the Karoo: Critical Legal and Environmental Perspectives*, edited by Jan Glazewski and Surina Esterhuyse, 266–79. Cape Town: Juta, 2016.

Green, Lesley, and David Green. *Knowing the Day, Knowing the World: Engaging Amerindian Thought in Public Archaeology*. Tucson: University of Arizona Press, 2013.

Green, Lesley, David W. Gammon, M. Timm Hoffman, Joshua Cohen, Amelia Hilgart, Robert G. Morrell, Helen Verran, and Nicola Wheat. "Plants, People and Health: Three Disciplines at Work in Namaqualand." *South African Journal of Science* 111, nos. 9–10 (2015): article 2014-0276. http://dx.doi.org /10.17159/sajs.2015/20140276.

Green, Lesley, Nikiwe Solomon, Leslie Petrik, and Jo Barnes. "Distortions, Distractions and Falsehoods in the City of Cape Town's Riposte to Kuils River Effluent Article." *Daily Maverick*, 10 April 2019.

Green, Lesley, Nikiwe Solomon, Leslie Petrik, and Jo Barnes. "Environmental Management Needs to Be Democratised." *Daily Maverick*, 6 March 2019, accessed 11 July 2019. https://www.dailymaverick.co.za/article/2019-03-06 -environmental-management-needs-to-be-democratised/.

Groenewald, Gerald. "Slaves and Free Blacks in VOC Cape Town, 1652–1795." *History Compass* 8, no. 9 (2010): 964–83.

Gudynas, Eduardo. "Extractivismos y corrupción en América del Sur: Estructuras, dinámicas y tendencias en una íntima relación." *RevIISE* (University of San Juan) 10, no. 10 (2017): 73–87.

Gudynas, Eduardo. "Value, Growth, Development: South American Lessons for a New Ecopolitics." *Capitalism Nature Socialism* 30, no. 2 (2019): 234–43.

Guèye, Mbaye. "The Slave Trade within the African Continent." In *The African Slave Trade from the Fifteenth to the Nineteenth Century: Reports and Papers of the Meeting of Experts*, 150–63. Paris: UNESCO, 1979.

Guicciardini, Niccolo. *Isaac Newton and Natural Philosophy*. London: Reaktion Books, 2018.

Gutwirth, Serge, and Isabelle Stengers. "The Law and the Commons." Paper presented at the third Global Thematic International Association for the Study of the Commons (IASC) Conference on the Knowledge Commons, Paris, 20–22 October 2016.

Haff, Peter. "Humans and Technology in the Anthropocene: Six Rules." *Anthropocene Review* 1, no. 2 (2014): 126–36.

Hahn, Theophilus. *Tsuni-||Goam: The Supreme Being of the Khoikhoi*. London: Trübner, 1881. https://ia800500.us.archive.org/0/items/tsunillgoamsupre00hahnuoft/tsunillgoamsupre00hahnuoft.pdf.

Haraway, Donna. "Animal Sociology and a Natural Economy of the Body Politic, Part II: The Past Is the Contested Zone: Human Nature and Theories of Production and Reproduction in Primate Behavior Studies." In "Women, Science, and Society." Spec. issue, *Signs* 4, no. 1 (Autumn 1978): 37–60.

Haraway, Donna. "Anthropocene, Capitalocene, Plantationocene, Chthulucene: Making Kin." *Environmental Humanities* 6 (2015): 159–65.

Haraway, Donna. *Primate Visions: Gender, Race and Nature in the World of Modern Science*. New York: Routledge, 1989.

Haraway, Donna. *When Species Meet*. Minneapolis: University of Minnesota Press, 2008.

Harney, Stefano, and Fred Moten. *The Undercommons: Fugitive Planning and Black Study*. Wivenhoe: Minor Compositions, 2013.

Harrison, Alette. "Exclusive: Desalination Company to Take City of Cape Town to Court for Breach of Contract." *News24*, 13 June 2019.

Haskins, Candice. "Cape Town's Sustainable Approach to Stormwater Management." Unpublished paper, Catchment, Stormwater and River Management Branch, City of Cape Town, January 2012.

Hauck, Maria. "Rethinking Small-Scale Fisheries Compliance." *Marine Policy* 32 (2008): 635–42.

Hey, Douglas. "The Control of Vertebrate Problem Animals in the Province of the Cape of Good Hope, Republic of South Africa." In *Proceedings of the 2nd Vertebrate Pest Control Conference*, paper 11. Lincoln: University of Nebraska, 1964. https://digitalcommons.unl.edu/cgi/viewcontent.cgi?article=1010&context=vpc2.

Hickel, Jason. *The Divide: A Brief Guide to Global Inequality and Its Solutions*. London: Penguin, 2017.

Hidetosi, Fukagawa, and Tony Rothman. *Sacred Mathematics: Japanese Temple Geometry*. Princeton, NJ: Princeton University Press, 2008.

Hilgart, Amelia. "Determination of a Robust Metabolic Barcoding Model for Chemotaxonomy in Aizoaceae Species: Expanding Morphological and Genetic Understanding." PhD diss., University of Cape Town, 2015.

Hoffman, Hillel J. "When Life Nearly Came to an End: The Permian Extinction." *National Geographic* 198, no. 3 (2000): 100–113.

Hoffman, T. S., and M. J. O'Riain. "Monkey Management: Using Spatial Ecology to Understand the Extent and Severity of Human-Baboon Conflict in the Cape Peninsula, South Africa." *Ecology and Society* 17, no. 3 (2012): article 13.

Hoffman, T. S., and M. J. O'Riain. "The Spatial Ecology of Chacma Baboons (*Papio ursinus*) in a Human Modified Environment." *International Journal of Primatology* 32 (2010): 308–28.

Hoffman, T. S., and M. J. O'Riain. "Troop Size and Human-Modified Habitat Affect the Ranging Patterns of a Chacma Baboon Population in the Cape Peninsula, South Africa." *American Journal of Primatology* 74, no. 9 (September 2012): 853–63.

Horstman, D. A., S. McGibbon, G. C. Pitcher, D. Calder, L. Hutchings, and P. Williams. "Red Tides in False Bay, 1959–1989, with Particular Reference to Recent Blooms of *Gymnodinium* sp." *Transactions of the Royal Society of South Africa* 47, nos. 4–5 (1991): 611–28.

Hugo, Pieter. *Permanent Error*. New York: Prestel, 2011.

Hui, Yuk. *The Question Concerning Technology in China: An Essay in Cosmotechnics*. Falmouth, UK: Urbanomic, 2016.

Huigen, Siegfried. *Knowledge and Colonialism: Eighteenth-Century Travellers in South Africa*. Leiden: Brill, 2009.

Human Wildlife Solutions. *Monthly Report: September 2014*. Tokai, South Africa: Human Wildlife Solutions, 2014.

Hund, Wulf D., Charles W. Mills, and Silvia Sebastiani, eds. *Simianization: Apes, Gender, Class, and Race*. Berlin: Lit, 2015.

Hungwe, Brian. "Letter from Africa: Thomas Mapfumo." *BBC News*, 2 May 2018. http://www.bbc.com/news/world-africa-43963049.

Ingold, Tim. *Lines: A Brief History*. London: Routledge, 2007.

Isaacs, Moeniba. "Small-Scale Fisheries Governance and Understanding the Snoek (*Thyrsites atun*) Supply Chain in the Ocean View Fishing Community, Western Cape, South Africa." *Ecology and Society* 18, no. 4 (2013): 17.

Jaffer, Naseegh, and Jackie Sunde. "Fishing Rights or Human Rights?" *SAMUDRA Report* 44 (July 2006): 20–24.

Jagati, Kamal. "Uttarakhand High Court Declares Gangotri, Yamunotri Glaciers as Living Entities." *Hindustan Times*, 31 March 2017. http://www.hindustantimes .com/india-news/uttarakhand-high-court-declares-gangotri-yamunotri -glaciers-as-living-entities/story-q1e7sjBnAGefEKT5cpezkO.html.

Jardine, Christian N., Brenda Boardman, Ayub Osman Julia Vowles, and Jane Palmer. "Climate Science of Methane." In *Methane UK*, 14–23. Oxford: University of Oxford, Environmental Change Institute, n.d. http://www.eci.ox.ac .uk/research/energy/downloads/methaneuk/chapter02.pdf.

Johnston, S. J., and D. S. Butterworth. "Rock Lobster Scientific Working Group Agreed Recreational Catch Estimates." Unpublished ms., Department of Mathematics and Applied Mathematics, University of Cape Town, 2010. https://open.uct.ac.za/handle/11427/18920.

Jordan, Bobby. "Vengeance Is a Mine, Sayeth the Chairman." *Sunday Times*, 29 November 2015. https://www.pressreader.com/south-africa/sunday-times /20151129/281732678403889.

Jullien, Francois. *The Propensity of Things: Toward a History of Efficacy in China.* Cambridge, MA: Zone Books, 1999.

Kamwangamalu, Nkonko M. "Ubuntu in South Africa: A Sociolinguistic Perspective to a Pan-African Concept." *Critical Arts: A South-North Journal of Cultural and Media Studies* 13, no. 2 (1999): 24–41.

Kansky, Ruth. *Baboons on the Cape Peninsula: A Guide for Residents and Tourists.* Cape Town: International Fund for Animal Welfare, 2002.

Kaplan, Bentley, Justin O'Riain, Rowen van Eeden, and Andrew J. King. "A Low-Cost Manipulation of Food Resources Reduces Spatial Overlap between Baboons." *International Journal of Primatology* 32, no. 6 (2011): 1397–1412.

Keegan, Tim. *Colonial South Africa: The Origins of Racial Order.* Cape Town: David Philip, 1996.

Klopper, Dirk. "Boer, Bushman, and Baboon: Human and Animal in Nineteenth-Century and Early Twentieth-Century South African Writings." *Safundi: The Journal of South African and American Studies* 11, nos. 1–2 (2010): 3–18.

Kolbe, Peter. *Capvt Bonae Spei hodiernvm, das ist: Vollständige Beschreibung des africanischen Vorgebürges der Guten Hofnung: Worinnen in dreyen Theilen abgehandelt wird, wie es heut zu Tage, nach seiner situation und Eigenschaft aussiehet; ingleichen was ein Natur-Forscher in den dreyen Reichen der Natur daselbst findet und antrifft: wie nicht weniger, was die eigenen Einwohner die Hottentotten, vor seltsame Sitten und Gebräuche haben: und endlich alles, was die europaeischen daselbst gestifteten Colonien anbetrift. Mit angefügter genugsamer Nachricht, wie es auf des Auctoris Hinein.* Nuremberg: Peter Conrad Monath, 1719.

Kravitz, E. A. "Serotonin and Aggression: Insights Gained from a Lobster Model System and Speculations on the Role of Amine Neurons in a Complex Behavior." *Journal of Comparative Physiology* 186 (2000): 221–38.

Kretzmann, Steve. "Pharmaceuticals and Industrial Chemicals Found in Fish Caught off Cape Town's Coast." GroundUp, 25 June 2019. https://www.groundup.org.za/article/were-eating-our-waste/.

Laland, K. N., N. Atton, and M. M. Webster. "From Fish to Fashion: Experimental and Theoretical Insights into the Evolution of Culture." *Philosophical Transactions of the Royal Society B: Biological Sciences* 366 (2011): 958–68.

Lalu, Premesh. *The Deaths of Hintsa: Post-apartheid South Africa and the Shape of Recurring Pasts.* Cape Town: HSRC Press, 2009.

Lange, Lis. "The Institutional Curriculum, Pedagogy and the Decolonisation of the South African University." In *Decolonisation in Universities: The Politics of Knowledge*, edited by Jonathan Jansen, 67–87. Johannesburg: Wits University Press, 2019.

Larson, Brendon M. H. "The War of the Roses: Demilitarizing Invasion Biology." *Frontiers in Ecology and Environment* 3, no. 9 (2005): 495–500.

Latour, Bruno. *An Inquiry into Modes of Existence.* Translated by Catherine Porter. Cambridge, MA: Harvard University Press, 2013.

Latour, Bruno. *Pandora's Hope: Essays on the Reality of Science Studies.* Cambridge, MA: Harvard University Press, 1999.

Latour, Bruno. *Politics of Nature: How to Bring the Sciences into Democracy.* Translated by Catherine Porter. Cambridge, MA: Harvard University Press, 2004.

Latour, Bruno. *Reassembling the Social: An Introduction to Actor-Network Theory.* Oxford: Oxford University Press, 2005.

Latour, Bruno. "The Recall of Modernity: Anthropological Approaches." Translated by Stephen Muecke. *Cultural Studies Review* 13, no. 1 (2007): 11–30.

Latour, Bruno. "Some Advantages of the Notion of 'Critical Zone' for Geopolitics." *Procedia Earth and Planetary Science* 10 (2014): 3–6.

Latour, Bruno. "'Thou Shalt Not Freeze Frame,' or, How Not to Misunderstand the Science and Religion Debate." In *On the Modern Cult of the Factish Gods,* 99–124. Durham, NC: Duke University Press, 2010.

Latour, Bruno. *We Have Never Been Modern.* Translated by Catherine Porter. Cambridge, MA: Harvard University Press, 1993.

Leibbrandt, H. C. V. *Precis of the Archives of the Cape of Good Hope, January 1656–December 1658, Part II: Riebeeck's Journal &c.* Cape Town: W. A. Richards and Sons, 1897.

Levertov, Denise. "Learning the Alphabet." In *Poems 1968–1972.* New York: New Directions, 1987.

Levinas, Emmanuel. "Ethics and Infinity." *CrossCurrents* 34, no. 2 (1984): 191–203.

Levinas, Emmanuel. *Otherwise Than Being.* Translated by A. Lingis. Dordrecht: Nijhoff, 1974.

Lien, Marianne, and John Law. "Emergent Aliens: On Salmon, Nature and Their Enactment." *Ethnos* 76, no. 1 (2010): 65–87.

Loewe, Mike. "Fracking Task Team Denied Opportunity to Address Meeting." *TimesLive,* 10 November 2015. Accessed 4 January 2016. http://www.timeslive.co.za/scitech/2015/11/10/Fracking-task-team-denied-opportunity-to-address-meeting.

Lorde, Audre. "The Master's Tools Will Never Dismantle the Master's House." In *Sister Outsider: Speeches and Essays,* 110–13. Berkeley: Crossing Press, 1984.

Maathai, Wangari. *Replenishing the Earth: Spiritual Values for Healing Ourselves and the World.* New York: Doubleday, 2010.

Maathai, Wangari. *Unbowed: A Memoir.* New York: Knopf, 2006.

Mabeza, Christopher Munyaradzi. *Water and Soil in Holy Matrimony? A Smallholder Farmer's Innovative Agricultural Practices for Adapting to Climate Change.* Bamenda and Buea, Cameroon: Langaa RPCIG, 2017.

MacKenny, Virginia, and Lesley Green. "African Artists in Post-colonial Colonial Landscapes: Re-membering the Discarded." Unpublished ms.

Makgoba, Malegapuru. "Wrath of Dethroned White Males." *Mail and Guardian,* 25 March 2005. http://mg.co.za/article/2005-03-25-wrath-of-dethroned-white-males.

Maldonado-Torres, Nelson. *Against War: Views from the Underside of Modernity.* Durham, NC: Duke University Press, 2008.

Marks, Shula E. "Southern Africa." In *Encyclopaedia Britannica Online.* Article published 1 July 2014. http://www.britannica.com/EBchecked/topic/556618 /Southern-Africa#toc234030.

Marshall, Michael. "The History of Ice on the Earth." *New Scientist Environment,* 24 May 2010. http://www.newscientist.com/article/dn18949-the-history-of -ice-on-earth.html.

Marx, Karl. *Capital.* Vols. 1–3. 1867–94. Translated by Ben Fowkes. London: Penguin Classics, 1976.

Marx, Karl. "Debates on the Law on Thefts of Wood." *Rheinische Zeitung,* 25 October–3 November 1842. Translated by Clemens Dutt. https://www .marxists.org/archive/marx/works/download/Marx_Rheinishe_Zeitung .pdf.

Master, Sharad. "Darwin as a Geologist in Africa—Dispelling the Myths and Unravelling a Confused Knot." *South African Journal of Science* 108, nos. 9–10 (2012): 45–49.

Mbembe, Achille. *Critique of Black Reason.* Durham, NC: Duke University Press, 2017.

Mbembe, Achille. "Necropolitics." Translated by Libby Meintjies. *Public Culture* 15, no. 1 (winter 2003): 11–40.

Meadows, Donnella H., Dennis L. Meadows, Jørgen Randers, and William W. Behrens III. *The Limits to Growth: A Report for the Club of Rome's Project on the Predicament of Mankind.* New York: Universe Books, 1972.

Mellet, Patric Tariq. "Camissa." Camissa People, accessed 30 December 2015, https://camissapeople.wordpress.com/camissa/.

Mellet, Patric Tariq. "Cape Indigenes." Camissa People, accessed 14 August 2019, https://camissapeople.wordpress.com/cape-indigenes-2/.

Memela, Mhlaba. "MEC's Call for Probe into Lightning." *Sowetan,* 4 January 2011. https://www.sowetanlive.co.za/news/2011-01-07-mec-dube-calls-for -lightning-probe/.

Minister of Environmental Affairs and Tourism v George and Others. 437/05, 437/05 (2006) ZASCA 57; 2007 (3) SA 62 (SCA) (18 May 2006), http://www.saflii .org/za/cases/ZASCA/2006/57.html.

Minnaar, Jolynn, dir. *Unearthed.* Cape Town: Stage 5 Films, 2014.

Mitchell, Tim. *Carbon Democracy: Political Power in the Age of Oil.* New York: Verso, 2013.

Mitford-Barberton, Ivan, and Violet White. *Some Frontier Families: Biographical Sketches of 100 Eastern Province Families before 1840.* Cape Town: Human and Rousseau, 1969.

Mkhize, Nomalanga. "Game Farm Conversions and the Land Question: Unpacking Present Contradictions and Historical Continuities in Farm Dwellers' Tenure Insecurity in Cradock." *Journal of Contemporary African Studies* 32, no. 2 (2014): 207–19.

Moller, Lucie A. *Of the Same Breath: Indigenous Animal and Place Names.* Bloemfontein: Sun Press, 2017.

Montgomery, David. *Dirt: The Erosion of Civilizations*. Berkeley: University of California Press, 2008.

Moore, Jason, ed. *Anthropocene or Capitalocene?* Oakland, CA: Kairos PM Press, 2016.

Moore, Jason. *Capitalism in the Web of Life: Ecology and the Accumulation of Capital*. London: Verso, 2015.

Moreton-Robinson, Aileen. *The White Possessive: Property, Power and Indigenous Sovereignty*. Minneapolis: University of Minnesota Press, 2015.

Morgan, Nathaniel John. "An Account of the Amakosae." *South African Quarterly Journal* 1, no. 3 (October 1833): S.1–12, 33–48, 65–71.

Morton, Tim. *Hyperobjects*. Minneapolis: University of Minnesota Press, 2013.

Motala, Siddique. "Higher Education Well-Being: Transcending Boundaries, Reframing Excellence." Paper presented at the conference of the Higher Education Learning and Teaching Association of South Africa, Durban, 21–24 November 2017.

Moukala, Edmund, ed. co-ord. *Tagore, Neruda, Césaire: For a Reconciled Universal*. Paris: UNESCO, 2011. http://unesdoc.unesco.org/images/0021/002116/211645e.pdf.

Murove, Munyaradzi. "Ubuntu." *Diogenes* 59, nos. 3–4 (2012): 36–47. https://doi.org/10.1177/0392192113493737.

Nadasdy, Paul. *Hunters and Bureaucrats: Power, Knowledge, and Aboriginal-State Relations in the Southwest Yukon*. Vancouver: University of British Colombia Press, 2003.

Nail, Thomas. *Lucretius: An Ontology of Motion*. Edinburgh: Edinburgh University Press, 2018.

Nancy, Jean-Luc. *The Birth to Presence*. Translated by Brian Holmes et al. Stanford, CA: Stanford University Press, 1993.

Nattrass, Nicoli. *The AIDS Conspiracy: Science Fights Back*. New York: Columbia University Press, 2012.

Negarastani, Reza. *Cyclonopedia: Complicity with Anonymous Materials*. Victoria, Melbourne, Australia: RePress, 2008.

Netshitenze, Joel. 2016. "The Black Man's Burden." *Sunday Times*, 21 February 2016. https://www.timeslive.co.za/sunday-times/opinion-and-analysis/2016-02-21-the-black-mans-burden-/.

Ngabonziza, Dan. "Rwanda to Plant 43,000 Hectares of Trees in Six Months." *KT Press*, 26 October 2018.

Nhemachena, Artwell. "Knowledge, Chivanhu and Struggles for Survival in Conflict-Torn Manicaland, Zimbabwe." PhD diss., University of Cape Town, 2014.

Norton, Marieke. "At the Interface: Marine Compliance Inspectors at Work in the Western Cape." PhD diss., University of Cape Town, 2014.

Nyamnjoh, Francis B. "Cameroonian Bushfalling: Negotiation of Identity and Belonging in Fiction and Ethnography." *American Ethnologist* 38, no. 4 (2011): 701–13.

Nyamnjoh, Francis B. "Fiction and Reality of Mobility in Africa." *Citizenship Studies* 17, nos. 6–7 (2013): 653–80.

Office of the ANC Chief Whip, Luthuli House, Johannesburg. "Specific Law to Criminalize Racism and Promotion of Apartheid Necessary." State-

ment, 5 January 2016. https://www.politicsweb.co.za/news-and-analysis/
specific-law-needed-to-criminalise-racism-and-prom.

Ojemaye, Cecilia Y., and Leslie Petrik. "Occurrences, Levels and Risk Assessment Studies of Emerging Pollutants (Pharmaceuticals, Perfluoroalkyl and Endocrine Disrupting Compounds) in Fish Samples from Kalk Bay Harbour, South Africa." *Environmental Pollution* 252A (2019): 562–72, doi: https://doi.org/10.1016/j.envpol.2019.05.091.

Ojemaye, Cecilia Y., and Leslie Petrik. "Pharmaceuticals in the Marine Environment: A Review." *Environmental Reviews* 27, no. 2 (2019): 151–65. https://doi.org/10.1139/er-2018-0054.

Okri, Ben. "The Joys of Storytelling III: Aphorisms and Fragments." In *The Joys of Storytelling*. London: Phoenix House, 1997.

Ommanney, Sarah. *Lacuna: Groote Schuur Zoo*. Cape Town: Centre for Curating the Archive, Michaelis School of Fine Art, University of Cape Town, 2012.

Parikka, Jussi. *The Anthrobscene*. Minneapolis: University of Minnesota Press, 2015.

Parliamentary Monitoring Group. "Marine Living Resources Amendment Bill [B30-2013]: Public Hearings Day 2." 16 October 2013. http://www.pmg.org.za/report/20131016-marine-living-resources-amendment-bill-b30-2013-public-hearings-day-2.

Pauw, Jacques. *The President's Keepers: Those Keeping Zuma in Power and out of Prison*. Cape Town: Tafelberg, 2017.

Penn, Nigel. *The Forgotten Frontier: Colonist and Khoisan on the Cape's Northern Frontier in the 18th Century*. Athens: Ohio University Press, 2005.

Petersen, Wally. *Cape of Good Hope Wildlife Guide*. Cape Town: Kommetjie Environmental Action Group, 2015.

Petrik, Leslie, Lesley Green, Adeola P. Abegunde, Melissa Zackon, Cecilia Y. Sanusi, and Jo Barnes. "Desalination and Seawater Quality in Green Point, Cape Town: A Study on the Effects of Marine Sewage Outfalls." *South African Journal of Science* 113, no. 11/12 (2017).

Pignarre, Philippe, and Isabelle Stengers. *Capitalist Sorcery: Breaking the Spell*. London: Palgrave Macmillan, 2007.

Plaatje, Sol. *Native Life in South Africa: Before and since the European War and the Boer Rebellion*. London: P. S. King, 1916.

Plumwood, Val. *Feminism and the Mastery of Nature*. London: Routledge, 1993.

Ponte, Stefano, Simon Roberts, and Lance van Sittert. "Black Economic Empowerment, Business and the State in South Africa." *Development and Change* 38, no. 5 (2007): 933–55.

Povinelli, Elizabeth. *Geontologies: A Requiem to Late Liberalism*. Durham, NC: Duke University Press, 2016.

Probyn, Elspeth. *Eating the Ocean*. Durham, NC: Duke University Press, 2016.

Raemaekers, Serge. "'Storied' Seafood and Community-Led Technology." TEDx Talks, published 23 January 2018. Video, 11:28. https://www.youtube.com/watch?v=dArhoEfNbwY.

Raemaekers, Serge, and Jackie Sunde. "Extending the Ripples." *Samudra Report* 75 (January 2017): 11–16. http://aquaticcommons.org/21180/1/017%20Extending%20the%20Ripples.pdf.

Raven-Hart, R. *Cape of Good Hope 1652–1702: The First Fifty Years of Dutch Coloni-sation as Seen by Callers*. Vol. 1. Cape Town: A. A. Balkema, 1971. http://www.dbnl.org/tekst/rave028cape01_01/rave028cape01_01.pdf.

Raworth, Kate. *Doughnut Economics: Seven Ways to Think Like a 21st-Century Economist*. London: Random House Business, 2017.

Reichel-Dolmatoff, Gerardo. *Rainforest Shamans: Essays on the Tukano Indians of the Northwest Amazon*. Devon: Themis, 1997.

Rhodes, Cecil. "Confessions of Faith." 2 June 1877. In *Sources of European History: Since 1900*. 2nd ed., edited by Marvin Perry, Matthew Berg, and James Kru-kones, 29. Boston: Wadsworth, 2009.

Ribe, Neil M. "Cartesian Optics and the Mastery of Nature." *Isis* 88 (1997): 42–61.

Ross, Robert. *Cape of Torments: Slavery and Resistance in South Africa*. London: Routledge and Kegan Paul, 1983.

Rousseau, Jean-Jacques. *Discours sur l'origine et les fondemens de l'inégalité parmi les hommes*. Amsterdam: Marc Michel Rey, 1755.

Rowell, Thelma. "The Concept of Social Dominance." *Behavioral Biology* 11 (1974): 131–54.

Roy, Arundhati. *Capitalism: A Ghost Story*. Chicago: Haymarket Books, 2014.

Roy, Arundhati. *The End of Imagination*. Chicago: Haymarket Books, 2016.

Roy, Arundhati. *Field Notes on Democracy: Listening to Grasshoppers*. Chicago: Haymarket Books, 2009.

Rundgren, Carl David. "Aspects of Pollution in False Bay, South Africa (with Special Reference to Subtidal Pollution)." Master's thesis, University of Cape Town, 1992.

Sacks, Oliver. "The Mental Life of Plants and Worms." *New York Review of Books*, 24 April 2014. http://www.nybooks.com/articles/archives/2014/apr/24/mental-life-plants-and-worms-among-others/.

Sanabria, Guillhermo Vega. "Science, Stigmatization and Afro-pessimism in the South African Debate on AIDS." *Vibrant* 13, no. 1 (2016): 22–51.

Sankara, Thomas. "Imperialism Is the Arsonist of Our Forests and Savannahs." In *Thomas Sankara Speaks*, 254–60. Atlanta: Pathfinder, 1988.

Saro-Wiwa, Ken. *Genocide in Nigeria: The Ogoni Tragedy*. Lagos, Nigeria: Saros International, 1992.

Sartre, Jean-Paul. "Black Orpheus." Translated by John MacCombie. *Massachusetts Review* 6, no. 1 (Autumn 1964–Winter 1965): 13–52.

Sassen, Saskia. *Expulsions: Brutality and Complexity in the Global Economy*. Cambridge, MA: Harvard University Press, 2014.

Sato, Daisuke, and Toshiki Nagayama. "Development of Agonistic Encounters in Dominance Hierarchy Formation in Juvenile Crayfish." *Journal of Experimental Biology* 215 (2012): 1210–17.

Schapera, I. *The Khoisan Peoples of South Africa*. London: Routledge, 1930.

Scheub, Harold. *A Dictionary of African Mythology: The Mythmaker as Storyteller*. Kindle ed. Oxford: Oxford University Press, 2000.

Schoeman, Karel. *Early Slavery at the Cape of Good Hope, 1652–1717*. Pretoria: Protea, 2007.

Schoeman, Karel. *Portrait of a Slave Society: The Cape of Good Hope, 1717–1795*. Pretoria: Protea Book House, 2012.

Schultz, Oliver. "Belonging on the West Coast: An Ethnography of St Helena Bay in the Context of Marine Resource Scarcity." Master's thesis, University of Cape Town, 2010.

Scott, Dianne, and Clive Barnett. "Something in the Air: Civic Science and Contentious Environmental Politics in Post-apartheid South Africa." *Geoforum* 40 (2009): 373–82.

Scully, W. C. *The White Hecatomb and Other Stories.* London: Methuen, 1897.

Segar, Sue. "Raw Deal: Desalinator's Job Is to Remove Salt, Not Sewage." *Noseweek* 236, 1 June 2019, https://www.noseweek.co.za/articles/4277/Raw-deal -desalinators-job-is-to-remove-salt,-not-sewage.

Sen, Amartya, Jean-Paul Fitoussi, and Joseph Stiglitz. *Mismeasuring Our Lives: Why GDP Doesn't Add Up.* New York: New Press, 2010.

Serres, Michel. *The Birth of Physics.* Translated by Jack Hawkes. Manchester, UK: Clinamen, 2000.

Serres, Michel. *The Five Senses: A Philosophy of Mingled Bodies.* Translated by Margaret Sankey and Peter Cowley. London: Continuum Books, 2008.

Serres, Michel. *Geometry: The Third Book of Foundations.* Translated by Randolph Burks. Kindle edition. New York: Bloomsbury, 2017.

Serres, Michel. *The Natural Contract.* Translated by Elizabeth MacArthur and William Paulson. Ann Arbor: University of Michigan Press, 1995.

Setiloane, Gabriel. "How the Traditional World-View Persists in the Christianity of the Sotho-Tswana." 1978. Michigan State University Libraries African e-Journals Project typescript, 27–42. http://pdfproc.lib.msu.edu/?file=/DMC /African%20Journals/pdfs/PULA/pula001001/pula001001003.pdf, original at http://digital.lib.msu.edu/projects/africanjournals/html/issue.cfm?colid =256.

Seymour, Colleen L., Suzanne J. Milton, Grant S. Joseph, W. Richard J. Dean, Tsholofelo Ditlhobolo, and Graeme S. Cumming. "Twenty Years of Rest Returns Grazing Potential, but Not Palatable Plant Diversity, to Karoo Rangeland, South Africa." *Journal of Applied Ecology* 47, no. 4 (August 2010): 859–67.

Shakespeare, William. *King Lear.* Edited by G. K. Hunter. London: Penguin Books, 1972.

Shakespeare, William. *A Midsummer Night's Dream.* Edited by Stanley Wells. London: Penguin Books, 1995.

Shepherd, Nick. "Archaeology Dreaming: Post-apartheid Urban Imaginaries and the Bones of the Prestwich Street Dead." *Journal of Social Archaeology* 7, no. 1 (2007): 3–28.

Singh, Jerome. "Project Coast: Eugenics in Apartheid South Africa." *Endeavour* 32, no. 1 (2008): 5–9. https://wikileaks.org/gifiles/attach/169/169033 _ProjectCoastEugenics.pdf.

"Sketch Map of South Africa Showing British Possessions, July 1885." The British Empire, website of Stephen Luscombe. Accessed 12 November 2016. http:// www.britishempire.co.uk/images3/southafrica1885map.jpg.

Skotnes, Pippa. *Claim to the Country: The Archive of Wilhelm Bleek and Lucy Lloyd.* Johannesburg: Jacana Media; Athens: Ohio University Press, 2007.

"Slaves and Negro Apprentices: April 26 1816, Proclamation: Offices for the Enregisterment of Slaves Established in Cape Town and the Districts." In *Proclamations, Advertisements and Other Official Notices Published by the Government of the Cape of Good Hope from January 10th 1806 to May 21st 1825*, compiled by Richard Plaskett and T. Miller, 361–63. Cape Town: Cape of Good Hope Government Press, 1827.

Sloterdijk, Peter. *Philosophical Temperaments: From Plato to Foucault*. New York: Columbia University Press, 2013.

Smith, Andrew B. "The Disruption of Khoi Society in the Seventeenth Century." In *Africa Seminar: Collected Papers*, 3:1–16. Cape Town: Centre for African Studies, University of Cape Town, 1983.

Smith, Andrew B., and R. Pheiffer. *The Khoikhoi at the Cape of Good Hope: Seventeenth-Century Drawings in the South African Library*. Cape Town: South African Library, 1993.

Smith, C. A. *Common Names of South African Plants*. Botanical Survey Memoir 35. Pretoria: Department of Agricultural Technical Services, 1966.

Smuts, Barbara. *Sex and Friendship in Baboons*. New Brunswick, NJ: Transaction, 1985.

Smuts, Eckard. "The 'Baboon Boy' of the Eastern Cape and the Making of the Human in South Africa." *Social Dynamics* 44, no. 1 (2018): 146–57.

Stapleton, Timothy J. "Hintsa (c. 1790–1835)." In *Encyclopedia of African Conflicts*, vol. 1, edited by Timothy J. Stapleton, 349. Santa Barbara, CA: ABC-CLIO, 2017.

Stengers, Isabelle. *Another Science Is Possible: A Manifesto for Slow Science*. Translated by Stephen Muecke. Cambridge, UK: Polity, 2018.

Stengers, Isabelle. "Book VII: The Curse of Tolerance." In *Cosmopolitics II*, translated by Robert Bononno, 303–416. Minneapolis: University of Minnesota Press, 2011.

Stengers, Isabelle. "The Cosmopolitical Proposal." In *Making Things Public*, edited by Bruno Latour and Peter Weibel, 994–1003. Cambridge, MA: MIT Press, 2005.

Stengers, Isabelle. *Cosmopolitics*. 2 vols. Translated by Robert Bononno. Minneapolis: University of Minnesota Press, 2010–11.

Stengers, Isabelle. "An Ecology of Practices." *Cultural Studies Review* 11, no. 1 (2005): 183–96.

Stengers, Isabelle. *In Catastrophic Times: Resisting the Coming Barbarism*. Translated by Andrew Goffey. London: Open Humanities Press, 2015.

Stengers, Isabelle. "The Intrusion of Gaia." Translated by Andrew Goffey. 29 December 2015. An Inquiry into Modes of Existence, http://modesofexistence.org/isabelle-stengers-the-intrusion-of-gaia/.

Stengers. Isabelle. "Thinking with Deleuze and Whitehead: A Double Test." In *Deleuze, Whitehead, Bergson: Rhizomatic Connections*, edited by Keith Robinson, 28–44. London: Palgrave Macmillan, 2009.

Stengers, Isabelle. *Thinking with Whitehead: A Free and Wild Creation of Concepts*. Translated by Michael Chase. Cambridge, MA: Harvard University Press, 2011.

Stengers, Isabelle, and Vinciane Despret. *Women Who Make a Fuss: The Unfaithful Daughters of Virginia Woolf.* Minneapolis: University of Minnesota Press, 2014.

Stetka, Brett. "By Land or by Sea: How Did Early Humans Access Key Brain-Building Nutrients?" *Scientific American,* 1 March 2016. https://www.scientificamerican.com/article/by-land-or-by-sea-how-did-early-humans-access-key-brain-building-nutrients/.

Stiegler, Bernard. *The Neganthropocene.* Translated by Daniel Ross. London: Open Humanities Press, 2018.

Strum, Shirley C. "Activists and Their Anthropomorphism Remain the Greatest Threat to Baboons." *Cape Times,* 23 July 2012.

Strum, Shirley C., and Linda Marie Fedigan, eds. *Primate Encounters: Models of Science, Gender and Society.* Chicago: University of Chicago Press, 2000.

Sullivan, Sian. "The Natural Capital Myth; or, Will Accounting Save the World? Preliminary Thoughts on Nature, Finance and Values." Leverhulme Centre for the Study of Value Working Papers #3, 2014. http://thestudyofvalue.org/wp-content/uploads/2013/11/WP3-Sullivan-2014-Natural-Capital-Myth.pdf.

Swanepoel, Janie. "Custodians of the Cape Peninsula: A Historical and Contemporary Ethnography of Urban Conservation in Cape Town." Master's thesis, Stellenbosch University, 2013.

Swartz, C. D., B. Genthe, J. Chamier, L. F. Petrik, J. O. Tijani, A. Adeleye, C. J. Coomans, A. Ohlin, D. Falk, and J. G. Menge. "Emerging Contaminants in Wastewater Treated for Direct Potable Re-use: The Human Health Risk Priorities in South Africa, Volume III: Occurrence, Fate, Removal and Health Risk Assessment of Chemicals of Emerging Concern in Reclaimed Water for Potable Reuse." Report to the Water Research Commission TT 742/3/18, March 2018. Gezina, South Africa: Water Research Commission.

Tobias, Philip V. "Homo erectus." In *Encyclopedia Britannica.* Article published in 2015. http://www.britannica.com/EBchecked/topic/270386/Homo-erectus.

Towers, Lucy. "South African Fisheries Win Major Victory in New York Court." *The Fish Site,* 8 July 2013. https://thefishsite.com/articles/south-african-fisheries-win-major-victory-in-new-york-court.

"Tragedy as Lightning Strikes More Than Twice in KZN." *Injobo* (KwaZulu-Natal Cooperative Governance and Traditional Affairs internal newsletter), February 2015. Accessed 15 March 2018. http://www.kzncogta.gov.za/Portals/0/Documents/newsletters/2015/injobo%20Newsletter%20February%20Edition%202015.pdf (link no longer active).

Tsing, Anna. *Friction: An Ethnography of Global Connection.* Princeton, NJ: Princeton University Press, 2005.

Tsing, Anna. "More-Than-Human Sociality: A Call for Critical Description." In *Anthropology and Nature,* edited by Kirsten Hastrup, 27–42. London: Routledge, 2013.

Tsing, Anna. *The Mushroom at the End of the World.* Princeton, NJ: Princeton University Press, 2015.

Turnbull, David. "Maps, Narratives and Trails: Performativity, Hodology and Distributed Knowledges in Complex Adaptive Systems—An Approach to Emergent Mapping." *Geographical Research* 45 (2007): 140–49.

Turnbull, David. "Movement, Boundaries, Rationality and the State: The Ngaan-yatyarra Land Claim, the Tordesillas Line and the West Australian Border." In *Moving Anthropology: Critical Indigenous Studies,* edited by T. Lea, E. Kowal, and G. Cowlishaw, 185–200. Darwin: Charles Darwin University Press, 2006.

Turrell, Robert. *Capital and Labour on the Kimberley Diamond Fields, 1871–1890.* Cambridge: Cambridge University Press, 1987.

Uwimbabazi, Penine. "Analysis of Umuganda: The Policy and Practice of Community Work in Rwanda." PhD diss., College of Humanities at the University of KwaZulu-Natal, Pietermaritzburg, South Africa, 2012.

Valentyn, François. *Oud en Nieuwe Oost-Indien.* Dordrecht: Johannes van Braam and Gerard Onder de Linden, 1726.

Van Breda, P. A. B., and S. A. Barnard. *100 Veld Plants of the Winter Rainfall Region.* Bulletin No. 422. Pretoria: Department of Agricultural Development, 1991.

Van Dooren, Thom. "Vultures and their People in India: Equity and Entanglement in a Time of Extinctions." *Australian Humanities Review* 50 (2011): 45–61.

Van Dooren, Thom, and Deborah Bird Rose. "Storied-Places in a Multispecies City." *Humanimalia* 3, no. 2 (2012): 1–27.

Van Rensburg, André. "The Secret Modus Operandi in Obtaining Slaves for the Cape: The Ship *Hasselt*, 1658." South Africa's Stamouers. Accessed 21 May 2018. n.d. https://www.stamouers.com/people-of-south-africa/slaves/678-obtaining-slaves.

Van Sittert, Lance. "The Fire and the Eye: Fishers' Knowledge, Echo-Sounding and the Invention of the Skipper in the St. Helena Bay Pelagic Fishery ca. 1930–1960." *Marine Policy* 60 (2015): 300–308.

Van Sittert, Lance. "The Supernatural State: Water Divining and the Cape Underground Water Rush 1891–1910." *Journal of Social History* 37, no. 4 (2004): 915–37.

Van Sittert, Lance. "'Those Who Cannot Remember the Past Are Condemned to Repeat It': Comparing Fisheries Reforms in South Africa." *Marine Policy* 26 (2002): 295–305.

Van Zyl, Marieke. "Ocean, Time and Value: Speaking about the Sea in Kassiesbaai." *Anthropology Southern Africa* 32, nos. 1–2 (2009): 48–58.

Vengosh, Avner, Robert B. Jackson, Nathaniel Warner, Thomas H. Darrah, and Andrew Kondash. "A Critical Review of the Risks to Water Resources from Unconventional Shale Gas Development and Hydraulic Fracturing in the United States." *Environmental Science and Technology* 48, no. 15 (2014): 8334–48.

Viveiros de Castro, Eduardo. "Perspectival Anthropology and the Method of Controlled Equivocation." *Tipití: Journal of the Society for the Anthropology of Lowland South America* 2, no. 1 (2004): 3–22.

Wade, David. *Li: Dynamic Form in Nature.* New York: Walker & Company/Wooden Books, 2003.

Wall, Kevin. "Water Supply: Reshaper of Cape Town's Local Government a Century Ago." In *Proceedings of the Water Distribution Systems Analysis Conference WDSA 2008, August 17–20, 2008, Kruger National Park, South*

Africa, edited by J. E. van Zyl, A. A. Ilemobade, and H. E. Jacobs. Reston, VA: American Society of Civil Engineers.

Wark, McKenzie. 2015. *Molecular Red: Theory for the Anthropocene*. London: Verso.

Watson, Letitia. "Insurance Claims Flood In." *Independent Online*, 27 July 2014. https://www.iol.co.za/personal-finance/insurance-claims-flood-in-1726042.

Webb, Denver Arnold. "Kraals of Guns and Redoubts of Authority: Military Conflict and Fortifications in the Wars of Dispossession and Resistance in South Africa's Eastern Cape, 1780–1894." PhD diss., University of Fort Hare, 2015.

Webster, Mike M., Nicola Atton, William J. E. Hoppitt, and Kevin N. Laland. "Environmental Complexity Influences Association Network Structure and Network-Based Diffusion of Foraging Information in Fish Shoals." *American Naturalist* 181, no. 2 (February 2013): 235–44.

Webster, Mike M., A. J. W. Ward, and P. J. B. Hart. "Shoal and Prey Patch Choice by Co-occurring Fishes and Prawns: Inter-taxa Use of Socially Transmitted Cues." *Proceedings of the Royal Society B: Biological Sciences* 275 (2008): 203–8.

Westrich, Jason R., Alina M. Ebling, William M. Landing, Jessica L. Joyner, Keri M. Kemp, Dale W. Griffin, and Erin K. Lipp. "Saharan Dust Nutrients Promote *Vibrio* Bloom Formation in Marine Surface Waters." *Proceedings of the National Academy of Sciences* 113, no. 21 (2016): 5964–69.

Wheat, Nicola. "An Ethnobotanical, Phytochemical and Metabolomics Investigation of Plants from the Paulshoek Communal Area, Namaqualand." PhD diss., University of Cape Town, 2014.

Wicks, Jeff. "'It's Just the Facts'—Penny Sparrow Breaks Her Silence." *News24*, 4 January 2016. https://www.news24.com/SouthAfrica/News/its-just-the-facts -penny-sparrow-breaks-her-silence-20160104.

Wiegand, Thorsten, Suzanne J. Milton, Karen J. Esler, and Guy F. Midgley. "Live Fast, Die Young: Estimating Size-Age Relations and Mortality Pattern of Shrubs Species in the Semi-Arid Karoo, South Africa." *Plant Ecology* 150 (2000): 115–31.

Wilder, Gary. *Freedom Time: Negritude, Decolonization and the Future of the World*. Durham, NC: Duke University Press, 2015.

Woolworths Holdings. *Annual Report 2012*. Cape Town: Woolworths. Accessed 9 March 2013. http://www.woolworthsholdings.co.za/investor/annual_reports /ar2012/sustainability/its_our_world/water.asp.

Worden, Nigel. "Cape Slaves in the Paper Empire of the VOC." *Kronos* 40, no. 1 (2014): 23–44.

Worden, Nigel. *Slavery in Dutch South Africa*. Cambridge: Cambridge University Press, 1985.

Worden, Nigel, and Clifton Crais, eds. *Breaking the Chains: Slavery and Emancipation in the Nineteenth Century Cape Colony*. Johannesburg: Witwatersrand University Press, 1994.

World Bank. *Overcoming Poverty and Inequality in South Africa: An Assessment of Drivers, Constraints and Opportunities*. Washington, DC: World Bank Publications, 2018.

Wynter, Sylvia, and Katherine McKittrick. "Unparalleled Catastrophe for Our Species? Or, to Give Humanness a Different Future: Conversations." In *Sylvia Wynter: On Being Human as Praxis*, edited by Katherine McKittrick, 9–89. Durham, NC: Duke University Press, 2015.

Yusoff, Kathryn. *A Billion Black Anthropocenes or None*. Minneapolis: University of Minnesota Press, 2019.

Ziervogel, Gina, Mark Pelling, Anton Cartwright, Eric Chu, Tanvi Deshpande, Leila Harris, Keith Hyams, et al. "Inserting Rights and Justice into Urban Resilience: A Focus on Everyday Risk." *Environment and Urbanization* 29, no. 1 (2017): 123–38.

Ziporyn, Brook. *Beyond Oneness and Difference: Li and Coherence in Chinese Thought and Its Antecedents*. Albany: SUNY Press, 2013.

Ziporyn, Brook. *Ironies of Oneness and Difference: Coherence in Chinese Thought; Prolegomena to the Study of Li*. Albany: SUNY Press, 2012.

INDEX

forced removals. *See* land

fortress conservation, 11, 161, 190. *See also* conservation; nature

fracking, 60–76; impact on water, 70–73. *See also* shale and shale gas

gas. *See* shale and shale gas

geological eras: Carboniferous, 4, 29, 62, 101, 211; Paleozoic (Karoo Ice Age), 29, 61. *See also* Anthropocene era; Capitalocene era

geology, 48, 60–63, 71–72, 101, 209–12, 224–26; glaciers, 61, 224; hydrogeology, 48, 57–59, 71. *See also* geological eras; shale and shale gas; Table Mountain (TMNP)

Glissant, Édouard, 20, 79–80, 218–20

government departments: Centre for Scientific and Industrial Research (CSIR), 174, 199; Department of Agriculture, Forestry and Fisheries (DAFF), 191–94; Department of Mineral Affairs (DMR), 71; Department of Trade and Industry, 173, 194; Environmental Management Department, 51

green whiteness, 106, 109–12, 120, 144, 205

Haraway, Donna, xvii, xix, xx, 142, 146, 148, 161, 164–66, 210, 253n24, 253n29, 254n38, 256n63, 69, 70, 71, 264n16

Heerengracht, 38, 47–48, 54

HIV and AIDS, 58, 84–93; ARVS, 84, 86, 92, 94; Treatment Action Campaign (TAC), 89, 92–93

Hoerikwaggo, 26, 38, 48–55; Hoerikwaggo Trail, 26. *See also* Table Mountain National Park (TMNP)

Homo erectus, 61

Homo oeconomicus, 134–36, 194, 207

Homo sapiens, 61, 134, 207

Human Wildlife Solutions (HWS). *See* baboon monitoring and management

hydrogeology. *See* geology

hydropolitics, 64–69

indigenous knowledge. *See* knowledge, indigenous/traditional

justice, 54, 57, 66–69, 131, 180, 187–229

Karoo desert and region, 60–76; restoration of, 65. *See also* geological eras, ice age

Khoe/Khoena, 9, 21–27, 29–48, 54–58, 61–64, 139–40, 227

knowledge: belief and, 27, 38–41, 50, 63, 89, 91–92, 95, 98–104, 218; scientific, 205–15, 226–27. *See also* concepts; feminist science studies; knowledge economy; knowledge, indigenous/traditional; modernity; ontology and ontologies; relations; science and scientists; scientism

knowledge economy, 9, 38, 50, 135–37, 179, 214–19, 226–27

knowledge, indigenous/traditional, 12, 16, 26–27, 84–95, 103–4. *See also* feminist science studies; plant medicine; Science Must Fall movement

land, 120–27; access and rights to, 22–24, 108–9; conflict over, 32–33, 108–9; forced removal of people from, 36, 64–65, 118; Native Land Act (1913), 118–19. *See also* maps and mapping, surveyors and cartographers; redistribution; soil; water

language and languages, 39–42

Latour, Bruno, xvii, 38, 74, 79, 86, 101–2, 234n8, 236n27, 236n33, 238n53, 242n34, 244n8, 242n10, 246n37, 246n39, 259n26, 260n34, 261n56, 265n41, 265n 46, 267n74

lobsters. *See* marine conservation and management

Lutzville, 8, 106–9, 206

maps and mapping, 17, 23–26, 34–38, 53, 115–16; cartography and surveying, 34–36. *See also* land

marine conservation and management, 9, 15, 136, 172–200, 227; algal blooms, 18, 175–85, 189, 196–99, 211; Blue Flag status, 178–79; lobsters, 11, 172, 176–78, 182–85, 189, 197, 209; Marine Protected Areas (MPAs), 179, 192–93; perlemoen (abalone), 12, 173, 177–78, 185–88; red tides, 172, 181; role of fishers in, 191–98. *See also* ecology, ecological regime shift; fishers; fishing and hunting; sewers and sewerage

war. *See* conflict

water, 18, 41–59; access and rights to, 31, 50–52; Cape Flats aquifer, 18, 51, 56–58, 109, 129–30, 206; desalination of, 174–75; drainage of, 52–54; drought, 51–59, 130, 173–75, 219; extraction of, 18, 44; ownership and management of, 31, 47, 54; rainfall, 26–29, 40–43, 51–53, 58, 64, 75–76, 121–22, 173, 243n43; springs, streams and rivers, 14–15, 26–33, 41–58, 211, 224, 238n61; from Table Mountain, 29–31, 44, 47; Vaarsche (the Fresh River), 29–31, 52. *See also* fracking; hydropolitics; land; marine conservation and management; marine conservation and management, Blue Flag status; marine conservation and management; sewers and sewerage; windmills

whiteness, 14, 79–80, 109–11, 120, 204. *See also* green whiteness

Wild Card (South African National Parks), 1–5, 14

wildlife, 10–13, 65, 110–11, 135, 147–48, 154–55, 165, 241n19

windmills, 18, 64, 75–76, 117

Xhosa, 113–15, 139–40

Xolobeni, 8, 106–9, 206